Evolve from Infrastructure to Innovation with SAP on AWS

Strategize Beyond Infrastructure for Extending your SAP applications, Data Management, IoT & AI/ML integration and IT Operations using AWS Services

Bidwan Baruah
Krishnakumar Ramadoss
Abarajith Vivekanandha

Apress®

Evolve from Infrastructure to Innovation with SAP on AWS: Strategize Beyond Infrastructure for Extending your SAP applications, Data Management, IoT & AI/ML integration and IT Operations using AWS Services

Bidwan Baruah
Sussex, WI, USA

Krishnakumar Ramadoss
Frisco, TX, USA

Abarajith Vivekanandha
Westerville, OH, USA

ISBN-13 (pbk): 979-8-8688-0889-0   ISBN-13 (electronic): 979-8-8688-0890-6
https://doi.org/10.1007/979-8-8688-0890-6

Managing Director, Apress Media LLC: Welmoed Spahr
Acquisitions Editor: Aditee Mirashi
Development Editor: Aditee Mirashi
Editorial Assistant: Kripa Joseph
Copy Editor: William McManus

Cover designed by eStudioCalamar

Cover image designed by Unsplash.com

Distributed to the book trade worldwide by Springer Science+Business Media New York, 1 New York Plaza, Suite 4600, New York, NY 10004-1562, USA. Phone 1-800-SPRINGER, fax (201) 348-4505, e-mail orders-ny@springer-sbm.com, or visit www.springeronline.com. Apress Media, LLC is a California LLC and the sole member (owner) is Springer Science + Business Media Finance Inc (SSBM Finance Inc). SSBM Finance Inc is a **Delaware** corporation.

For information on translations, please e-mail booktranslations@springernature.com; for reprint, paperback, or audio rights, please e-mail bookpermissions@springernature.com.

Apress titles may be purchased in bulk for academic, corporate, or promotional use. eBook versions and licenses are also available for most titles. For more information, reference our Print and eBook Bulk Sales web page at https://www.apress.com/bulk-sales.

Any source code or other supplementary material referenced by the authors in this book is available to readers on GitHub. For more detailed information, please visit https://www.apress.com/gp/services/source-code.

If disposing of this product, please recycle the paper

Table of Contents

About the Authors...xi

About the Technical Reviewers xv

Introduction xvii

Chapter 1: Introduction: Why Evolve from Infrastructure to Innovation with SAP on AWS? 1

What Is "Evolving from Infrastructure to Innovation"?........................2

Change of Mindset and Culture...5

    Decoupling and Why It Matters...6

    Automate Everything 8

    Fail Fast, Learn Early, and Pivot Quickly 8

    Establish Organizational Practices 9

Role of SAP Enterprise Architects 11

    AWS Cloud Adoption Framework 12

    SAP Enterprise Architecture Framework 13

Pillars of SAP on AWS Innovation 15

    Pillar 1: Apps and APIs..16

    Pillar 2: IoT...20

    Pillar 3: Data and Analytics 25

    Pillar 4: AI and ML...42

    Pillar 5: DevOps and SysOps.......................................54

AWS Solutions Library..69

Summary...70

Chapter 2: Fundamentals of Modern Architecture 73

APIs and Microservices..74

Why APIs Are Important from a Customer Angle.............................75

Evolution of SAP APIs and Integration..76

Protocols in the Context of SAP APIs..78

Statefulness vs. Statelessness 80

Can SAP ERP Be the API Layer? 81

Can SAP API Management Be the API Layer? 84

Decoupling with Microservices...89

The Evolution of Software Architecture...90

    Monolithic Architecture 91

    Service-Oriented Architecture 93

    Microservices 95

    Serverless...98

Coupling: The Integration Magic Word 100

Integration Patterns 101

    Synchronous (Request-Response) Model.................................101

    Asynchronous Point-to-Point Model (Queue)............................104

    Asynchronous Point-to-Point Model (Router) 107

    Asynchronous Point-to-Point Model (Message Bus) 108

Event-Driven Architecture...111

    What Is an Event?..112

    Event-Driven Architecture with SAP S/4HANA, SAP BTP,
    and AWS Services..114

Summary..116

Chapter 3: Modern Data Strategy **117**

Data Archiving..119

    Types of Data to Archive 121

    Data Management Evolution: Archiving, SAP ILM, and Data Aging in
    SAP HANA 122

    What to Consider Before Archiving SAP Data 124

    SAP Data Archiving Options with AWS Services........................127

Data Lakes: A Modern Approach to Data Management...................130

    How to Build Data Lakes on AWS 132

    Integrating SAP Data into an AWS Data Lake 136

    SAP Data Extraction Patterns 140

Data Virtualization and Data Federation..150

    Data Virtualization 150

    Data Federation: Integrating Disparate Data Sources 151

    Comparison of Data Federation and Data Virtualization............153

End-to-End Enterprise Analytics 154

    Streamlining with AWS Lake Formation 156

    Applying Machine Learning 156

    Performing Analytics 156

    User Access 157

    Achieving End-to-End Enterprise Analytics 157

Data Mesh: A Paradigm Shift in Data Architecture........................158

    Importance of Data Mesh from a Data Lake Perspective 158

    Achieving Data Mesh Architecture for SAP Data Using AWS Services.........159

Summary...162

Chapter 4: Extending SAP Business Processes 163

Strategic Innovation Approach 164

How to Use the Evaluation Framework..................................165

Leveraging the Right Technology Stack 168

SAP Business Technology Platform 168

AWS Technology Stack 171

Methodology for Choosing a Technology Stack...................172

Joint Reference Architectures from SAP and AWS.........................174

Data-to-Value Architecture 175

Integration and App Development 180

Platform Foundation 183

Extending SAP Business Processes with the AWS SDK for ABAP 186

Use Case: Modernize Document Processing in SAP....................187

Use Case: Increasing Efficiency in Financial Consolidation by 90%...........193

Use Case: Improving Location Accuracy 196

Integration of AWS B2B Data Interchange into SAP Business Processes 198

Use Cases of AWS B2B Data Interchange specific to SAP..........................202

Summary...206

Chapter 5: Smart Factory 209

What Is a Smart Factory?..210

Edge and Edge Devices 210

Importance of Smart Factories...212

Digital Manufacturing and Its Key Tenets....................................214

Industrial Data Lakes and Analytics 215

Predictive Maintenance 232

Predictive Quality...240

Digital Twin 244

Voice Technology 248

AWS Solutions for Manufacturing and Industry 251

    Asset Maintenance & Reliability 252

    Asset Performance Management 252

    Automation Software Management 252

    Cloud Manufacturing Execution System (MES) 253

    Connected Worker 253

    Composable Operations Applications 254

    Industrial Data Fabric 255

    Lean Daily Management 256

    Computer Vision for Quality Insights 257

    Operational Technology (OT) Cybersecurity 257

    Predictive Quality 257

SAP Solutions to Enable Digital Manufacturing 258

    SAP Solutions That Complement AWS Offerings 265

Secrets to a Successful Implementation 266

Summary 268

Chapter 6: Business Transformation with AI/ML 271

Why AI and ML Matter Now 272

AI and ML Offerings from AWS and Use Cases 273

Generative AI 276

Foundation Models 278

    The Need for Amazon Bedrock 279

    Amazon Bedrock Agents 280

How LLMs Work and How to Make Them Smarter 281

    Vector Embeddings 282

    What Is a Transformer? 285

    Customization of LLMs 288

Amazon Q...307

Patterns to Use Generative AI 309

Gen AI Use Cases for SAP..314

    Technical Use Cases..316

    Business Use Cases...327

Summary...336

Chapter 7: DevOps and SysOps for SAP...........................339

DevOps in Action 340

    Putting DevOps into Practice...340

    Building a DevOps Culture...341

Building a CI/CD Pipeline with AWS 351

    AWS CodeCommit...351

    AWS CodeBuild 352

    AWS CodeDeploy 352

    AWS CodePipeline 353

DevOps for SAP 353

    Plan and Setup Phase...354

    Develop Phase 360

    Test Phase 367

    Monitor and Operate Phase...368

SysOps for SAP 386

    Automated Operations..386

    Security and Compliance...397

    Business Continuity 403

Summary...413

Chapter 8: Starting your SAP on AWS Modernization Initiatives......417

What's Worth Solving? 418

Strategy Is Everything 418

Ready-to-Use Solutions for Your Business Challenges.....................422

AWS Solutions and SAP Missions...422

AWS CAF in Action...427

Creating an SAP on AWS Modernization Roadmap 430

Modernization Process: An Iterative Approach 434

How to Get Business Buy-in..436

Building a Strong Business Case...437

Presenting Your Case for Innovation...437

Creating a Stakeholder Engagement Plan 438

Demonstrating Proof of Concept 441

Governance 445

Enable Governance with AWS Services 447

Transition to Operations 449

Summary..451

Correction to: Evolve from Infrastructure to Innovation with SAP on AWS C1

Index...453

About the Authors

The authors of this book have helped multiple Fortune 500 SAP customers globally to transform their SAP ecosystem.

 **Bidwan Baruah** brings 18 years of IT experience as a cloud architect and technology advisor. He has led various implementation, upgrade, and migration projects of SAP solutions for customers across the globe. He has also been an SAP customer and, as the SAP architect, led the successful execution of a large-scale SAP digital transformation project for a leading organization. Currently serving as a Senior Solutions Architect at Amazon Web Services (AWS), Bidwan specializes in guiding enterprises through the complexities of cloud migration and digital transformation, particularly in integrating SAP environments with AWS services. In this role, he has assisted a number of Fortune 100 and other leading enterprises in North America with migrating and modernizing their SAP workloads on AWS, making him a sought-after advisor for leveraging cloud technologies to drive business innovation. Bidwan is passionate about educating and empowering professionals, sharing his insights through speaking engagements, technical blogs, and this comprehensive guide on modernizing SAP with AWS.

Krishnakumar Ramadoss is a distinguished IT professional and Senior SAP Innovation Solutions Architect at AWS. With over two decades of experience in the technology sector, he has been pivotal in helping global enterprises transition to cloud environments, focusing on SAP migrations and building cloud-native solutions. His deep expertise in cloud infrastructure, data analytics, and enterprise software integration has established him as a leader in driving technological innovation and operational excellence. Krishnakumar focuses on building innovative solutions through an "extend" strategy that enables customers to develop advanced capabilities on top of their SAP migrations. These capabilities span various domains, including data analytics, artificial intelligence/machine learning, applications and integration, and the Internet of Things (IoT). Through these cutting-edge solutions, he empowers organizations to unlock new levels of efficiency, scalability, and competitive advantage. Krishnakumar played a pivotal role in developing a critical low-code data integration service, enabling enterprises across different verticals like manufacturing, automotive, biotech, aerospace, and retail to build scalable SAP data lakes on AWS and derive value from their mission-critical data. He also contributed to the development of AWS Supply Chain, a cloud-based service that unifies data and provides ML-powered actionable insights. This application helps customers improve supply lead times, better forecast demand, and reduce their carbon footprint. Known for his strategic insights and practical approach, Krishnakumar is committed to mentoring IT leaders and sharing his expertise through industry conferences and webinars. In this authoritative book, he provides insights on leveraging AWS to modernize SAP systems, pushing the boundaries of what's possible in cloud computing, and driving innovation in a rapidly evolving digital landscape.

 **Abarajith Vivekanandha** is a seasoned cloud architect and IT strategist with extensive expertise in SAP on AWS solutions. As a Senior Solutions Architect at AWS, he has been instrumental in guiding numerous organizations, including Fortune 500 companies, through their cloud migration journeys, optimizing their SAP landscapes for enhanced performance and agility. With a strong background in enterprise architecture, data management, and cloud-native technologies, Abarajith is passionate about driving digital transformation and innovation. He is a sought-after speaker at industry events and an advocate for best practices in cloud adoption, dedicated to empowering businesses to achieve their strategic objectives through cutting-edge technology solutions.

About the Technical Reviewers

Rajendra Narikimelli currently serves as an SAP Solutions Architect at AWS, where he aids SAP customers in harnessing the power of AWS to transition into cloud-native architectures. Prior to this role, he was a Services Delivery Architect in the MaxAttention CoE at SAP America, conducting advisory workshops for premium and large SAP customers. With 22 years of IT experience, Rajendra has provided thought leadership to several multinational organizations globally, focusing on the management of complex SAP landscapes.

With 18 years of hands-on experience in the dynamic realm of SAP, **Rozal Singh** brings a wealth of expertise encompassing the full spectrum of SAP system implementation, management, and operations. His proficiency extends to both cloud-based and on-premises SAP environments, positioning him as a versatile and seasoned professional in the SAP space. Before joining AWS, Rozal distinguished himself as an SAP Technical Lead Architect and a migration specialist at a consulting company. His career is marked by a steadfast dedication to technology and a keen focus on developing robust architectural foundations to bring his customers' strategic visions to fruition.

Introduction

If you have always done it that way, it is probably wrong.

—Charles Kettering

This powerful insight from Charles Kettering challenges us to rethink our traditional methods and embrace innovation with open arms. In an era where "slow is the new broken" and "data is the new gold," the ability to adapt and innovate is not just an advantage but a necessity. As businesses transcend beyond mere infrastructure, they must leverage cutting-edge technologies to seize new opportunities and drive transformation across all aspects of their operations.

As we embarked on writing this book, we drew inspiration from the transformative journeys of countless customers who are reshaping their business models and IT landscapes through digital transformation. In today's volatile world, businesses face relentless disruptions driven by social, economic, and technological changes, compelling them to adapt swiftly and enhance agility. Those who fail to evolve risk being outpaced by more nimble competitors. This urgency is driving CEOs to prioritize digital transformation, challenging IT teams to deliver innovative solutions that amplify flexibility, agility, and competitive advantage.

SAP stands as a cornerstone in digital transformation initiatives. As a mission-critical ERP system, SAP is pivotal for unlocking the immense value embedded in enterprise data. By integrating various business processes and functions, SAP empowers organizations to streamline operations, boost efficiency, and gain real-time insights.

The original version of the book has been revised. A correction to this book can be found at https://doi.org/10.1007/979-8-8688-0890-6_9

Its robust capabilities in finance, supply chain, sales and distribution, manufacturing, and production make it an indispensable tool for businesses striving for excellence.

SAP S/4HANA, the next-generation enterprise resource planning (ERP) suite, is designed to help businesses run simple in a digital and networked world. Built on the advanced in-memory platform SAP HANA, SAP S/4HANA offers a dramatically simplified data model and a user-friendly interface, providing powerful analytics and real-time decision-making capabilities. As SAP continues to evolve with the transition to SAP S/4HANA, it ensures that businesses stay at the forefront of digital transformation, armed with the necessary tools to thrive in a dynamic and competitive environment.

Two key enablers of SAP's digital transformation capabilities are the SAP Business Technology Platform (SAP BTP) and AWS services, also known as AWS native services.

SAP BTP is a comprehensive platform that integrates data and analytics, application development, automation, and AI technologies. It provides businesses with the flexibility to customize and extend their SAP applications to meet specific needs, facilitating innovation and enhancing business agility. BTP's robust features allow organizations to harness the power of their data, streamline processes, and accelerate digital transformation initiatives, making it an essential component of a modern SAP ecosystem.

The advent of cloud computing, especially AWS, revolutionized the SAP landscape. AWS offers a scalable, reliable, and secure infrastructure platform, enabling businesses to swiftly adapt to changing demands. By leveraging AWS, companies can extend their SAP processes with cutting-edge technologies such as analytics, AI/ML, and IoT via AWS services. This transformation empowers businesses to innovate rapidly, deploy new applications and services, and extract profound insights from their data. AWS's relentless pursuit of innovation, coupled with its extensive ecosystem of partners, empowers SAP customers to achieve unparalleled agility, efficiency, and competitive edge in their digital transformation journeys.

Consider these insights from recent research studies:

- A survey conducted by SAPInsider found that 72% of organizations implementing SAP S/4HANA experienced significant improvements in process efficiency, while 61% reported enhanced end-user and business satisfaction following their migration.[1]

- According to a McKinsey report, "Up to $3 trillion is up for grabs for Forbes Global 2000 companies that go beyond cloud adoption and venture into innovation and pioneering.[2]

This book serves as a guide for SAP customers embarking on their digital transformation journey with AWS. By tackling the dual challenges of identifying the right business use cases and implementing them effectively, we aim to empower organizations not only to modernize their SAP landscapes but also to leverage AWS native services and SAP BTP services powered by AWS to innovate and extend their SAP applications. This approach ensures that businesses remain adaptable and competitive in an ever-evolving landscape.

---

[1] SAPInsider, "SAP S/4HANA Migration Benchmark Report," https://sapinsider.org/sap-s-4hana-migration-benchmark-report/

[2] McKinsey Digital, "The State of Cloud Computing in Europe: Increasing Adoption, Low Returns, Huge Potential," https://www.mckinsey.com/capabilities/mckinsey-digital/our-insights/the-state-of-cloud-computing-in-europe-increasing-adoption-low-returns-huge-potential

CHAPTER 1

Introduction: Why Evolve from Infrastructure to Innovation with SAP on AWS?

Migrating to AWS unlocks quick wins, while innovating with its capabilities shapes a modern enterprise, prepared to tackle today's challenges and tomorrow's opportunities.

In this chapter, we will introduce the concept of evolving beyond infrastructure hosting to achieve innovation with SAP on Amazon Web Services (AWS). Many organizations are migrating their SAP workloads to AWS as a part of their digital transformation strategy, and they are already seeing substantial business benefits in terms of reduced operating costs, increased flexibility, and improved resiliency. But is this "real" digital transformation? Migrating your SAP workloads to AWS is just the starting step. Digital transformation is not just about hosting on AWS to get infrastructure flexibility; it's all about changing the core business model, redefining business processes, and changing IT from just a cost center to the driver of innovation for the business. You can achieve all of these objectives by taking advantage of native AWS services and SAP Business Technology Platform (BTP) services powered by AWS.

Embracing innovation and transformation often entails change and disruption. The famous author and educator Peter Drucker once said, "The greatest danger in times of turbulence is not the turbulence; it is to act with yesterday's logic." We will explore how organizations can transform their culture and mindset by leveraging enterprise architecture frameworks offered by AWS and SAP. These frameworks facilitate the alignment of business and IT by integrating AWS and SAP technologies with business objectives, enabling more strategic planning of current and future IT investments and cloud transformations. We'll also present the SAP on AWS innovation pillars along with the AWS services, extending beyond compute, storage, and network functionalities. The Pillars of SAP on AWS Innovation will serve as the foundation of this book and will be utilized throughout the chapters.

The following topics will be covered in this chapter:

- What is "evolving from infrastructure to innovation"?

- Change of mindset and culture

- Role of SAP enterprise architects

- Pillars of SAP on AWS Innovation

- AWS Solutions Library

What Is "Evolving from Infrastructure to Innovation"?

The world of SAP is experiencing a substantial shift right now. With the imminent end of support for SAP ERP Central Component (SAP ECC) in 2027 and SAP's focus on the RISE with SAP program, nearly all enterprises running SAP as an ERP solution are seeking to migrate either their entire

SAP infrastructure to the cloud as a part of a data center exit initiative or at least specific components to the cloud, such as starting with a small footprint like only SAP S/4HANA on the cloud.

But *why use SAP on the cloud*? Customers are moving their SAP workloads to the cloud primarily due to factors such as cost effectiveness, the flexibility provided by a diverse range of infrastructure options, resiliency, security, and, most importantly, the ease of embracing the innovative services consistently introduced by cloud providers. A traditional data center is inherently unable to offer these benefits due to the significant advantage cloud providers possess in terms of *economies of scale*, which refers to the cost benefits that arise as an increasing number of customers adopt, consume, or purchase a service or a product.

Unlike a traditional data center, a cloud platform can swiftly provision infrastructure with no lead time and seamlessly transition that infrastructure to the latest offering the very next moment. Additionally, the cloud stands out in terms of resiliency, achieved through a combination of architectural principles and redundancy facilitated by a global network of data centers, which a conventional data center cannot support. But the biggest differentiator is that a cloud platform enables you to adopt innovative services and modernize the way you run your business.

SAP on the cloud transformation is not a destination, *it's a journey*. The decisions made at various intersections shape this journey, and most often, the choices available later are influenced by the decisions made early on in the journey. Additionally, your choice of vehicle for the journey plays a pivotal role. It's a matter of choosing a brand with innovation ingrained in its DNA throughout its history, or opting for one that assimilates features inspired by others. All cloud providers bring different strengths to the table. They might offer more or less the same kind of services, but the underlying capabilities of the services and the culture of the organization define what you are going to get as a partner. These are the reasons why customers trust AWS as the cloud provider of choice to run their mission-critical SAP workloads.

As previously mentioned, many organizations who have already
started on this journey to SAP on AWS as a part of their digital
transformation initiative are already seeing big business benefits, primarily
in terms of reduced operating costs, increased flexibility, and improved
resiliency. But what is "real" digital transformation? Keeping your existing
business processes as is with SAP on AWS is not digital transformation.
Making copies of your processes is completely different from taking
advantage of the digital technology and changing your business model.
Gone are the days when you look for technology tools and software to
fit your existing manual business processes. Now, you transform your
processes by leveraging technology, which helps you to stay ahead of your
competitors.

Companies such as Volkswagen, Invista, Moderna and Zalando are
using AWS beyond hosting to optimize their business processes thereby
changing IT from just a cost center to the driver of innovation for the
business.

Volkswagen has created a cloud-native solution that integrates its
back-office processes on SAP with its smart factories across multiple
plants, including automating the stock replenishment process in which
Internet of Things (IoT) devices perform automatic stock rechecks at
the shop floor, which allows SAP to start the replenishment process
without human intervention and even create the purchase requisition
in the SAP system for the missing stock (see `https://www.youtube.com/
watch?v=IPkekBzClTk`).

Similarly, Invista (a subsidiary of Koch Industries) leverages AWS
artificial intelligence and machine learning (AI/ML) capabilities to
perform visual inspections, effectively reducing defects. Additionally,
they implement predictive maintenance strategies to minimize downtime
(`https://aws.amazon.com/solutions/case-studies/invista-
case-study/`).

During the Covid pandemic, Moderna used AI/ML to sequence and release its first batch of vaccine in just 27 days. Moderna scaled its SAP S/4HANA system to support global distribution of 300M+ doses (`https://aws.amazon.com/solutions/case-studies/moderna-therapeutics/`).

Finally, Zalando has integrated its SAP systems with 36 AWS technologies and created a hybrid data architecture, thereby lowering the cost of ownership for its SAP data architecture by 30% (`https://aws.amazon.com/solutions/case-studies/zalando-sap/`).

For many organizations, the challenge to digitally transform is that it requires moving the goal post continuously and not having a constant target state. As you can imagine, a typical "IT project" always has a known target state. But with cloud transformation, the target state evolves continuously. This demands a change of culture and mindset.

Change of Mindset and Culture

Based on a study by Americas' SAP Users' Group (ASUG), the top three barriers to innovation are

- Lack of innovation strategy

- Change management issues

- Strained resources

All of these barriers are related to mindset and culture.

While engaging with SAP customers considering a move to AWS, a significant number express a preference for a lift and shift approach, also known as *rehosting*, wherein there are no changes to the workload, including the application, database, and operating system. While it's understandable that this method is chosen to minimize business disruption or expedite the migration project timeline, opting for a lift and shift strategy significantly limits the potential benefits offered by AWS and foregoes the opportunity to re-architect your SAP applications to make them cloud native. Having collaborated with numerous SAP customers throughout the years, a recurring observation we've made is that many of them approach AWS usage similar to on-premises. So, what is *cloud native*? Cloud native means architecting your applications in such a way that they *don't just run in the cloud but are built for it*. That is, the applications take advantage of the features and technologies offered by the cloud. *Cloud-native applications are built for the cloud—designed to thrive*. Thus, transitioning to AWS requires a shift in mindset—a change of mindset in the way you architect, operate, and innovate.

With regard to architecting, first and foremost, rightsize your SAP systems—don't plan for three years ahead, but instead plan for today and select the right AWS Savings Plan so that you can change to the instance of your choice tomorrow. To optimize your SAP landscape, it's advisable to architect it in a cloud-native manner, leveraging the innovative services offered by AWS. Common use cases are using Amazon Relational Database Service (Amazon RDS) wherever possible, using Amazon Elastic File System (Amazon EFS) for SAP Content Server, using Amazon Simple Storage Service (Amazon S3) for SAP Archiving, and so forth. These are some of the cloud-native capabilities that can be easily leveraged.

Decoupling and Why It Matters

Another common on-prem mindset that we have seen customers implement is installing multiple SAP systems on a single Amazon Elastic Cloud Compute (Amazon EC2) instance. Implementing this architecture increases the *blast radius*, meaning that if there is a service disruption to one EC2 instance, multiple applications will be impacted. In fact, consolidation is the opposite of digital decoupling. So, unlike on-prem, on AWS, you spread out the applications across several small EC2 instances and do not consolidate them in a large single instance. Some people may argue that this approach would result in increased operational overhead and cost. This argument disregards the crucial role of automating operational activities, and AWS provides a seamless platform for achieving this without the necessity of investing in third-party tools. Cost is a nonfactor given that AWS has SAP-certified instances of every size you possibly need.

AWS is not merely an infrastructure platform, but a driving force for architectural evolution. Enterprises running SAP can utilize the extensive range of AWS services to modernize and enhance their business processes, achieving unprecedented levels of agility, scalability, and innovation. No longer constrained by the limitations of monolithic architectures, enterprises can now adopt modern architectural principles such as microservices, event-driven integrations, and an API-first approach. This allows them to break down silos, develop extensions outside the ERP core, and orchestrate complex business processes effortlessly.

By adopting microservices and event-driven architectures, enterprises can break down monolithic SAP application extensions into modular, independently deployable units, enabling rapid development cycles and seamless scalability. An API-first approach allows organizations to expose SAP functionalities as reusable services, facilitating integration with external systems and empowering developers to create innovative applications at scale. All these newer architecture patterns provide the backbone for decoupling SAP systems from external dependencies, enabling real-time data processing and ensuring seamless communication across distributed architectures. This paradigm shift empowers enterprises to adapt to evolving business requirements, respond to market dynamics, and drive continuous innovation.

Automate Everything

From an operational perspective, the primary challenge to innovation is
that too much time is spent on manual, repetitive tasks, taking customers
away from higher value tasks. Streamline time-consuming and error-prone
system administration tasks by implementing automation. This empowers
your IT workforce to spearhead business innovation. Implement
standardized best practices for core SAP operations to streamline your IT
processes, boosting business agility. We talk about this topic in detail in
Chapter 7.

Fail Fast, Learn Early, and Pivot Quickly

Another reason organizations are not able to innovate is because of the
lack of a platform to experiment and fail fast. AWS provides you that
platform. You develop expertise and begin leveraging AWS services
tailored to your specific needs.

In instances where a particular AWS service proves insufficient,
users can seamlessly transition to the next one, without any long-term
commitment, all under a flexible pay-as-you-go model. Both AWS and SAP
are rolling out new features, services, and capabilities at an unprecedented
rate. From an innovation perspective, the mindset should be to
continuously adapt based on technology advancements. A comprehensive
project spanning three to six months to implement a solution for a three-
year use case no longer applies in today's world. Enterprises must adopt an
approach that facilitates the swift implementation of innovative use cases,
allowing for continuous adaptation and improvement of solutions based
on ongoing AWS innovations.

For example, you learn how to use the AWS SDK for SAP ABAP to
implement your first use case and then, as and when new releases are
available to support integration with new AWS services, implementing
your new use cases is a seamless experience. One customer that we

know used the AWS SDK for SAP ABAP as soon as it was released by AWS to automate the customer's check-processing business process within SAP. Later, when a subsequent release of the SDK enabled integration with Amazon Bedrock and SAP, the customer was able to effortlessly explore other use cases related to generative AI.

Establish Organizational Practices

While embarking on the cloud journey, organizations often take different routes. Some organizations create a team, commonly known as a *Cloud Center of Excellence (CCoE)*, that is dedicated to establishing the foundation of cloud best practices and frameworks. Others opt to train their entire IT department on cloud technologies. Regardless of the chosen approach, a recommended strategy is to initiate the process swiftly without overly restrictive guardrails that could impede progress. Once operations have stabilized, it is advisable to reassess and redefine governance based on the insights and lessons learned during the initial stages. A prescriptive guidance is published by AWS at `https://docs.aws.amazon.com/prescriptive-guidance/latest/cloud-center-of-excellence/introduction.html`.

For organizations to reap the power of a CCoE, they should have a congenial organizational structure. This can be achieved by establishing a Cloud Adoption Office (CAO) to adopt and implement the CCoE tenets. The following are the two key components within a CAO:

- The Cloud Business Office:

    - Aligns the offerings provided by the Cloud Platform Engineering Team with the requirements of enterprise customers and leadership

    - Offers ongoing support for onboarding, training, and change management to ensure a smooth transition and successful adoption of cloud technology across the organization

- The Cloud Platform Engineering Team:

  - Configures and standardizes the AWS platform
    according to enterprise guidelines for architecture,
    operations, security, governance, and financial
    management

  - Packages and iteratively enhances these standards
    as deployable products and services that can be
    easily accessed and utilized

The diagram in Figure 1-1 illustrates how all the components (teams)
integrate. This may appear overwhelming at first. As with all cloud
initiatives, think big but start small. Establish a Cloud Adoption Office
and a few development teams to initiate the process. Scale as your cloud
transformation journey progresses and expands.

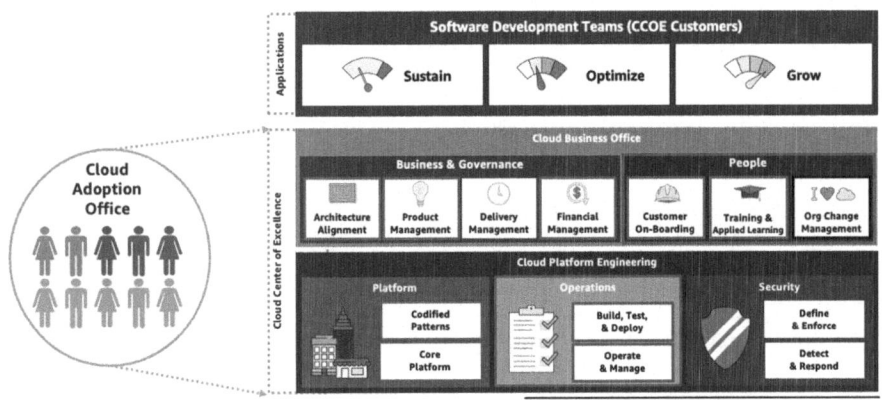

Figure 1-1. *AWS Cloud Center of Excellence (CCoE)*

A CCoE is often considered to be a part of the broader SAP Enterprise
Architecture (EA) Framework within an organization. Over the past few
decades, organizations have evolved from SAP systems to SAP ecosystems.
This evolution necessitates a holistic approach to transformation,

considering both business and IT aspects. Achieving this holistic perspective requires the implementation of an enterprise architecture practice, which will be our focus in the next section.

Role of SAP Enterprise Architects

Survival and success in today's world is not guaranteed by being the oldest or most established organization; rather, it is the organizations that prove most adaptable to change that endure and flourish. And, enterprise architecture serves as the driving force for enterprises to achieve adaptability and embrace change by translating business needs to solution capabilities and technical components. With the ongoing shift toward cloud migration and the continuous innovation within cloud computing, the role of enterprise architects (EAs) has undergone some transformations as well.

Enterprise architects now need to develop a cloud enterprise. They need to help in making the enterprise aware of how cloud transformation will impact the business as well as how it will change the landscape of IT applications and services. EAs need to take into account the business and IT strategy and existing investments, which is essential to make sure that innovative solutions do not become isolated from the perspective of business value, processes, and technology. EAs articulate the drivers and business benefits (like automation of business processes). This will be discussed more in Chapter 4, which explores extending SAP business processes of the cloud transformation to the business leaders.

Although there are multiple frameworks for enterprise architects, we are going to discuss two of them in the context of SAP on AWS transformation: the AWS Cloud Adoption Framework and the SAP Enterprise Architecture Framework.

AWS Cloud Adoption Framework

To help the enterprise architects, AWS has published the AWS Cloud
Adoption Framework (AWS CAF), which is a set of guidelines provided
by AWS to help organizations digitally transform and accelerate their
business outcomes through innovative use of AWS. These guidelines
are categorized into six perspectives: Business, People, Governance,
Platform, Security, and Operations. Each perspective comprises a set
of capabilities that stakeholders own or manage in an enterprise's
cloud transformation journey. Achieving success in an organization's
transformation requires envisioning the desired target state, gauging the
cloud readiness, and embracing an agile approach to address the gaps.
Incremental transformation enables the enterprise to showcase value
swiftly while minimizing the necessity for extensive predictions. An
iterative approach helps maintain momentum and refine the roadmap
based on experiential learning, which ties back to the change of mindset
mentality we highlighted earlier. This is what AWS CAF is all about. It helps
organizations to

- Evaluate and improve their cloud readiness

- Identify and prioritize transformation opportunities

- Iteratively evolve their transformation roadmap

The AWS CAF value chain, as shown in Figure 1-2, illustrates how
organizational change powered by a set of foundational capabilities
can accelerate business outcomes. The value chain is simply the
transformation domains, where process transformation is powered
by technological transformation. This process transformation enables
organizational transformation, which enables product transformation.
The business outcomes include reducing risk; improved environmental,
social, and governance (ESG) performance; and increased revenue and
operational efficiency. AWS CAF 3.0 broadens the scope of AWS CAF 2.0

from facilitating "cloud adoption" to accelerating "cloud-powered digital transformation," places a greater emphasis on data and analytics, and expands the number of foundational capabilities from 31 to 47. We discuss the role of AWS CAF in detail in Chapter 8.

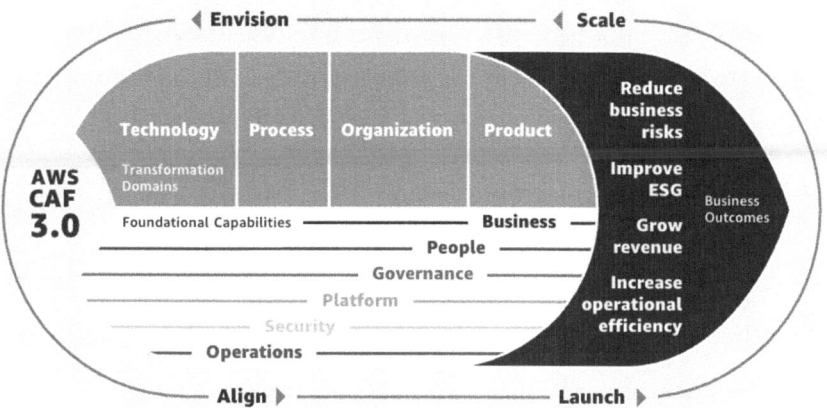

Figure 1-2. *AWS Cloud Adoption Framework value chain*

SAP Enterprise Architecture Framework

SAP has been making efforts to strengthen its capabilities in the enterprise architecture space in the last decade, recognizing that almost all of its customers are undergoing significant transformation. Also, SAP no longer exists in a "box", as enterprises are now more diversified in terms of types of deployments for different applications, like on-prem, software as a service (SaaS), platform as a service (PaaS), hybrid, and so on, with strong integration between each other. So, the entire enterprise architecture needs to be evaluated for any transformation.

SAP recognizes the increasing importance of enterprise architects in helping organizations drive this transformation. In addition to acquiring SAP Signavio (for business process management) and LeanIX (for enterprise architecture management), SAP has also published the SAP

Enterprise Architecture Methodology, all of which together is intended to help in translating business requirements to architectural principles to SAP's technology portfolio.

The SAP EA Methodology (see Figure 1-3) guides enterprise architects in defining a target business architecture to facilitate the achievement of business goals. Subsequently, it involves building an SAP Solution Architecture based on the business architecture and architecture vision, culminating in the creation of a comprehensive technology architecture.

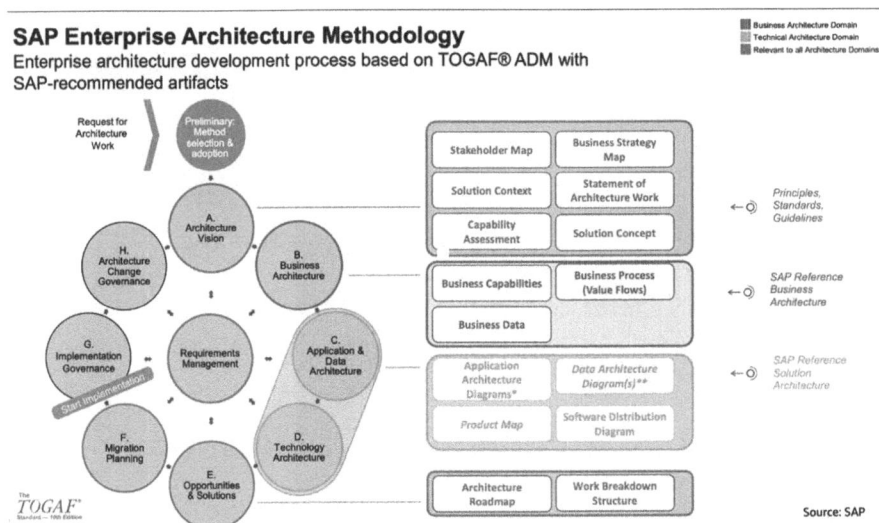

Figure 1-3. *SAP Enterprise Architecture Methodology*

The SAP Enterprise Architecture Framework (https://groups. community.sap.com/t5/enterprise-architecture-blog-posts/sap-enterprise-architecture-framework/ba-p/124037) is centered on the following four cornerstones:

- **Methodology**: SAP EA Framework is based on TOGAF and industry standards-based EA methodologies.

- **Reference Architecture content**: SAP Reference Business and Solution Architecture maps business challenges with SAP IT solutions.

- **Tools**: SAP Signavio and LeanIX are the two primary tools, but other tools are available that can help you identify the target state, its value proposition, and the assessment of peripheral systems' impact based on the activity. For example, for upgrading, there are tools available as listed at `https://support.sap.com/en/tools/upgrade-transformation-tools.html` (login required).

- **Services**: Standardized SAP EA services to support customer transformation.

SAP Signavio is a business process management (BPM) and process intelligence platform tool that monitors business processes' performance and conformance, detects blockers, provides actionable recommendations based on SAP best practices, and implements process improvements. Enterprise Architecture Management (EAM) by LeanIX can help to visualize the entire IT application landscape, identify applications nearing their end of life, facilitate the design of a target state, and assist in planning new architectural roadmaps. When used together, BPM and EAM provide a comprehensive overview of the interconnections between SAP and related applications, which helps you to reduce uncertainty, continuously innovate, and accelerate the value of your SAP transformation. This SAP transformation from an AWS perspective is driven through five pillars that we call the "Pillars of SAP on AWS Innovation," discussed next.

Pillars of SAP on AWS Innovation

Leveraging AWS provides a robust and resilient foundation in terms of compute, network, and storage. What SAP on AWS customers are truly interested in is how they can drive innovation in areas such as extension of SAP applications (the first pillar, which we refer to as Apps and APIs, for simplicity),

15

IoT (second pillar), data and analytics (third pillar), AI and ML (fourth pillar), and DevOps and SysOps (fifth pillar)—all the capabilities that AWS offers. These are five key focal points where customers aim to sustain continuous innovation by harnessing the potential in these areas. Let's look at each of these Pillars of SAP on AWS Innovation one by one.

Pillar 1: Apps and APIs

Decades ago, the integration of third-party applications with SAP was considerably less complex and more straightforward. The introduction of the Web necessitated integrating applications with SAP, and the subsequent mobile revolution placed a priority on developing mobile-first apps. Consumer expectations shifted, with a demand for the same flexibility and sophistication in enterprise applications as experienced in consumer applications. Presently, there is a significant focus on investments in AI-enabled applications, and SAP is asking customers to separate their SAP extensions from the core SAP system so that the core SAP system can be upgraded faster without any adaptation and test efforts on the extensions. Extensions do not break upgrades and upgrades do not break extensions. This is the concept of maintaining a *clean core* for seamless upgrades. In this approach, extensions are separated from SAP code on the core system yet integrated using APIs.

Figure 1-4 shows the different options to create new extensions/ customizations with SAP S/4HANA 2022. Classic ABAP development no longer is recommended unless you have no other option. One of the criteria for deciding which option to select is whether the extension requirements need proximity to and coupling with SAP S/4HANA data, transactions, and apps.

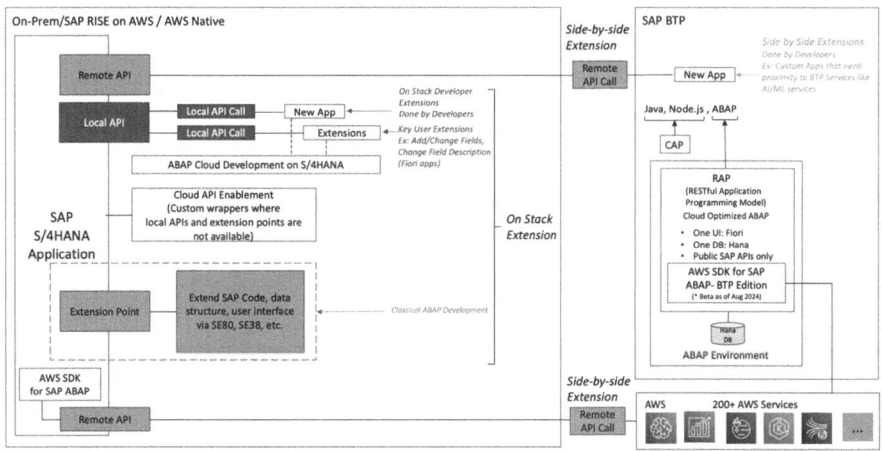

Figure 1-4. *SAP S/4HANA extensibility options*

Here we distinguish between side-by-side (or loosely coupled) extensions and on-stack (or tightly coupled) extensions. Organizations should go with a side-by-side or loosely coupled extension if any of these following apply:

- It needs to have close proximity with SAP BTP services or AWS native services.

- It encompasses native mobile apps, apps requiring freestyle UI development for consumer-grade interfaces, or apps designed for individuals without a user ID in SAP S/4HANA.

- They want to keep it separate from the core SAP S/4HANA infrastructure and operations.

- It is a stand-alone application with not much usage of SAP S/4HANA data.

- The data is duplicated to SAP BTP or retrieved through remote API calls from the core SAP S/4HANA system.

17

- Custom data is not changed together with core data (no transactional consistency required).

- The application requires data sourced from data hubs or integration hubs, which consolidate, gather, or distribute data from various systems throughout the organization.

An organization should opt for an on-stack or tightly coupled extension if:

- They are extending the look and feel of the app with minor cosmetic changes (also known as key user extensions)

- If there is a requirement of frequent reading or modifying of SAP data, involving multiple round trips.

- If there is a requirement of reading of SAP data in complex SQL queries like joins and with high data volume.

- It needs transactional consistency.

- It needs to use the extension point (also known as classic ABAP development)

For new development needs where the app requires close proximity with SAP BTP services, side-by-side extension is recommended.

APIs play a pivotal role, serving as an agile, flexible, secure, and scalable layer for integrating various business applications. APIs provide prebuilt integration packages that cater to a wide range of interface and integration requirements. Therefore, the success of all extension projects relies heavily on having a robust set of APIs in place. Integration no longer is about integrating SAP with a few non-SAP systems from an application level. Every customer's landscape is now heterogenous, all driven by process-centric composition. No longer does a business process run end to end only on SAP. There are SAP SaaS products and third-party solutions involved in almost all business processes. That is why SAP S/4HANA exposes many APIs to integrate with all these products. These APIs can be consumed from all these platforms.

SAP BTP serves as one of the platforms to develop these SAP cloud solutions. As detailed in the SAP Business Technology Platform Regions and Service Portfolio (`https://discovery-center.cloud.sap/ serviceCatalog`), AWS is the only cloud provider that offers all SAP BTP services available for deployment on hyperscalers. As of July 2024, all 83 SAP BTP services currently run on AWS and 22 run exclusively on AWS, including impactful services such as SAP Build Apps, SAP Build Code, SAP Master Data Governance, cloud edition, and so on, emphasizing global availability, scalability, and elasticity. Leveraging SAP BTP on AWS presents unparalleled opportunities for developing SAP extensions on the foremost cloud infrastructure provider. SAP continuously delivers new APIs with high priority via its SAP Business Accelerator Hub (`https:// api.sap.com`). If an API is not available, a customer can request it via a dedicated *Customer Influence Session* or build its own API. SAP Integration Suite is an integration platform as a service (iPaaS) that combines tools and prepackaged content that address various integration scenarios across hybrid landscape. The SAP Integration Suite includes Cloud Integration, which facilitates the modeling and orchestration of integration flows for both application-to-application (A2A) and business-to-business (B2B) scenarios. As shown in Figure 1-5, the suite also incorporates open connectors and AWS Adapter, offering a unified interface for seamless integration with various SaaS providers. Furthermore, the Integration Suite features an Event Mesh, enabling event-driven data flows between SAP and target consumers in both directions. The suite also provides an API Management layer, allowing the smooth exposure of digital assets in a hybrid setting.

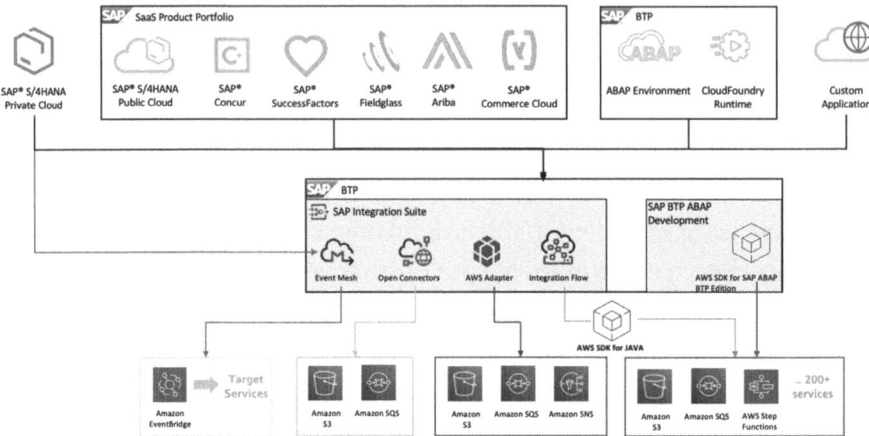

Figure 1-5. *Integration options between SAP BTP and AWS services*

We will discuss more details about APIs and event-driven architecture in Chapter 2.

Pillar 2: IoT

Industrial customers using SAP on AWS to run their enterprise are using AWS services to preform analytics of industrial data. They are leveraging AWS to harness their operational and machine data to achieve a spectrum of outcomes such as reducing energy costs, identifying and resolving equipment issues, pinpointing inefficiencies in manufacturing lines, enhancing product quality, and optimizing production output. The benefits of analyzing industrial data can be categorized into five categories:

- **Availability**: Reduction in unplanned downtime

- **Productivity**: Enhancement of labor productivity and reduction in overtime

- **Quality**: Decrease in scrap, material overuse, and related factors

- **Carbon footprint**: Lowering of energy
  consumption costs

- **Safety**: Reduction in safety incidents

These customers gain visibility into operational technology (OT) data
from machines, programmable logic controllers (PLCs), and supervisory
control and data acquisition (SCADA) systems. This visibility allows them
to conduct root cause analysis (RCA) when a production line or machine
experiences downtime, enhance production throughput, and establish
tighter integration between OT data-driven analytics, IT systems like SAP,
and manufacturing execution systems (MESs). At a very high level, the
entire process can be divided into five stages:

1. Ingest data from equipment to AWS.

2. Store, organize, and manage data on AWS.

3. Specify performance metrics for your equipment
   and processes.

4. Visualize the historical and live data.

5. Drive business outcomes by deploying AI/ML
   applications, which optimize factory output,
   improve product quality, maximize asset utilization,
   and identify equipment maintenance issues.

Before we talk about ingesting and analyzing data from different
industrial data sources, let's first understand the concept of "things" in IoT
and how they connect to AWS.

What Are "Things" in IoT and What Is AWS IoT Core?

An *IoT thing* is a representation and record of your physical device in
AWS. A physical device needs a thing record in order to work with AWS IoT
(https://aws.amazon.com/iot/), which provides a registry that helps you

21

manage things. A thing is a representation of a specific device or a logical entity. AWS IoT Core is a managed service that lets these IoT things talk to various AWS services. Figure 1-6 shows how real-world things are mapped to AWS things.

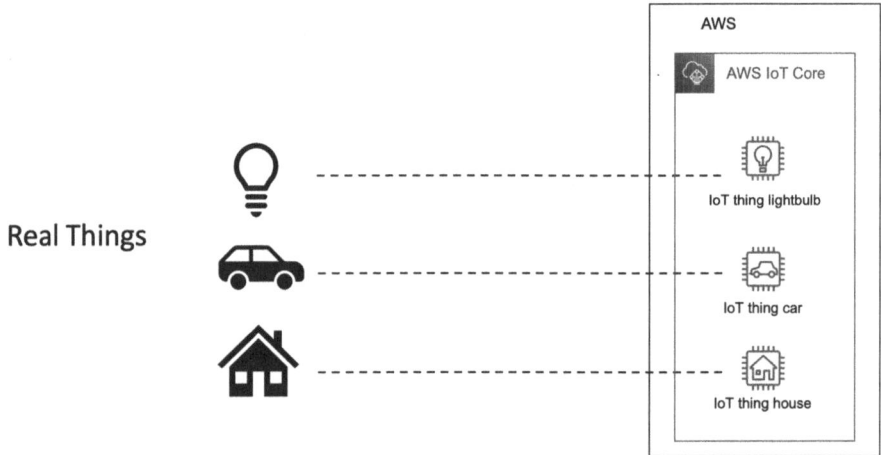

Figure 1-6. *Real-world things mapped to IoT things*

How Does a Device Talk to AWS IoT Core or Other Services?

AWS IoT Core serves as the endpoint for device connections to the AWS Cloud through the MQTT/HTTP/LoRaWAN network protocols, as depicted in Figure 1-7. MQTT is a lightweight, publish–subscribe, machine-to-machine network protocol for message queuing service. It is designed for connections with remote locations that have devices with resource constraints or limited network bandwidth. Long Range Wide Area Network (LoRaWAN) is a protocol that operates on a wireless radio frequency technology called LoRa, which helps devices connect to the Internet.

After devices connect to AWS IoT Core via these protocols, IoT Core then facilitates the forwarding of data to various other AWS services. Once enterprises have the capability to utilize AWS's suite of 200+ services, the potential opportunities become limitless.

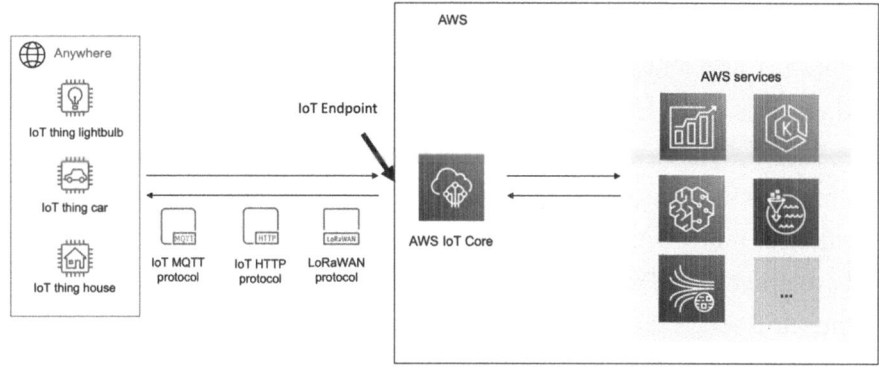

Figure 1-7. *IoT things and AWS IoT Core*

Now, on the floor of an industrial plant or in similar scenarios, not all devices are "smart," meaning not all of them can publish to IoT Core via MQTT messages. For those devices, AWS offers AWS IoT SiteWise, which efficiently collects data from a variety of equipment, including machines, sensors, PLCs, SCADA systems, and data historians, through a local gateway. AWS IoT SiteWise then creates virtual representations of industrial assets, organizes and labels the data, and generates real-time key performance indicators (KPIs) and metrics. Additionally, IoT SiteWise builds fully managed dashboards that facilitate the visualization and monitoring of data against the predefined KPIs. This comprehensive process aims to assist organizations in making more informed, data-driven decisions.

AWS IoT SiteWise data can be made available in AWS IoT Core easily to take advantage of the features of AWS IoT Core.

Predictive Maintenance

One of the AWS services that is widely used to provide predictive maintenance capabilities is Amazon Monitron. Built on AWS IoT and machine learning (ML) technologies, Amazon Monitron is essentially an end-to-end system that is fundamentally composed of sensors that capture temperature and vibration data from equipment, gateways for transferring this data to the AWS Cloud, a GUI/App, and a dedicated service to analyze the collected data. Amazon Monitron has the capability to export data to an Amazon Kinesis stream and Amazon S3, thereby ensuring that the data is accessible for downstream SAP integration. Figure 1-8 shows the high-level architecture. At regular intervals, the sensor takes a reading for a few seconds and sends the data to the gateway, which in turn sends it to AWS for analysis.

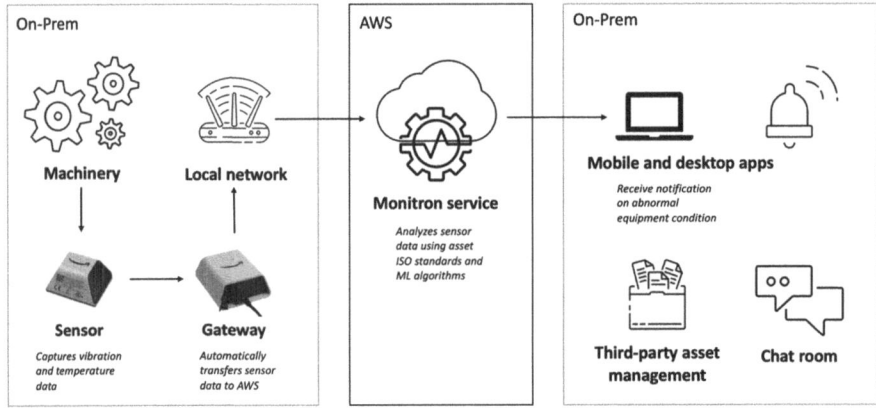

Figure 1-8. *End-to-end architecture of Amazon Monitron*

Amazon Monitron is ideal for monitoring simple and/or small equipment. However, for larger or more complex assemblies, predicting failure is not as straightforward as comparing sensor values against predefined thresholds. Such prediction often requires unsupervised machine learning models. Amazon Lookout for Equipment offers a solution that eliminates the need for creating a time-consuming and costly comprehensive ML solution. It analyzes data from various sensors and systems, such as pressure, flow rate, RPMs, temperature, and power. Based on this data, it automatically trains a model specific to your equipment. Using this customized ML model, Lookout for Equipment analyzes incoming sensor data in real time and detects early warning signs of potential machine failures.

AWS provides a comprehensive suite of IoT services and solutions designed to connect and manage millions of devices, bringing intelligence to edge devices across various sectors, including industrial, consumer, commercial, and automotive workloads. These services enable the collection, storage, and analysis of IoT data, empowering organizations to derive valuable insights from the vast amounts of data generated by connected devices. In Chapter 5, we are going to discuss in more detail the topic of industrial data analytics along with predictive maintenance and additional topics like predictive quality, digital twin, and so on. However, this serves as an excellent segue to introduce our next pillar, *Data and Analytics*.

Pillar 3: Data and Analytics

A 2023 Statista report (https://www.statista.com/statistics/871513/worldwide-data-created/) states that approximately 328.77 million terabytes of data are created each day across the globe. A data strategy stands as the primary asset for securing a competitive edge over other companies. Neglecting or undervaluing data will never allow enterprises to stay ahead in the competitive landscape.

Many SAP on AWS customers are leveraging AWS's advanced analytics tools to derive valuable insights from their data lakes, aiding in data-driven decision-making, resulting in creation of "intelligent enterprises." For example, Invista established an AWS data lake, leading to a reduction in inventory costs and improved supply chain efficiency (https://aws.amazon.com/solutions/case-studies/invista-case-study/). Previously, Invista depended on storage that compartmentalized data at each plant, with manual and reactive maintenance processes. The solution involved harnessing the AWS Cloud, resulting in annual savings of $2 million for Invista and the capacity to automate visual inspection data. As another example, an SAP exploration and production customer in Australia integrated data from numerous IoT sensors in a plant with SAP asset data, consolidating the data into an AWS data lake. Through the application of AI/ML algorithms across this dataset, the customer was able to identify patterns and initiate predictive maintenance processes within SAP. This integration resulted in the development of what the customer referred to as "intelligent assets."

The core purpose of analytics is for business units to define their learning objectives from the data. Subsequently, they collaborate with IT to determine data sources and access controls, conduct the analysis, and present the results in a user-friendly visual format. It's not just SAP data but also data from non-SAP systems, SaaS applications, and emerging data sources such as social media and IoT devices that needs to be analyzed to derive useful business insights.

In recent years, there has been an increase in the demand for consolidating data in cost-effective storage solutions, coupled with cloud-native analytics and AI/ML capabilities. This is driving customers to transition to AWS data lakes on its object store called Amazon Simple Storage Service (S3) due to its unparalleled durability, availability, scalability, security, compliance, and audit capabilities. Data lakes built on Amazon S3 provide the foundation for better record management, advanced analytics, and innovation through data science with AWS native services with built-in ML integration or via AWS Analytics solutions developed by partners. AWS data lakes are designed to offer access to structured, semi-structured, and unstructured data at scale. The rationale is to centralize all data in a single repository, catalog it effectively, and ensure seamless access for users across diverse data types and sources.

Modern Data Warehouse

There are use cases that require data to be stored in a data warehouse. Data warehousing today does a lot more for customers than it did even three to four years back. A modern cloud-based data warehouse must provide uncompromised analytics capabilities. Users expect the ability to analyze their entire dataset without data movement, support for widely used open standard file formats (such as Parquet, Hudi, or JSON), and seamless access to data across both the data warehouse and operational databases. The provisioning, operation, and workload management of the data warehouse should be handled without requiring users' active involvement. Given the unpredictable and diverse nature of customer workloads, the system should autonomously manage these variations without manual intervention.

Users desire a system that remains within their budgetary constraints and avoids incurring costs during idle periods. Additionally, they seek top-notch price performance to ensure affordability in the face of escalating data volumes. Lastly, there is a demand for real-time capabilities to accommodate the simultaneous access of thousands of users through business intelligence tools. To meet these customer challenges and address their modern needs, AWS released Amazon Redshift. It helps analyze all the data—the log files, the transactional data, the click-stream data, semi-structured data, and nested data—without movement or transformation. Amazon Redshift is the most affordable cloud data warehouse, delivering up to three times better price performance relative to competition. It requires zero administration and reliably delivers insights from the data in seconds in the most secure way.

In 2022, Amazon introduced the General Availability (GA) of Amazon Redshift Serverless. When considering whether to opt for Amazon Redshift or Redshift Serverless, it's important to note that Amazon Redshift is the preferred choice for stable usage workloads such as SAP. On the other hand, Amazon Redshift Serverless is well suited for unpredictable variable and/or periodic workloads characterized by significant idle time and spikes in usage. It is also recommended for addressing ad hoc analytics that need to get started quickly and for test and development environments.

For some other customers, the preference is to perform a shadow replication of the database without a predetermined plan for data usage. In such cases, where an object store like Amazon S3 is not the preferred option, AWS offers Amazon Relational Database Service (Amazon RDS) as a relational store alternative. Amazon RDS is a completely managed and scalable database service offered by AWS, making it versatile for various requirements. Amazon RDS offers a comprehensive suite of database engines, each tailored to distinct data schema needs—whether it's row-based, columnar, in-memory, document, graph, time series, or ledger databases—providing the flexibility and scalability required to efficiently manage diverse workloads under a fully managed service. This ensures that you can choose the optimal database solution for your specific use case, all while benefiting from automated management, high availability, and robust security.

The Journey of Data

The journey of data commences with its ingestion into the data lake, where it is gathered and defined for future discovery. Subsequently, the raw data undergoes a process of cleansing and making it available for exploration. Once the data is appropriately prepared for discovery, it becomes a valuable resource for finding answers to business questions or identifying patterns to aid in predicting experiences. With these extensive capabilities, data is able to tell stories. As depicted in Figure 1-9, AWS offers a platform and special services to facilitate the entire journey of data, providing optimization tailored to each and every use case.

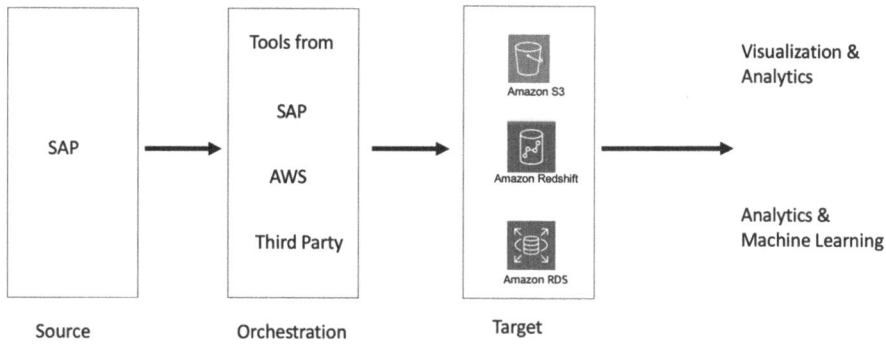

Figure 1-9. *The journey of data*

Now, let's look at this journey in a bit more detail. Back in 2017, Amazon launched AWS Glue as the revolutionary fully managed serverless service in the extract, transform, and load (ETL) space. AWS Glue automates a significant portion of the repetitive and labor-intensive tasks associated with discovering, categorizing, cleaning, enriching, and moving data. This automation allows organizations to allocate more time to the analysis of their data. AWS Glue performs automated crawls of data sources, identifies data formats, and provides suggestions for schemas and transformations. This eliminates the need for manual coding of data flows, streamlining the data processing workflow. AWS Glue has three core components:

- **Data Catalog**: It is the central metadata repository.
  Tables and databases containing metadata are stored
  here, but not actual data from a data store.

- **Crawler**: It discovers the data and associated metadata
  from various data sources (source or target) such as
  Amazon S3, Amazon RDS, Amazon Redshift, and so
  on. Crawlers automatically keep the AWS Glue Data
  Catalog up to date as and when new data arrives or the
  data evolves.

- **ETL job**: It is the business logic that is required to
  perform data processing comprising a transformation
  script (for which you need to write code), data sources,
  and data targets.

An AWS Glue ETL job is initiated in any of the following ways:

- Synchronously as a step within AWS Step Functions,
  which is a serverless orchestration service that makes it
  is easy to build a workflow to orchestrate various event-
  driven different steps that uses different AWS services.
  Think of AWS Step Functions as a service that can help
  you build an orchestration workflow.

- Asynchronously by an Amazon EventBridge event.
  EventBridge is used to orchestrate events in an event-
  driven workflow by allowing users to create rules.
  EventBridge evaluates each event against the rules,
  and if there is a match, it sends a copy of the event to
  the specified target in the rule. For example, let's say
  some new data lands in an Amazon S3 bucket, which

is nothing but an event. Using EventBridge, you can
configure a rule to start an AWS Glue workflow (which
consists of the AWS Glue job) every time new data
lands in the S3 bucket. EventBridge was formerly called
CloudWatch Events.

- Manually using the AWS CLI or AWS SDK.

Later in 2020, AWS introduced AWS Glue Studio to create, run, and
monitor the ETL jobs in an easy way, as well as AWS Glue DataBrew to
help clean, prepare, normalize, and transform the data without writing any
code, especially for team members in a customer's data team (such as data
analysts) who lacked the skillset of writing code. As of December 2023,
AWS Glue now comes with integrated support for SAP HANA. AWS Glue
Studio provides a visual interface to connect to SAP HANA, author data
integration jobs, and run them on the AWS Glue Studio serverless Spark
runtime. Figure 1-10 shows step by step the process of extracting data from
an SAP system using AWS Glue.

Figure 1-10. *Extracting data from SAP using AWS Glue*

Fully Managed Data Ingestion and Transfer Service

Fast forwarding a few years, AWS introduced Amazon AppFlow, which is a
fully managed service that enables you to (with just a few clicks) securely
transfer data between SAP, SaaS applications such as Salesforce, Marketo,
Slack, and ServiceNow, and AWS services such as Amazon S3 and Amazon
Redshift (see Figure 1-11). Amazon AppFlow gives you a simple no-code
user interface, allowing you to connect to different sources and ingest data

into AWS. Although you need to write code for the AWS Glue job, you don't have to do that while using Amazon AppFlow because "under the hood" Amazon AppFlow leverages several AWS services, including AWS Glue, Amazon CloudWatch, AWS Lambda, and Amazon EventBridge.

Amazon AppFlow streamlines data preparation tasks by offering capabilities such as transformations, partitioning, and aggregation. It automates the process of preparing and registering schemas through integration with the AWS Glue Data Catalog. This seamless integration enables effortless discovery and sharing of data across AWS analytics and machine learning services. As of August 2024, for smaller systems or systems with low change rate, Amazon AppFlow is the preferred way to ingest SAP data into an AWS data lake. However, for larger systems, third-party solutions like BryteFlow or SNP are recommended. Refer to "Guidance for SAP Data Integration and Management on AWS" (`https://aws.amazon.com/solutions/guidance/sap-data-integration-and-management-on-aws/`) in the AWS Solutions Library for more information. We are going to introduce the AWS Solutions Library in the last section of this chapter and discuss it at length in Chapter 8.

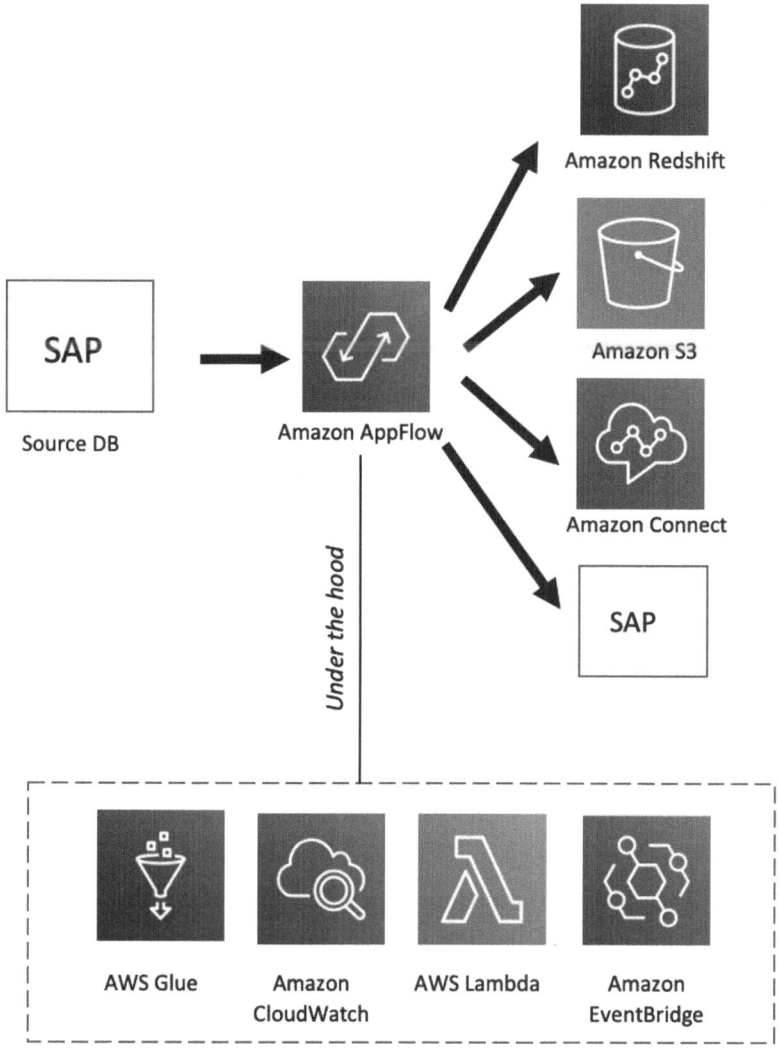

Figure 1-11. *Underlying architecture of Amazon AppFlow*

Generative AI Capabilities

With the emergence of generative AI in 2023, AWS launched Amazon Q in November 2023. Amazon Q is a generative AI–powered assistant designed for engaging in conversations, resolving issues, creating content, extracting insights, and executing tasks. This is achieved by establishing connections with an organization's information repositories, code, data, and enterprise systems. Capabilities of Amazon Q were extended to data integration within AWS Glue. Amazon Q data integration is a generative AI–powered AWS Glue capability that enables users to build data integration pipelines effortlessly using natural language. Through conversations with Amazon Q, you can perform various tasks such as authoring jobs, troubleshooting issues, and obtaining answers related to AWS Glue and data integration. Describe your data integration workload, and Amazon Q data integration will generate a comprehensive AWS Glue script tailored to your requirements. Amazon Q data integration facilitates connections to common AWS sources like Amazon S3, Amazon Redshift, and Amazon DynamoDB. You can troubleshoot job-related errors by requesting explanations and proposed solutions from Amazon Q data integration. Throughout the entire data integration workflow, Amazon Q provides detailed guidance to assist you. Additionally, Amazon Q aids in learning and constructing data integration jobs within AWS Glue.

Visualization

Figure 1-12 shows the services offered by AWS for data visualization within SAP data analytics along with other capabilities like ingestion, storage and compute. For ad hoc universal querying of data from an Amazon S3 bucket, AWS offers Amazon Athena. Amazon Athena is an interactive query service that makes it easy to analyze data directly in Amazon S3 using standard SQL. Amazon QuickSight is a comprehensive business intelligence (BI) service designed to help users easily create and visualize interactive dashboards for their data, now with intuitive and powerful natural language experiences, available via large language models (LLMs) capabilities, which AWS calls "Generative BI" or Amazon Q for QuickSight.

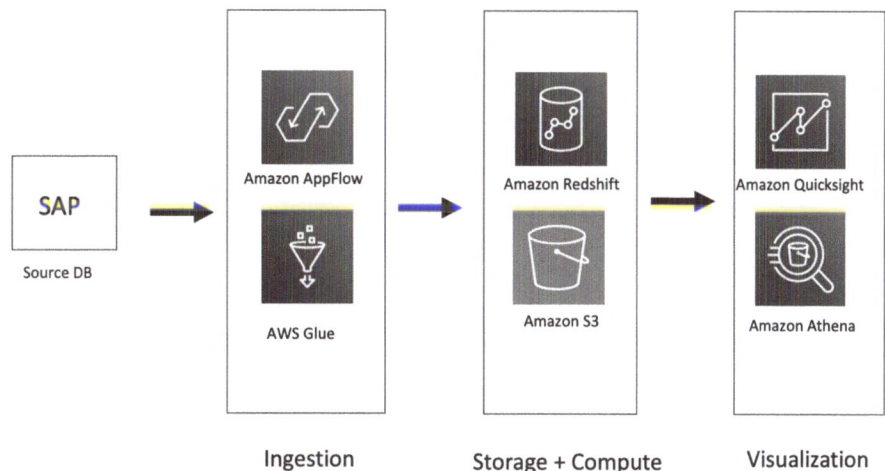

Figure 1-12. *AWS services and their capabilities for SAP data analytics*

Putting It All Together

Figure 1-13 shows what an end-to-end solution would look like. The various AWS services and their capabilities that we have discussed so far, have been put together to give you a visual end-to-end architecture pattern.

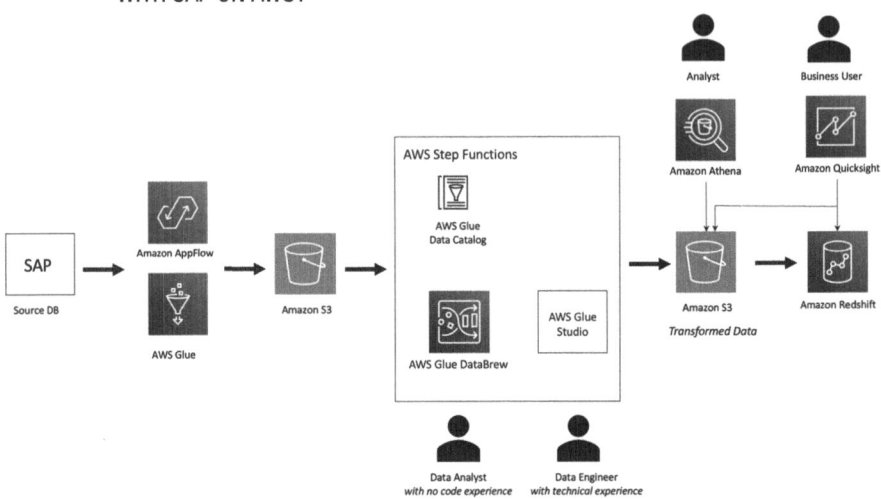

Figure 1-13. *Sample end-to-end architecture pattern of SAP data analytics using AWS services*

Exchanging Data the Easy Way

Data serves as the catalyst for growth and innovation, yet data teams allocate 80% of their time to source and prepare data, leaving only 20% for actual data analysis. The labor-intensive and time-consuming nature of finding, acquiring, and preparing the required data makes this process challenging.

The simplification and democratization witnessed in machine learning and application development through cloud technology should be applies similarly to the exchange of data. The process of discovering, preparing, and utilizing data in the cloud, where a growing volume of data is generated and stored, should be easy. Subscribers should have a place where they can go to procure all of the data that they need, in whatever format they need it in. They shouldn't need to jump through hoops to set up subscriptions, contracts, and billing. And they shouldn't need to invest months of effort building pipelines to prepare and load data into their production systems each time they license new data. That is where AWS Data Exchange comes into the picture to bridge the gap between providers and subscribers exchanging data, helping customers lower costs, become more agile, and innovate faster.

AWS Data Exchange helps organizations find the data they're searching for with AWS's comprehensive catalog, including over 3,500 public products from over 350 data providers delivered—through files, APIs, or queries—directly to the data lakes, applications, analytics, and ML models that use it. AWS Data Exchange also supports custom and private products for situations in which customers want data that's not currently available in the public catalog.

Customers have the flexibility to ingest third-party data files directly into Amazon S3, enabling them to swiftly commence data preparation and analysis using integration, analytics, and ML tools within minutes of subscribing. Additionally, customers can request data delivery via Amazon Redshift tables, relieving providers of the tasks involved in cleansing, validating, and transforming the data into production-ready tables. This allows subscribers to promptly query, analyze, and integrate the data with their production systems upon subscription. Moreover, customers have the option to request data delivery via APIs, empowering their developers to seamlessly integrate the data into production applications regardless of their location.

AWS Data Exchange stands out as the only platform that enables customers to discover and license files, tables, and APIs within a single product, offering complete automation of data ingestion and usage with preferred tools. Furthermore, for providers, global distribution of their data business through AWS Data Exchange is easily achievable with our user-friendly API and console experience, requiring just a few clicks.

How to Start Building SAP Data Lakes on AWS

The easiest way to start building SAP data lakes on AWS is to use AWS Lake Formation, a fully managed serverless service that allows you to build clean and secure data lakes in days. It has several features, but the top four features at a high level are

- It helps to easily build a data ingestion pipeline from your data source to Amazon S3 in a few clicks.

- It optimizes the way data is stored in the S3 data lake.

- It simplifies complex access and permission management for data lakes.

- It enables data sharing across different AWS Accounts and AWS Organization that an enterprise might use.

AWS also provides accelerators for key business process areas within SAP, which allows customers to create pipelines that connect to SAP, stage and transform the data in a "schema aware" way, provide target schemas in a data warehouse, and provide operational dashboards for data visualization that provide turnkey reporting and insights. These accelerators are called AWS Analytics Fabric for SAP (`https://github.com/aws-samples/aws-analytics-fabric-for-sap`), which is a set of prebuilt accelerators for SAP-specific business processes like Order-to-Cash (O2C) that helps to adopt data extraction/analytics insights from the customer data. For details, refer to this blog article: `https://aws.amazon.com/blogs/awsforsap/extend-rise-with-sap-on-aws-with-analytics-fabric-for-sap-accelerators/`.

AWS Analytics Fabric for SAP gives customers the ability to accelerate data transformation initiatives, modernize aging and self-managed integrations, and ensure data security and compliance that supports their business. Additionally, it allows them to connect applications, services,

and processes to automate workflows cost-effectively, ultimately deriving
actionable insights from their data in a way that suits their scenario
and budget.

Data and Analytics Solutions from SAP

Over time, as organizations endeavor to extract additional business
insights from data, there is a growing demand to enhance data
accessibility, driven by an increasing number of individuals seeking
access to information. However, a common dynamic exists wherein the
business users typically find themselves on the receiving end, with IT
being responsible for preparing and making data ready for consumption.
There is a natural tension between the business's need for self-service,
with decentralized access, and IT's desire to centralize governance. SAP
Datasphere aims to reconcile this conflict by offering self-service data
modeling capabilities, along with data cataloging, data warehousing,
data federation and data virtualization features. SAP Datasphere is SAP's
next-generation cloud data warehouse solution. SAP Datasphere was first
introduced on AWS, and as of November 2023, SAP Datasphere runs in
three times more regions than the next-leading cloud service provider.

Just like SAP Datasphere, other BTP services or products from SAP
can complement AWS services in your SAP data analytics use case. SAP
Analytics Cloud, also first launched on AWS, is another SAP BTP service
that can take your enterprise planning to the next level with an optimized,
vertically integrated consumption layer for planning and analytics
workloads. Instead of maintaining your database on your own, if you
are looking for database as a service (DBaaS) in the SAP realm, then SAP
HANA Cloud, powered by AWS Graviton, is your answer. Serving as your
operational database, SAP HANA Cloud provides a platform for developing
applications that harness analytical capabilities such as machine learning,
graph analysis, and spatial data processing. HANA Cloud enables the
integration of historical and real-time data, facilitating comprehensive
insights for your business needs.

The previous diagram from Figure 1-12 depicting the role of different AWS services in the data analytics journey is updated in Figure 1-14 to show some of the currently relevant complementary offerings from SAP.

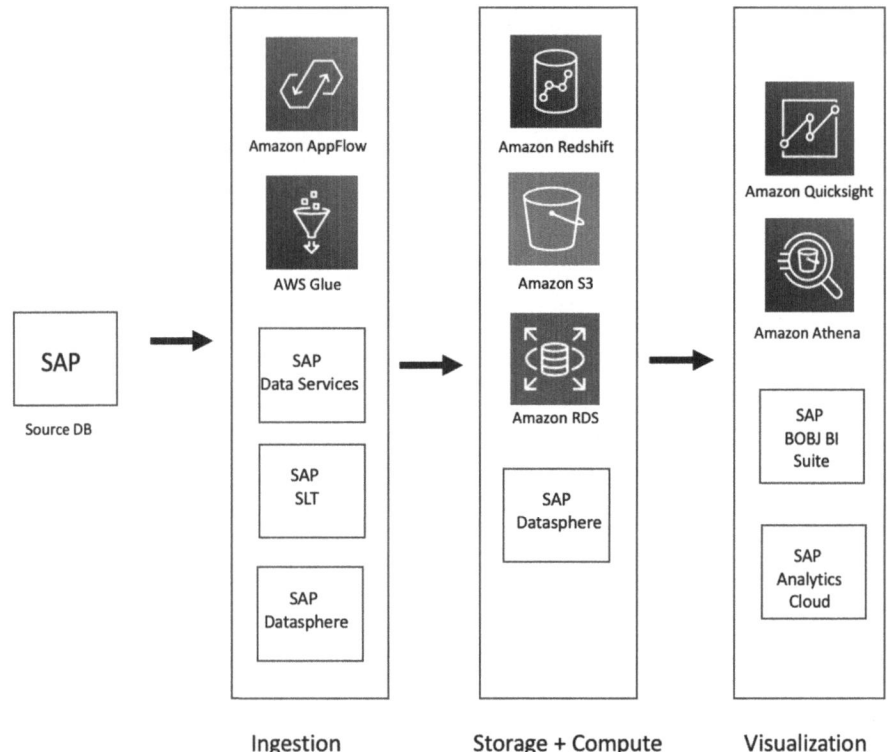

Figure 1-14. *AWS and SAP solutions for SAP data analytics*

Third-Party Data and Analytics Solutions

AWS and SAP solutions for SAP Analytics do not solve every use case for every customer. That's where third-party solutions play an important role. These solutions provide specialized analytics features, advanced data processing, and visualization options that may not be fully covered by SAP and AWS services alone. The updated diagram shown in Figure 1-15 now includes examples of third-party solutions and their capabilities; however, this does not represent an exhaustive list of available third-party solutions.

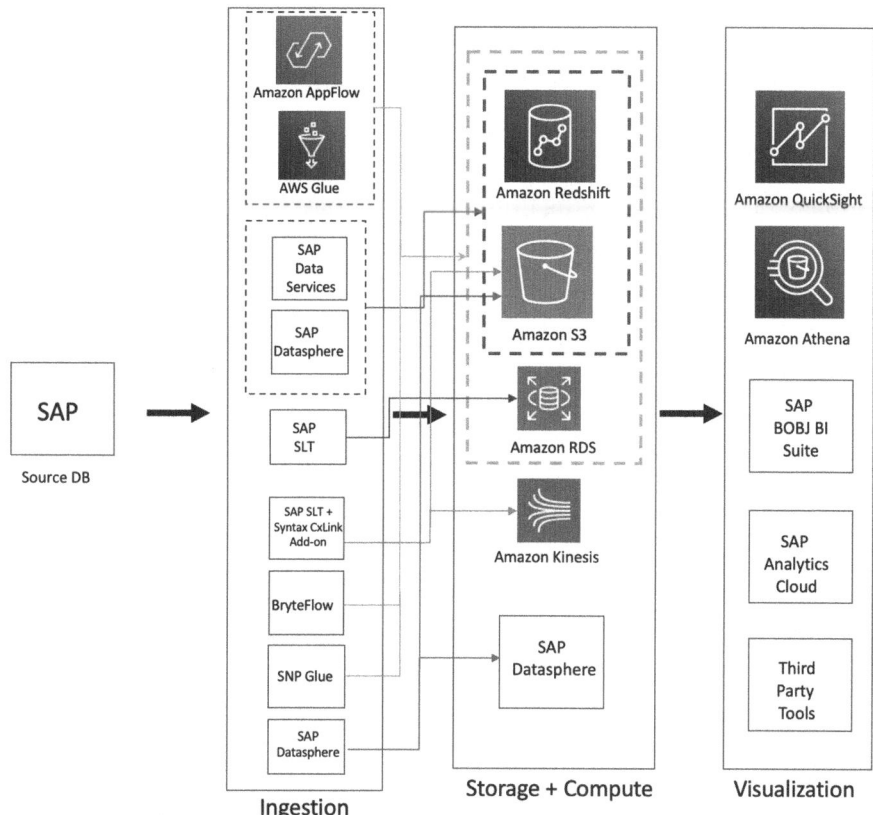

Figure 1-15. *Incorporating third-party data and analytics solutions*

We will discuss the topic of data and analytics in much more detail in Chapter 3.

Pillar 4: AI and ML

As shown in Figure 1-9 depicting the journey of data, once you have your data foundation, it's all about using this data for data-driven decision-making, and to use it for machine learning (ML). First of all, we need to establish working definitions for artificial intelligence, machine learning, generative AI, and other related terms.

Artificial intelligence (AI) is the ability of computer systems to complete tasks by mimicking human intelligence, by using machine learning, natural language processing, computer vision, and other technologies. It is the ability of computer systems to discover new information, infer something that is not explicitly stated, and reason by putting multiple pieces of information together.

Machine learning (ML) is a subset of AI that uses large amounts of data to build statistical models that learns from the data to make predictions or decisions. It is the ability to make predictions on new data based on learnings from data that is already available.

Deep learning (DL) is a subset of ML inspired by the human brain, which is a neural network with three or more layers. Because of the involvement of multiple layers, it is called "deep" learning. Generative AI is a subset of DL that is able to create new content like text and images by using foundation models (FMs).

Figure 1-16 represents the hierarchy of AI, ML, DL, and generative AI. Let's start by discussing DL generative AI first and progress up the stack, culminating in the exploration of the broader AI layer.

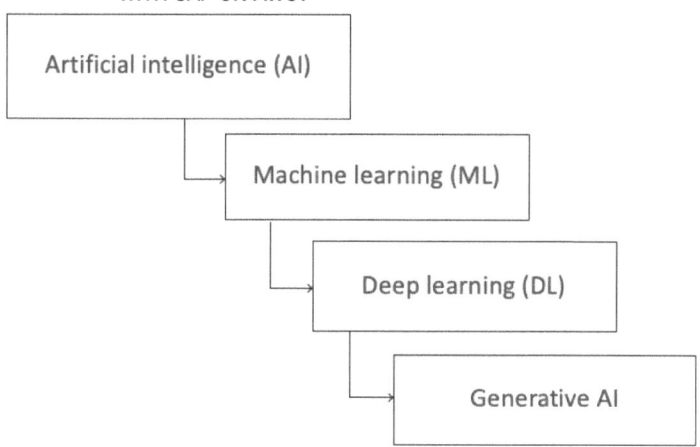

Figure 1-16. *Relationship between AI, ML, DL, and generative AI*

How Deep Learning Generative AI Differs from Traditional ML

As previously mentioned, generative AI is a subset of DL that creates new content by using foundation models. How are the FMs different from ML models? Let's say we need to create a system that differentiates between apples and oranges. If done using ML, we would need to feed the machine the features by which the two can be differentiated, like color, shape, and so forth. With DL, on the other hand, we skip the manual step of specifying the features, which are picked up automatically by the neural network. Foundation models are significantly larger and more complex than traditional ML models. FMs are trained on vast amounts of data and parameters, enabling them to learn a broad range of tasks and concepts. On the other hand, traditional ML models require labeled data for supervised learning. As shown on the left in Figure 1-17, traditional ML models are confined to narrow scope and applicability, so to get a bigger

task done, you would need multiple ML models. As shown on the right in
Figure 1-17, foundation models have the potential to impact a wide range
of applications and industries due to their versatility and scalability. So,
often one foundation model is enough.

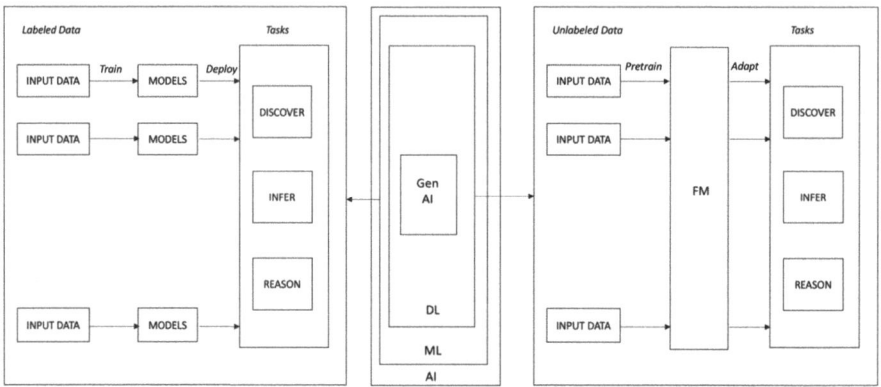

Figure 1-17. *Difference between traditional ML model (left) and*
foundation model (right)

As evident from Figure 1-18, traditional ML uses architecture that
requires the complex and time-consuming process of data preparation,
data labeling, and model training, and each model is meant for each
single specific task. Deep learning/generative AI, on the other hand, uses
unlabeled data, pretrained models, and one model for multiple tasks.

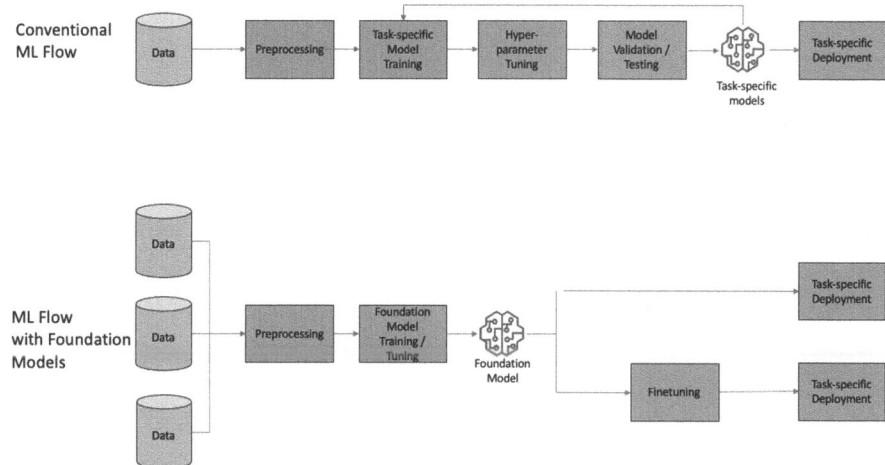

Figure 1-18. *Foundation models have transformed the ML workflow*

How AWS Can Help in Machine Learning

ML relies on data as its fuel; the quality of data directly influences
the quality of ML models. Consequently, enhancing data quality and
employing appropriate feature engineering techniques are vital for the
development of precise ML models. ML practitioners often undergo
meticulous cycles of training models, tuning parameters, deploying
models, and managing the intricacies of ML processes in the pursuit of
optimal models that effectively generalize on real-world data and deliver
the desired outcomes. Amazon SageMaker simplifies this entire process all
the way from data preparation to model deployment and management by
providing fully managed infrastructure, tools, and workflows. As depicted
in Figure 1-19, Amazon SageMaker is a suite of several capabilities, some of
which we are going to discuss now.

Amazon SageMaker Data Wrangler helps you with 300+ built-in
transformations to prepare data without writing code. Amazon SageMaker
Autopilot helps to accelerate the process of building and deploying ML
models. With Amazon SageMaker Canvas, you can explore and visualize

45

your data and quickly generate accurate ML predictions, helping you to gain advanced insights into your data before building your ML models. SageMaker Canvas is primarily used by business analysts. Data engineers and data scientists typically use Amazon SageMaker Studio, which is the IDE to help in the entire process. The Amazon SageMaker suite of tools makes collaboration between data engineers and business analysts seamless.

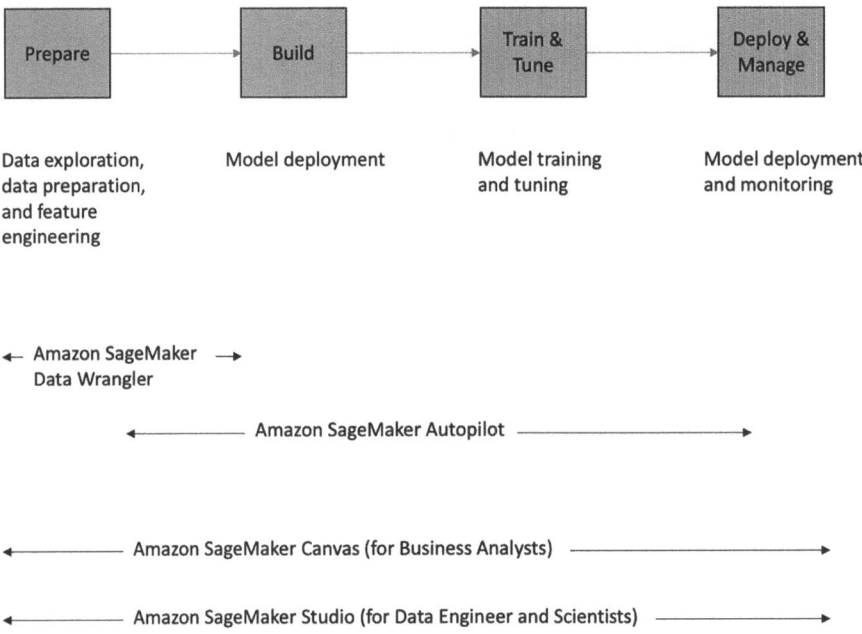

Figure 1-19. *Capabilities of Amazon SageMaker in the ML workflow*

Now, going back to our data lake diagram (see Figure 1-14) from the last section, SageMaker Canvas can work directly with Amazon S3 data only if the data is in.csv format (as of November 2023); otherwise, you need to use Amazon Athena. Usage of SageMaker Studio and SageMaker Canvas in this context looks like the diagram shown in Figure 1-20. If you are wondering how the access control for business analysts and data scientists is controlled, specifically on what they should and should not see, it's all via AWS LakeFormation.

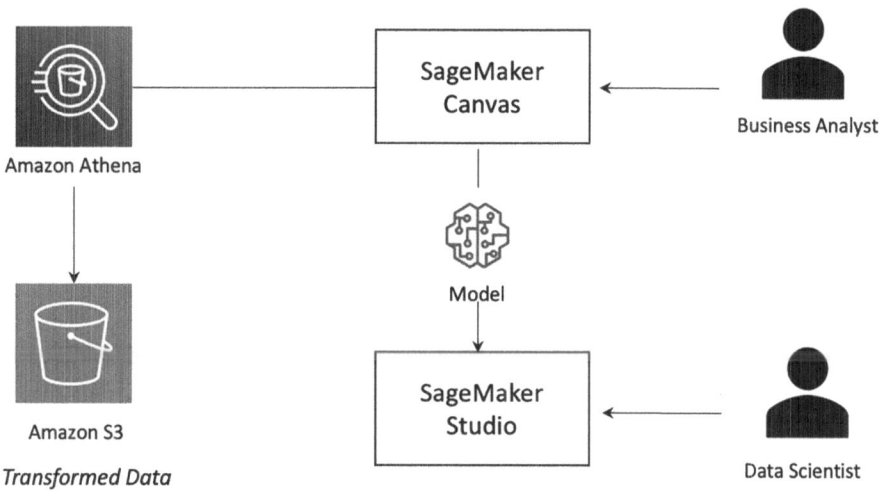

Figure 1-20. *Building/connecting ML model to AWS data lake using Amazon SageMaker*

Another capability within Amazon SageMaker is Amazon SageMaker JumpStart, which provides access to pretrained, open source models for a wide range of problem types to help enterprises get started with ML. Organizations can then incrementally train and tune these models before deployment.

Importantly, to use Amazon SageMaker, you need to have ML expertise within your organization. If you don't have in-house ML skills, you can use pretrained AI/ML models to start your journey directly from inference without the pain of going all the way through preparation, building, training, and deployment. Using AWS AI services (`https://aws.amazon.com/machine-learning/ai-services/`) is a good place to start in this scenario, together with solution accelerators called AWS AI/ML Solutions (`https://aws.amazon.com/solutions/ai-ml/`), which use AWS AI services under the hood. One such AWS AI service is Amazon Bedrock, discussed next.

How Amazon Bedrock Works

With the surge of generative AI in mid-2023, AWS released Amazon Bedrock, effectively elevating the capabilities offered by SageMaker JumpStart by providing access to the pretrained foundation models through APIs, eliminating the need of machine learning expertise in enterprises. Amazon Bedrock offers a managed service experience with models from providers such as AI21, Anthropic, Stability AI, and others.

The choice between Amazon SageMaker JumpStart and Amazon Bedrock depends on an organization's specific needs and priorities. Amazon Bedrock serves as a more straightforward option that eliminates the necessity of handling infrastructure, whereas SageMaker JumpStart offers hands-on access and customization capabilities, catering to experienced ML teams working with a wider array of models. With Amazon Bedrock, customers do not need to manage the model deployment or scalability; they just choose the model, customize it privately (if needed), and interact with it through APIs. Amazon SageMaker JumpStart offers increased flexibility for ML practitioners, enabling them to explore, fine-tune, and deploy both proprietary and open source frameworks that might not be accessible on Bedrock. Additionally, SageMaker JumpStart provides customers with greater control over the ML infrastructure when training and overseeing their models.

The choice between the two Amazon generative AI tools depends on whether prioritizing ease of use or model selection holds greater significance for a given AI project. Table 1-1 summarizes the high-level differences between the two.

Table 1-1. *Comparison of Amazon Bedrock and Amazon SageMaker*

	Amazon Bedrock	Amazon SageMaker
High Level Difference	Bedrock provides APIs for developers.	SageMaker is a platform for ML engineers.
Model Availability	Provides a curated selection of several top-tier generative models, including some proprietary ones.	Offers access to an extensive range of open-source models, numbering in the hundreds of thousands, covering both generative and non-generative types.
Inference Infrastructure	Fully handles and abstracts the underlying inference infrastructure, requiring no deep learning (DL) expertise.	Offers control over the selection of accelerated and non-accelerated instances, along with diverse deployment options.
Customization & Fine-Tuning	Enables customization/fine-tuning of certain models without needing DL expertise.	Facilitates comprehensive management of the entire ML workflow, including tasks like data labeling, model versioning, etc.
Governance & Guardrails	Provides prompt-based guardrails for content generation.	Supports advanced ML governance, including service control policies, integration with IAM/Organizations, bias detection, etc.
Pricing Model	Charged based on the number of input/output tokens and images generated.	Charged based on the type and duration of the compute instances used.

The diagram in Figure 1-21 illustrates the high-level architecture between Bedrock and an organization's application. The AWS account positioned on the left side represents the sole customer AWS account, while the remaining accounts belong to AWS—encompassing Bedrock service accounts and the Model account. Steps 1 through 4 demonstrate how the Bedrock API is invoked by the customer's application. When the application initiates an inference, the request is routed through the API endpoint in the Inference Service account. The inference service then retrieves the fine-tuned model to execute the inference process.

This application may be an ABAP program, in which case, access to the Bedrock API is facilitated through the AWS SDK for ABAP, or it may be a Fiori application created using SAP Business Application Studio and SAP Build Work Zone (formerly Fiori Launchpad). AWS PrivateLink ensures that customers establish private connectivity directly to Amazon Bedrock service accounts through the network. This eliminates the need for an Internet gateway in the customer's virtual private cloud (VPC), and there's no requirement to expose the customer's VPC to incoming traffic.

The diagram in Figure 1-21 also demonstrates the training process, via steps a to f. The training process is executed via the Bedrock Orchestration account/API endpoint. This account initiates the Fine-Tuning Job, extracting Training Data from the customer accounts and preparing the Fine-Tuned Model.

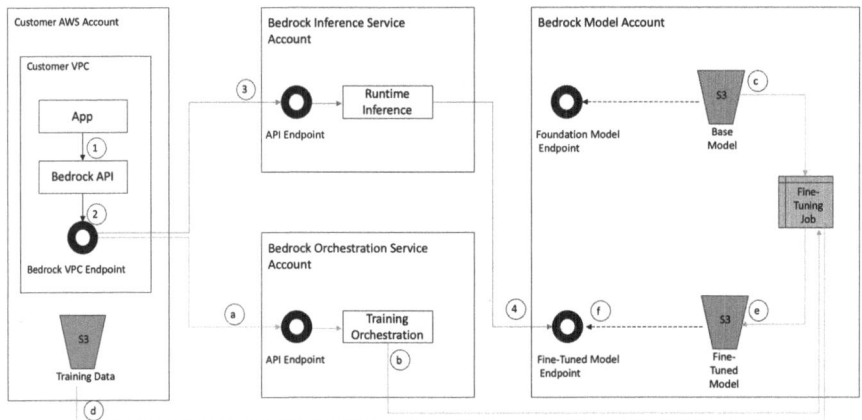

Figure 1-21. *Amazon Bedrock architecture diagram*

AI Services from AWS

Now that you are familiar with the ML tools and services for consuming data and using that data to train and fine-tune models, let's go a level up to the AI layer. Let's explore how AWS AI services can integrate into the SAP ecosystem. SAP customers have already made substantial investments of both money and time to establish end-to-end integrated systems tailored to their business needs. Integrating AWS AI services into these existing business processes has the potential to bring about significant transformations, enhancing efficiency, accuracy, and cost-effectiveness.

Consider, for instance, the scenario of quality inspection. Imagine using computer vision to ensure that all products pass through an automated gate check, thereby increasing customer satisfaction and reducing returns. Similarly, employing visual equipment inspection can minimize unscheduled plant downtime. Invista has successfully implemented these two use cases, showcasing the practical benefits of integrating AWS AI ML services into established SAP ecosystems. Invista has harnessed AWS AI/ML capabilities to conduct visual inspections and implement predictive maintenance, effectively reducing costly unplanned downtime. Furthermore, Invista has revolutionized parts restocking by developing a fully automated system using AWS services, seamlessly integrated into its SAP ERP system. This system enables Invista to predict, order, and store only the necessary parts, optimizing inventory management.

Various other SAP customers have automated their check and invoice processing business processes in SAP, resulting in the ability to process 80% of the checks/invoices received automatically with no human intervention. They have greater visibility and control over their billing process, improved cash flows from faster processing time, and cost savings from reduced labor. This is where AWS AI services (`https://aws.amazon.com/machine-learning/ai-services/`) such as Amazon Lookout for Vision, Amazon Lookout for Equipment, Amazon Rekognition, Amazon Textract, and Amazon Forecast come into play.

AI and ML Offerings from SAP

SAP started embedding predictive and ML capabilities in SAP HANA almost a decade back. Being an in-memory database, SAP HANA moved the data-intensive calculation of typical application logic into the database layer by using SQLScript, which is an extension of SQL. It lets developers define complex application logic inside database procedures. However, describing predictive analysis logic with procedures is difficult, so SAP came up with the Predictive Analysis Library (PAL), which consists of functions that can be called from within SQLScript procedures to perform analytic algorithms. The algorithms also include machine learning algorithms for training a model. The SAP Automated Predictive Library (APL), on the other hand,

bundles together all the algorithms of SAP Predictive Analytics, which
is a data mining solution that enables you to build predictive models to
discover hidden insights and relationships in your data. The SAP APL
installation package can be downloaded from the SAP Support Portal and
installed on the SAP HANA database instance using the SAP HANA resident
lifecycle management tool `hdblcm` or `hdblcmgui`.

Fast forwarding to 2024, SAP is approaching the AI revolution by
keeping in mind its core strength—business knowledge via business
processes. So, SAP is calling its AI offering SAP Business AI. Aligned with
its overall strategy, SAP is building its AI-related technical elements on the
Business Technology Platform (BTP), naming it the AI Foundation. The
AI Foundation serves as a one-stop shop for developers, allowing them to
create AI- and generative AI–powered extensions and applications on the
SAP BTP. Additionally, SAP is developing ready-to-use services known as
SAP AI Services or SAP Business AI Services. These services are built upon
the SAP AI Core, the fundamental AI runtime that facilitates the training,
deployment, and monitoring of ML models. You can find more info at
`https://community.sap.com/topics/artificial-intelligence`. One
such service, which is comparable to one of the use cases described earlier,
is the Document Information Extraction service. Customers can integrate
these services with SAP S/4HANA using Intelligent Scenario Lifecycle
Management (ISLM), which is a framework that helps integrate ML
capabilities into business applications. ISLM is delivered via the SAP Basis
Component starting with SAP S/4HANA 2020 FPS0 and is part of standard
licensing. It offers easy operations and management of ML solutions
called *ML scenarios* in SAP S/4HANA. ML scenarios are available in two
flavors: embedded or via BTP. The embedded ML scenarios are intended
for moderate ML capabilities like forecasting, classification, and so forth,
which are delivered via SAP HANA ML libraries. The more complex ML
scenarios like natural language processing and deep learning are offered
through BTP.

SAP is also powering some of its BTP services, such as SAP
Datasphere and SAP HANA Cloud, and SaaS solutions, such as SAP
SuccessFactors, with built-in AI capabilities. SAP Joule is SAP's generative
AI–based digital assistant that helps all SAP users with their everyday
tasks. It is embedded into SAP applications, delivering immediate
value without the need for additional customization/integration. SAP
Joule can, for example, help a developer to write a piece of code or
help a business user to create a job requisition on the user's behalf. It
is named after the famous English physicist James Joule, after whom
the unit of energy is named as well. Similar to how energy exists in
various forms, with Joule offering a standardized measure across
all forms, data is distributed across diverse business divisions and
platforms, and this new AI assistant Joule extracts intelligence from the
data across all platforms and provides unified information as needed.
All of these AI capabilities provided by SAP is shown in Figure 1-22.

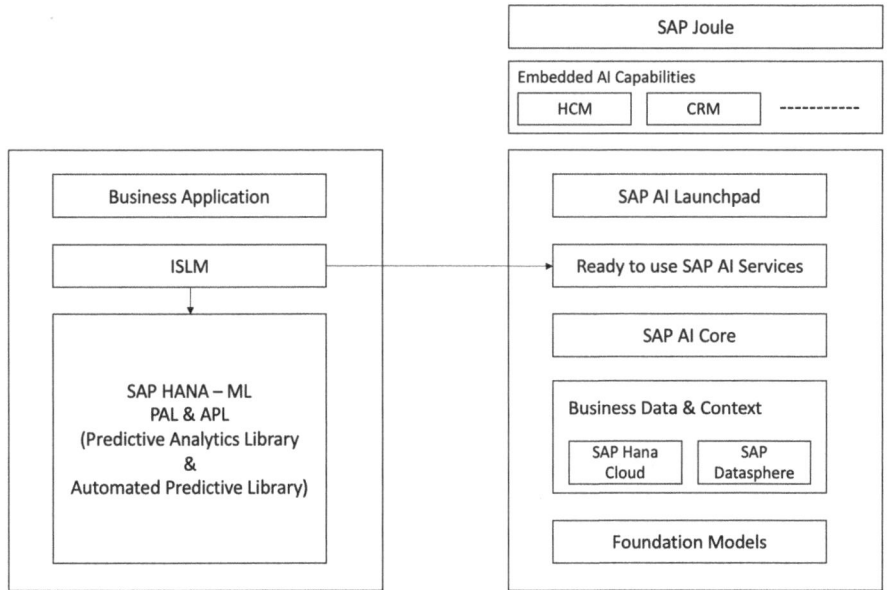

Figure 1-22. *AI capabilities provided by SAP*

We are going to discuss a lot of these AI/ML use cases in detail in Chapter 6.

Pillar 5: DevOps and SysOps

DevOps

DevOps is about bringing the Development team and Operations team together by implementing the appropriate cultural philosophies, practices, and tools that enable them to accelerate the delivery of applications and services, resulting in increased innovation and agility as well as improvement in security, quality, and cost optimization. Traditional ways of software development have proven slow and inefficient, and fail to support the teams' efforts to quickly deliver stable, high-quality applications. When we say "traditional ways," we mean a setup where infrastructure is provisioned manually, developers write the code, and then sys admins deploy the code at certain intervals, often weekly and sometimes even monthly.

There is not a universal standard of what perfect DevOps looks like; in many more ways than not, DevOps is a philosophy, a framework, and a methodology that organizations can use to modernize their end-to-end software delivery process. According to the Google Cloud/DORA report *Accelerate State of DevOps 2019* (`https://services.google.com/fh/files/misc/state-of-devops-2019.pdf`), every organization, irrespective of size and industry, can benefit from DevOps. Figure 1-23 illustrates the comparison between the traditional and DevOps approaches to working through visual representation.

Traditional way of working:

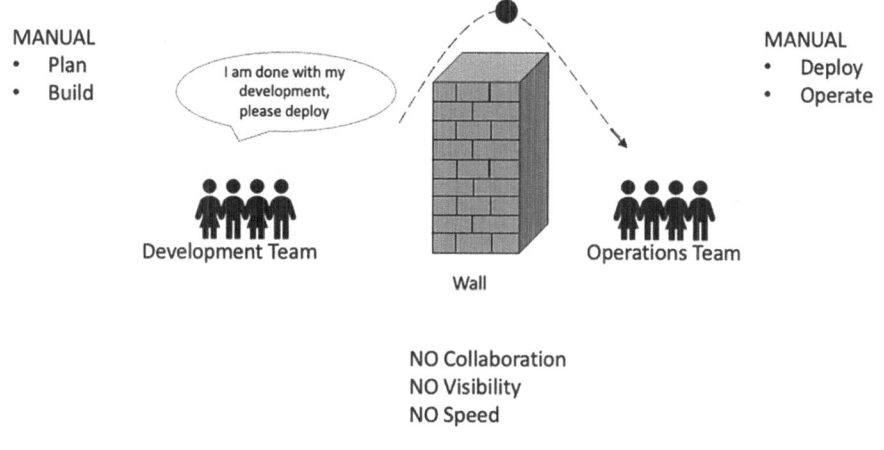

DevOps way of working:

Figure 1-23. *Traditional way vs. DevOps way of working*

As shown in Figure 1-24, DevOps is all about establishing a culture, driven by processes and enabled by tools. At the heart of the processes is automation, automating the *plan, build, deploy,* and *operate* phases so that organizations can deliver application changes and new applications faster and with more confidence. At the core of DevOps is CI/CD. *Continuous integration (CI)* involves the integration of various frequent changes from multiple developers into a single main line of code, while *continuous deployment (CD)* pertains to immediate deployment of these changes into production once they are successfully tested. Although both processes can be executed manually, the true value for organizations lies in automation achieved through a set of steps, also known as pipelines. DevOps is about moving things quickly from one end to the other with short feedback cycles; hence the word *continuous*. And to move fast, organizations need to adopt as much automation as possible, for which they need technology and tools.

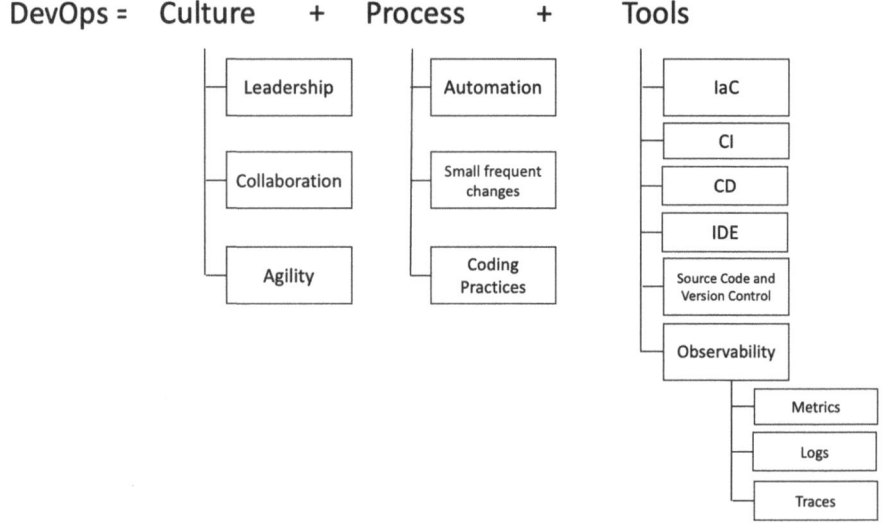

Figure 1-24. *DevOps overview*

DevOps for SAP

Let's first explore the end-to-end life cycle of SAP artifacts through a DevOps perspective, starting from provisioning infrastructure for SAP to daily operations of SAP. Figure 1-25 outlines the challenges at each phase and presents corresponding solutions and tools.

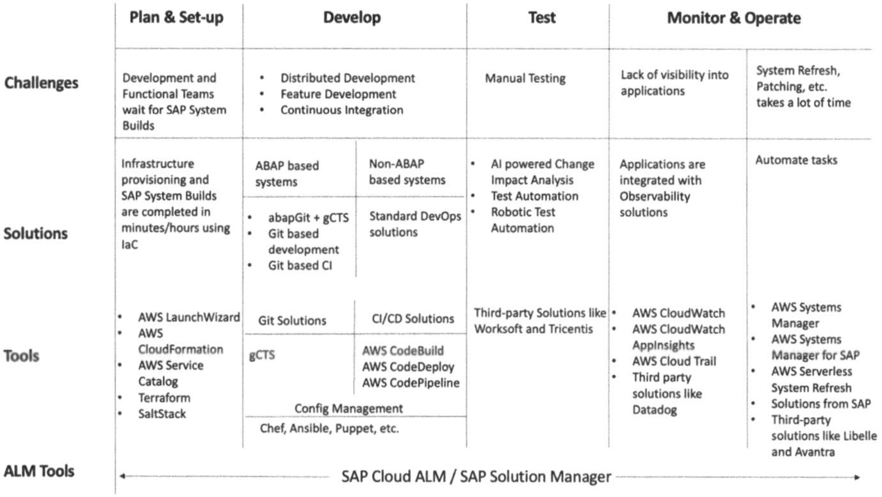

	Plan & Set-up	Develop		Test	Monitor & Operate	
Challenges	Development and Functional Teams wait for SAP System Builds	• Distributed Development • Feature Development • Continuous Integration		Manual Testing	Lack of visibility into applications	System Refresh, Patching, etc. takes a lot of time
Solutions	Infrastructure provisioning and SAP System Builds are completed in minutes/hours using IaC	ABAP based systems • abapGit + gCTS • Git based development • Git based CI	Non-ABAP based systems Standard DevOps solutions	• AI powered Change Impact Analysis • Test Automation • Robotic Test Automation	Applications are integrated with Observability solutions	Automate tasks
Tools	• AWS LaunchWizard • AWS CloudFormation • AWS Service Catalog • Terraform • SaltStack	Git Solutions gCTS Config Management Chef, Ansible, Puppet, etc.	CI/CD Solutions AWS CodeBuild AWS CodeDeploy AWS CodePipeline	Third-party Solutions like Worksoft and Tricentis	• AWS CloudWatch • AWS CloudWatch AppInsights • AWS Cloud Trail • Third party solutions like Datadog	• AWS Systems Manager • AWS Systems Manager for SAP • AWS Serverless System Refresh Solutions from SAP • Third-party solutions like Libelle and Avantra
ALM Tools	←		SAP Cloud ALM / SAP Solution Manager			→

Figure 1-25. *DevOps for SAP: challenges, solutions, and tools*

The first phase focuses on automating the build of SAP systems. With infrastructure as code (IaC), the entire build of an SAP system, which has traditionally been a manual process, can now be automated. When we say the "entire build of an SAP system," we mean not only defining the infrastructure for the database server and SAP application servers with storage, network, and compute but also the ability to install the SAP application, the database, and the cluster configuration for high availability.

Infrastructure as code (IaC) is a way of provisioning infrastructure, which includes servers, operating systems, storage, databases, and applications, via code rather than manually, with best practices for security, performance, and so on already baked in. So, what does the code contain? The code contains specifications of the infrastructure to deploy along with details needed to install databases and applications. It helps to manage AWS and third-party resources together in a template for future usage, making it easier to scale and to provision the same exact environment every time. Most importantly, IaC is a key driver for preventative governance, because you can enforce compliance effectively prior to or as a part of provisioning, which is very important specifically in regulated industries.

IaC extends beyond just provisioning; it also encompasses ongoing daily operations, including tasks such as patching and maintenance. SAP Basis teams, which used to do these manual activities in collaboration with Infrastructure teams entailing back-and-forth handovers during maintenance windows, have now evolved into SAP Cloud Platform teams. SAP Cloud Platform teams have automated these activities via IaC, which neither need requesting the business for an extended downtime window nor needs involvement of multiple teams, in a weekend.

As depicted in Figure 1-26, various tools are available for IaC, including AWS Cloud Formation, AWS Cloud Development Kit (CDK), AWS Serverless Application Model (SAM), AWS Launch Wizard for SAP, AWS Service Catalog, SSM Documents, Terraform, and others. One SAP on AWS customer decreased the time to build an SAP system from 2 weeks to 20 minutes by adopting an IaC approach. The customer integrated its AWS Service Catalog products with its existing IT service management (ITSM) solution, ServiceNow. Using self-service, the customer's users can now spin up new SAP environments via ServiceNow requests within minutes.

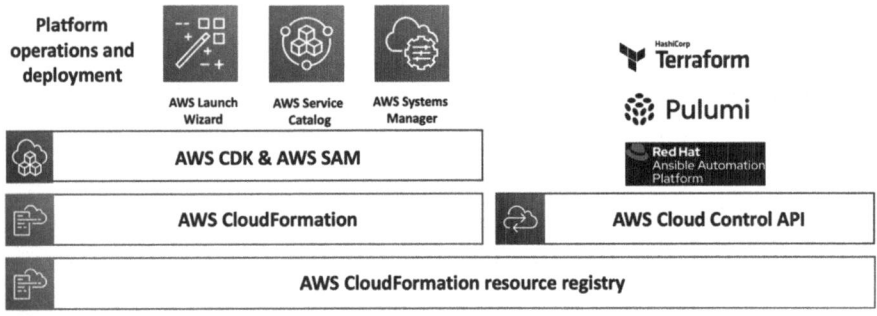

Figure 1-26. *AWS IaC portfolio for SAP*

Now that we have discussed provisioning and set-up aspects of SAP systems, lets move to testing. Testing in SAP is usually a prolonged and intricate process, primarily because even a minor change to a technical artifact results in a substantial overlap between various business processes within SAP. This necessitates testing across multiple business processes and obtaining sign-off from respective Business Process Owners.

Due to their expertise in SAP business processes, business users find themselves burdened with extensive testing responsibilities. However, this need not be the case, as testing can be automated. Third-party products from Tricentis, Worksoft, Basis Technologies, and others offer solutions for automating the testing process, reducing the involvement of Business Process Owners to cases where new test scenarios need to be recorded.

When dealing with changes impacting multiple business processes, it's essential to note that not all processes require testing. These third-party tools come equipped with AI-driven features that help identify and focus on the most at-risk processes, streamlining the testing effort rather than testing all impacted business processes indiscriminately.

Once the changes are tested and deployed, AWS makes continuous monitoring very easy. Although the terms *monitoring* and *observability* often are used interchangeably, monitoring is an activity that makes a system observable, alongside other activities like tracing and logging.

So, what is *observability*? In his engineering paper "On the General Theory of Control Systems," published in 1960, R.E. Kalman invented the term observability and defined it as follows: "A system is observable if its state can be inferred from measurements. If that system can be controlled and observed, it can be optimized." In modern terms, observability describes how well you can understand what is happening in a system, often by instrumenting it to collect metrics, logs, or traces. Hence, you'll often see monitoring, tracing, and logging described as the *three pillars of observability*.

Amazon CloudWatch provides complete visibility into your cloud
resources and applications. CloudWatch agents collect metrics and logs
from all your AWS resources and services that run on AWS (as well as on-
premises servers), which you can visualize with CloudWatch dashboards.
You can correlate logs and metrics side by side to troubleshoot and
set alerts with CloudWatch alarms, as well as automate response to
operational changes with CloudWatch Events, which has now evolved
to Amazon EventBridge. Amazon CloudWatch alarms allow you to set a
threshold on metrics and trigger an action. You can create high-resolution
alarms, set a percentile as the statistic, and either specify an action or
ignore as appropriate. Amazon CloudWatch composite alarms allow you
to combine multiple alarms and reduce alarm noise. Amazon CloudWatch
Application Insights (CWAI) elevates monitoring to a new level by
overseeing SAP NetWeaver applications, HANA, and Sybase databases.
It introduces automatic dashboards and insights that highlight potential
root causes, minimizing the need for SAP Basis administrators to dedicate
extensive time to troubleshooting. Amazon CWAI actively monitors the
health of AWS resources, the database (HANA and Sybase, as of December
2023), and SAP application resources, including high-availability clusters,
and offers the capability to automate corrective actions, such as initiating
the start of the database or SAP application as needed. This integrated
approach streamlines the monitoring and management processes,
enhancing the efficiency of SAP system administrators.

DevOps for ABAP-Based Systems

When we talk about the core ABAP-based SAP systems, SAP still remains a
monolithic application. Certain aspects, such as "object lock" in SAP ABAP
during development, deploying changes through transports, are distinctive
to SAP. Typical DevOps tools don't work for this monolithic architecture.
All these factors limit what we can establish as a process to implement
DevOps. The main impediment to realizing the fundamental principle

of DevOps, characterized by continuous integration/continuous delivery
(CI/CD) models—where code undergoes a continuous, seamless cycle
of writing, reviewing, testing, and releasing updates—are the constraints
posed by the concept of object lock in SAP ABAP during development.

Consequently, ABAP teams continue to collaborate within a single
monolithic code base, making distributed development impossible.
To overcome this obstacle, SAP introduced Git-enabled Change and
Transport System (gCTS) together with enabling developers to manage
their ABAP source code in Git repositories, which is used to manage
development artifacts irrespective of language and technology. This
allows developers to work in parallel on the ABAP code bases, collaborate,
monitor changes, and essentially manage their ABAP development in a
manner similar to non-SAP software development processes. gCTS was
first introduced in SAP S/4HANA 1909 but is available for both customizing
and workbench requests with SAP S/4HANA 2020. abapGit (`https://
abapgit.org/`) is typically the Git client used for ABAP development.

The process for ABAP developers remains the same during object
development, with a notable shift occurring when the transport request
is released, deviating from the classical ABAP change management
approach. Upon release of the transport request, a new commit is
generated. It is also possible to generate a new commit while releasing the
task. The pipeline monitors the Feature branch for any changes and, upon
detecting a new commit, initiates the deployment of the modification to
the QA/Test system and kicks off the unit test process. In the event of unit
test failures, gCTS can revert the changes, marking a significant change
from the classical ABAP change and release management process. As and
when the new development artifact is ready after successful tests, a merge
is executed to the Master branch, activating another pipeline to deploy
the object to the Production environment. The entirety of this process is
visually presented step by step in Figure 1-27.

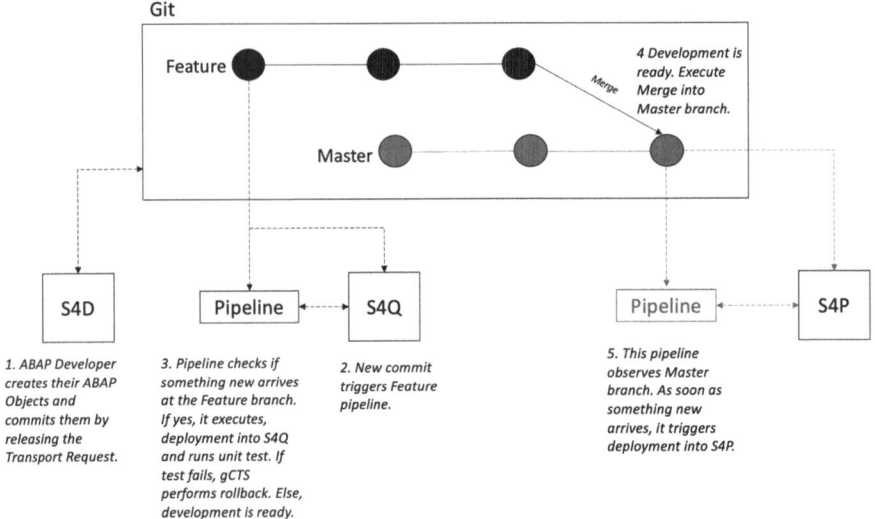

Figure 1-27. *SAP ABAP CI/CD pipeline using gCTS*

Now, let's take a look at DevOps for non-ABAP-based applications, which are primarily cloud native via BTP.

DevOps in BTP

SAP BTP facilitates the seamless implementation of DevOps for SAP applications running on the platform as part of your digital transformation. In your side-by-side extension scenarios, deploying DevOps is made straightforward. Figure 1-28 illustrates SAP's comprehensive approach to enable DevOps in BTP, showcasing various services at your disposal. As an initial step, setting up SAP CI, a pipeline as a service with no customer-maintained infrastructure, is recommended. Alternatively, experienced CI/CD users may opt for third-party solutions or open source offerings like Project Piper (https://www.project-piper.io/), which is primarily for advanced CI/CD customers wherein you get the building blocks for your pipeline but run it on your AWS infrastructure.

SAP BTP supports a diverse range of tools and programming languages that significantly enhance developers' flexibility and freedom in application creation, which can run on different environments (platform as a service) of your choice. Choices include the SAP Build portfolio, tailored for nonprofessional developers, which comprises SAP Build Apps, SAP Build Process Automation, and SAP Build Work Zone. Additionally, ABAP Development Tools is available, designed for ABAP, IDE for Kyma, and SAP Business Application Studio for Cloud Foundry. The SAP Business Application Studio, chosen for its rich offerings in enhancing developer productivity, is featured in Figure 1-28 as the development environment. It integrates seamlessly with SAP CI. Transporting changes can be achieved using SAP Cloud Transport Management, a standardized change management service that can synchronize on-premises changes. Alternatively, CD or CTS+ can be utilized, with integration options for orchestration tools such as SAP Cloud ALM, SAP Solution Manager, or ServiceNow.

For event subscription from the platform, the SAP Alert Notification service is available. Third-party solutions can be considered as well. From a monitoring and operation standpoint, SAP Automation Pilot service serves as a low-code, no-code automation service for DevOps actions and alert remediation. Alternatively, third-party solutions, shell scripts, and similar tools can be employed for operations automation.

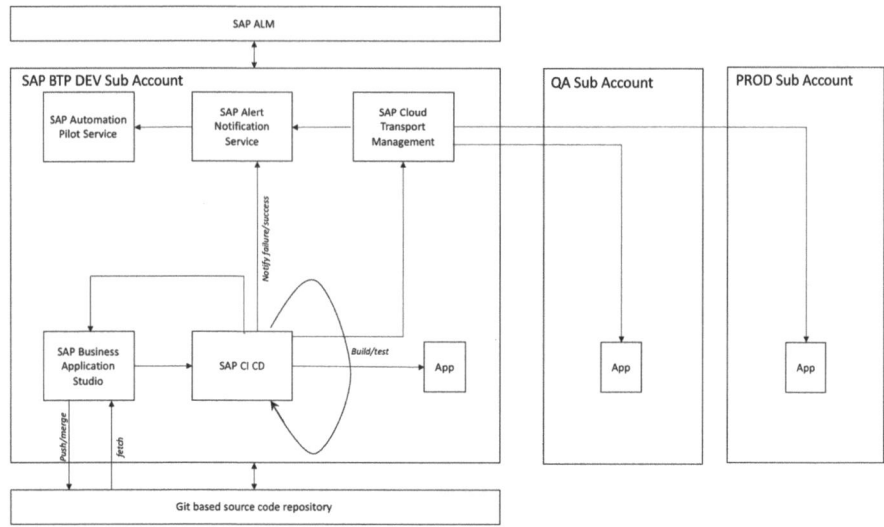

Figure 1-28. *DevOps for applications on BTP*

SysOps for SAP

While observability is one aspect of operational considerations, managing and operating your hybrid infrastructure securely at scale presents its own set of challenges. Questions arise about grouping on-prem and AWS resources by application (e.g., SAP), selectively viewing operational data for specific resource sets (e.g., all SAP production systems), and efficiently detecting and resolving operational issues. Additionally, the need to perform automated actions on groups of resources, like automating the OS patching and HANA database patching of all SAP sandbox systems simultaneously, is a consideration. AWS Systems Manager addresses all these concerns, providing a comprehensive solution for secure and scalable infrastructure operation and management. You define in an AWS Systems Manager document (SSM document) the actions that you would like to perform on the Amazon EC2 instances as a part of the automation. AWS provides a number of preconfigured SSM documents, in JSON or

YAML format, that organizations can use right out of the box. AWS Systems
Manager's Application Manager, a feature designed for DevOps engineers
to assist in the troubleshooting and resolution of issues related to AWS
resources within the specific context of a particular application, supports
SAP HANA databases, allowing you to discover and register your SAP
HANA systems and back up your SAP HANA databases using AWS Backup.

There is also an SAP-specific service of AWS Systems Manager called
AWS Systems Manager for SAP (SSM4SAP). Relatively still new as of this
writing, it is available to use with AWS APIs. AWS Systems Manager's
Application Manager operates using these APIs in the background.
It provides a seamless integration between AWS services and SAP
applications running on AWS and automation capabilities for managing
the SAP workloads on AWS.

From an operational perspective, some of the tasks that can be
automated are OS patching as well as SAP HANA DB patching, starting/
stopping SAP, autoscaling SAP application servers, and SAP system
refresh. All these tasks are automated by building solutions using native
AWS services like AWS Step Functions, AWS Lambda, and so on. These
solutions are helping customers save a lot of time and cost. For example,
the automated SAP system refresh solution is helping customers reduce
the SAP refresh process from two to three weeks to less than one day
with a downtime of fewer than 30 mins, essentially resulting in near-zero
downtime for the SAP QA system. At Zalora, the time required for SAP
system refresh has been significantly reduced, decreasing from five days
to under two days (`https://aws.amazon.com/solutions/case-studies/
zalora-sap/`). This improvement is attributed to automated mechanisms,
resulting in enhanced refresh quality. As a result, testing cycles can now
iterate within a shorter time window. For another SAP on AWS customer,
automated OS patching completed in less than one hour for 150+ SAP EC2
instances, compared to six to eight hours when it was done manually.

We dive deep into this topic of SysOps in Chapter 7.

Modern Disaster Recovery

In the modern era, the concept of disaster recovery (DR) has evolved beyond natural disasters to encompass a range of cyber threats such as ransomware, malware attacks, data breaches, hardware failures, and accidental deletions. As a result, modern disaster recovery plans (DRPs) are now predominantly cloud-based, offering near-instant restoration of business data, applications, and connectivity. This shift toward cloud-based solutions, like AWS Elastic Disaster Recovery (AWS DRS), AWS Backup Vault Lock, and Amazon S3 Object Lock for ransomware protection, is driven by several factors, including the cost-effectiveness of cloud infrastructure, automation capabilities, better security posture, faster setup and recovery times, and geographical advantages.

The rationale behind this transition to DR on AWS lies in the recognition that downtime is intolerable for both customers and businesses, as it can lead to significant financial losses and reputational damage. AWS DRS safeguards customers' SAP applications against various IT disruptions, while simultaneously reducing the total cost of ownership (TCO) associated with disaster recovery. With AWS DRS, AWS is the disaster recovery site, and the source environment may or may not be on AWS—it can be on premises or on another cloud provider.

As shown in Table 1-2, AWS DRS offers several advantages over traditional DR solutions. For instance, it allows organizations to easily scale their DR infrastructure by adding or removing small, low-cost replicating servers as needed, thereby avoiding large upfront investments in new hardware. Additionally, AWS DRS provides easy, repeatable, and nondisruptive testing capabilities along with easy capabilities to fail back to the primary site if an actual DR is invoked. Most importantly, with AWS DRS, customers can achieve a near-zero recovery point objective (RPO) and a recovery time objective (RTO) of minutes. Except in certain scenarios where native asynchronous database (DB) replication is deemed

more appropriate, AWS DRS can be utilized for both the SAP database
and SAP application layers. Several SAP on AWS customers, using a
combination of native DB replication and AWS DRS, have run several DR
drills wherein they are able to recover multiple SAP production systems on
the AWS DR region with less than 15 minutes of data loss and an RTO of
one hour or less.

Table 1-2. *Comparison of Traditional On-Prem DR and Modern
DR on AWS*

Traditional On-Prem DR	Modern DR on AWS
Massive upfront and ongoing hardware cost.	AWS DRS: Small, low-cost replication servers are installed on AWS DR region. Native DB replication: Smaller servers can be utilized for ongoing database replication, with the ability to swiftly switch to larger instances during DR invocation for increased workload demands.
Management and infrastructure overhead.	Lower IT overhead.
Testing is complicated and often needs a lot of effort.	Easy and repeatable DR drills are possible.
Failback is not easy.	Reverse replication and failback to primary site is seamless while using AWS DRS.
Higher RPO and RTO.	AWS DRS: Near-zero RPO and RTO in minutes. Native DB replication: RPO of ~15 mins with both primary and DR regions on the same continent. RTO can be significantly reduced with several automations.

SAP Cloud ALM

SAP Cloud ALM is SAP's next-generation cloud-native ALM solution to help organizations operate hybrid SAP landscapes; it is offered as a part of SAP Enterprise Support. It seems likely that at some point in the future SAP will replace SAP Solution Manger with SAP Cloud ALM.

As of this writing, only specific SAP Cloud ALM monitoring and operational capabilities are currently supported. If you are presently utilizing SAP Solution Manager on AWS, particularly for use cases like SAP Change Request Management (ChaRM), you may continue to use SAP Solution Manager. For SAP on AWS customers leveraging SAP Focused Run for advanced operational scenarios, it's important to note that SAP Cloud ALM is not a replacement but can serve as a complementary solution.

SAP Cloud ALM is positioned to empower customers by offering a comprehensive understanding of the health of SAP business solutions. As depicted in Figure 1-29, it provides full-stack monitoring and alerting that span business processes, integration, users, jobs and tasks, applications, and the overall health of services and systems. It introduces intelligent event processing as a central routing infrastructure, enabling effective event correlation and intelligent alerting, along with the capability to conduct root cause analysis on different levels through embedded dashboards. It also enables operations automation by seamlessly integrating with existing automation solutions.

In the case of customers utilizing SAP Solution Manager solely for monitoring and operational purposes or those seeking monitoring capabilities for SAP BTP services, planning a gradual transition from SAP Solution Manager to SAP Cloud ALM may be beneficial. SAP provides a tool called SAP Readiness Check for SAP Cloud ALM. It's a self-service tool to analyze an existing SAP Solution Manager system in preparation for a transition to SAP Cloud ALM. It provides visibility to the application lifecycle management capabilities used in an SAP Solution Manager

system along with the equivalent capabilities in SAP Cloud ALM. Based on
the SAP Readiness Check results, the transition can be started immediately
by setting up a "greenfield" SAP Cloud ALM and activating the specific
use cases of interest. Some of the monitoring capabilities available with
SAP Cloud ALM are Integration and Exception Monitoring, Real User
Monitoring, Job and Automation Monitoring and Health Monitoring for
applications built with SAP BTP Neo, SAP BTP ABAP and SAP BTP Cloud
Foundry. Please refer to latest documentation from SAP for more specific
information.

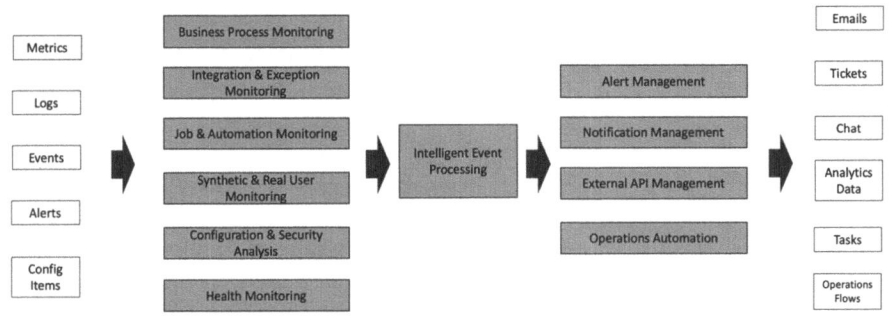

Figure 1-29. *SAP Cloud ALM operations overview*

We dive deep into other facets of DevOps and SysOps in Chapter 7.

AWS Solutions Library

Today, organizations are actively seeking reliable solutions and
architectural guidance to swiftly address their business challenges.
Whether customers opt for prepackaged deployments or tailored
architectures, the AWS Solutions Library (https://aws.amazon.com/
solutions/) offers a comprehensive collection of solutions developed by
AWS and AWS Partners to cater to a wide array of industry and technology
use cases.

The AWS Solutions Library is a great place to start evaluating the different options you might have. Figure 1-30 shows by way of example a limited set of the many options that you might have. You can customize and tailor the reference architectures to your needs. That's why AWS services are often compared to LEGO bricks. Your creativity sets the limit on what you can construct with LEGO bricks or AWS services and the methods you employ. However, similar to building with LEGO bricks, constructing a solid structure is crucial; otherwise, your creation may collapse.

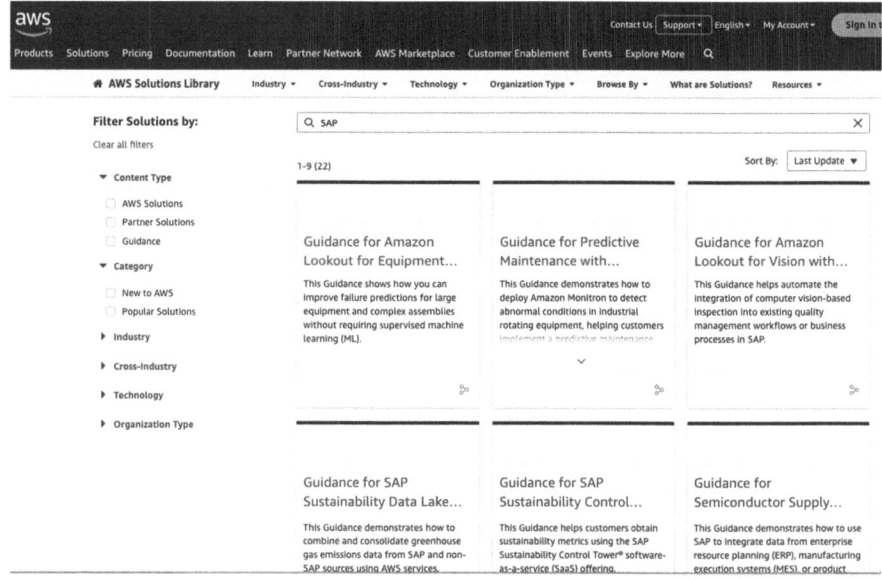

Figure 1-30. *AWS Solutions Library filtered by SAP use cases*

Summary

In this chapter, you have learned what it means to innovate beyond just hosting your SAP workloads on AWS. You learned about the importance of changing your mindset and culture as you transition from on-prem to

AWS. You also learned how you can use some of the proven frameworks provided by AWS and SAP to set up organizational practices you can build to foster a culture of innovation, not just for today but on an ongoing basis.

This chapter introduced to you to the five Pillars of SAP on AWS Innovation, which is the foundation of this book and on which subsequent chapters are based. You were introduced to the AWS services and offerings from SAP that are relevant to each pillar and that will be used throughout the book. The section covering the first pillar, Apps and APIs, focused on why an API-first approach is becoming more and more important. Through the exploration of the second pillar, IoT, you gained insights into how industrial customers leverage AWS IoT and ML-based services to effectively utilize their operational and machine data to achieve a spectrum of business outcomes. (In Chapter 5, we'll dive deep into this topic with a focus on integration of these services with SAP.) In the section covering the third pillar, Data and Analytics, you were introduced to a multitude of AWS services designed to assist you in constructing a data lake on AWS in a systematic manner, guiding you through each step of the process. In the coverage of the fourth pillar, AI and ML, you gained a foundational understanding of the distinct concepts (but often used interchangeably) within the evolving landscape of artificial intelligence (AI) and machine learning (ML). This coverage included delineating the differences between various terms and their applications. Additionally, a variety of AWS AI and ML services were unveiled to you along with their functionalities and operational mechanisms. (This set a good foundation for you to explore more about this hot topic in Chapter 6.) The discussion of the last pillar, DevOps and SysOps for SAP, emphasized the advantages of the DevOps way of work and what it truly means. It shed light on the challenges from an SAP perspective and the solution and tools you have at your disposal to overcome those. From a SysOps perspective, you learned about the AWS services and automation options available for modernizing activities like patching and disaster recovery. (Chapter 7 will focus purely on this aspect.)

In Chapter 2, you will embark on a journey into the core principle of modern architecture. It emphasizes the advantages of breaking down monolithic SAP extensions into microservices, allowing for independent scaling and faster development cycles.

CHAPTER 2

Fundamentals of Modern Architecture

Modern architecture, powered by APIs, microservices, serverless, and event-driven technologies, transforms complexity into simplicity.

In the rapidly evolving landscape of enterprise technology, the intersection of enterprise resource planning (ERP) applications such as SAP and solutions offered by cloud providers like Amazon Web Services (AWS) presents a unique opportunity for enterprises to go beyond traditional boundaries and embrace the full spectrum of modernization. Beyond merely hosting and managing SAP systems, enterprises now have the power to redefine their business processes, innovate at scale, and chart new pathways to success.

At the core of this transformative journey lies the recognition that AWS is not just a platform for infrastructure but a catalyst for architectural evolution. Enterprises running SAP can harness the vast array of AWS services to modernize and extend their business processes, unlocking unprecedented levels of agility, scalability, and innovation.

Enterprises no longer are confined by the limitations of monolithic architectures. Instead, they can leverage modern architectural principles such as microservices, event-driven integrations, and an API-first approach to break down silos, build extensions outside the ERP core, and orchestrate complex business processes with ease.

© Bidwan Baruah, Krishnakumar Ramadoss and Abarajith Vivekanandha 2024
B. Baruah et al., *Evolve from Infrastructure to Innovation with SAP on AWS*,
https://doi.org/10.1007/979-8-8688-0890-6_2

By embracing microservices, enterprises can decompose monolithic SAP application extensions into modular, independently deployable units, enabling rapid development cycles and seamless scalability. With an API-first approach, organizations can expose SAP functionalities as reusable services, facilitating integration with external systems and empowering developers to build innovative applications at scale.

Event-driven integrations provide the foundation for decoupling SAP systems from external dependencies, enabling real-time data processing, and ensuring seamless communication across distributed architectures. This paradigm shift empowers enterprises to adapt to changing business requirements, respond to market dynamics, and drive continuous innovation.

In this chapter, we delve into the fundamentals of modern architecture and uncover the transformative potential of SAP applications hosted on AWS under the following topics:

- APIs and microservices

- Software architecture patterns: monolith, service-oriented architecture (SOA), microservices, and serverless

- Integration patterns:

  - Synchronous

  - Asynchronous point-to-point model: queue, router, and message bus

- Event-driven architecture

APIs and Microservices

Twenty-five years ago, it was totally fine for enterprises doing SAP ERP implementations to use one big monolithic box and build everything on top of it. Then came the web revolution, and enterprises started building websites, which they had to integrate with SAP. Then came the mobile revolution, and now enterprises have started to build mobile apps. Today, enterprise stakeholders expect enterprise apps to offer the same flexibility, features, and rich experience they get in consumer apps. In response, enterprises are investing a lot more in voice-enabled applications and conversational applications.

Enterprises no longer are seeking to customize the core application, as they are looking at their journey toward SAP ERP migrations such as S/4HANA. They want to keep the core intact to lower the upgrade costs, and they are now seeking to build business extensions and innovation capabilities around SAP applications using cloud-native solutions.

Why APIs Are Important from a Customer Angle

The key stakeholders for an enterprise include its customers, partners, and employees. They want to be able to access their business processes and applications seamlessly across multiple devices. According to conservative estimates, by 2027, 40% of web transactions for consumers will happen through something that doesn't have a screen, like voice-controlled apps and chatbots. Customers are growing accustomed to these conversational, screen-free experiences. Now, imagine your own customers start demanding that same level of interaction with your enterprise systems, like your SAP application. Would your existing software be ready to deliver that? How could you adapt to meet these evolving needs?

That's where APIs come in, which enable you to expose those critical business functionalities safely and securely as consumable APIs and start using them through all the different channels depicted in Figure 2-1.

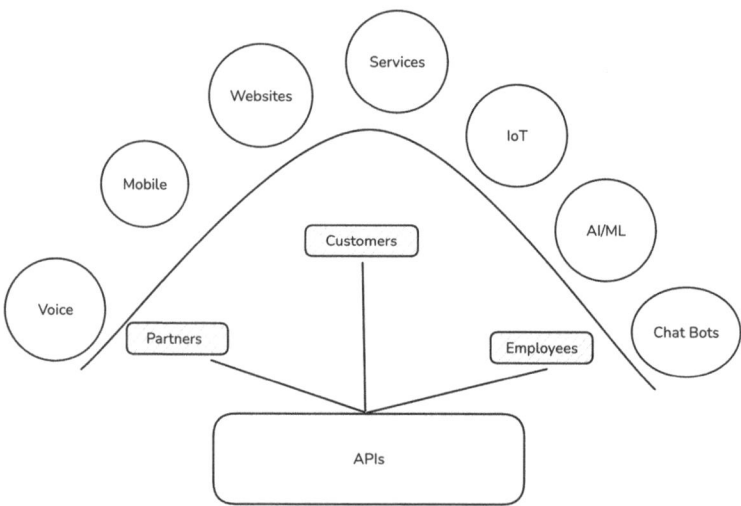

Figure 2-1. *APIs for a connected experience*

APIs are critical for this connected experience. APIs provide an agile, flexible, secure, and scalable layer for integrating various business applications, as well as an easy-to-consume service for all of your user interface and integration needs.

Evolution of SAP APIs and Integration

SAP ERP systems traditionally were designed as monolithic applications with tightly coupled components and a limited ability to expose functionality externally. As the need for integration and connectivity grew, SAP introduced various technologies and approaches to expose data and business processes through APIs. This evolution is depicted in Figure 2-2 and summarized in the following list:

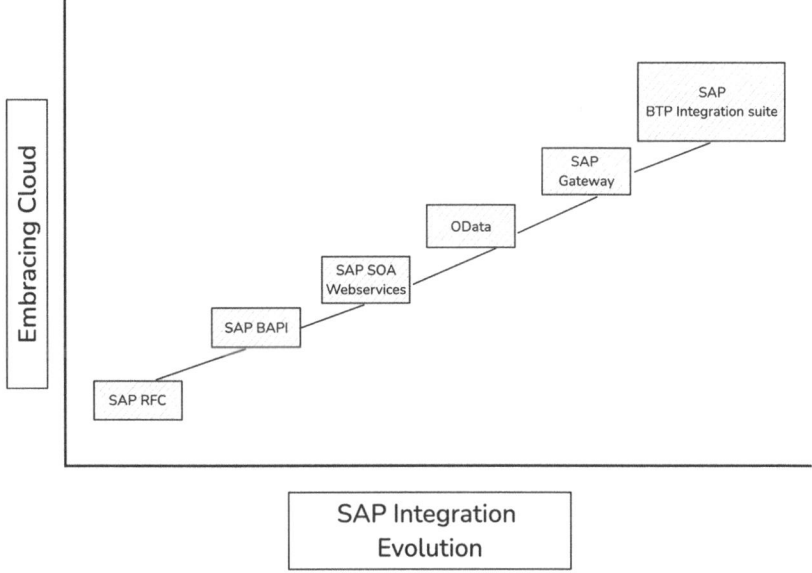

Figure 2-2. *Evolution of integration in the world of SAP*

- **SAP Remote Function Call (RFC):** One of the earliest methods for integrating with SAP systems was through RFC, which is an SAP-proprietary protocol that allows remote procedure calls to execute ABAP functions on the SAP system.

- **SAP Business Application Programming Interface (BAPI):** SAP introduced BAPI as a standardized way to access and manipulate business data and processes in SAP systems. BAPI provide a higher level of abstraction compared to RFC.

- **SAP Enterprise Services (ES):** With the introduction of service-oriented architecture (SOA), SAP developed Enterprise Services as a way to expose business objects and processes as web services, using standards like Simple Object Access Protocol (SOAP) and Extensible Markup Language (XML).

- **OData services:** SAP embraced the OData protocol, an open standard for building RESTful APIs, to provide a more modern and lightweight approach to exposing data and services from SAP systems.

- **SAP Gateway:** Introduced with SAP NetWeaver 7.0, SAP Gateway offers a unified platform for exposing and consuming APIs, including OData services, from SAP systems.

- **SAP Integration Suite:** The SAP Business Technology Platform (BTP) is a comprehensive platform-as-a-service (PaaS) offering. It provides a unified and integrated environment for enterprises to build, extend, and integrate their applications, including seamless integration with SAP systems. As SAP shifted toward a cloud-first strategy, the SAP Integration Suite became the primary platform on the BTP for developing, deploying, and managing APIs and integrations with SAP and non-SAP systems. Today, SAP's API strategy revolves around the Integration Suite, which includes SAP API Management, offering tools and services for building, securing, and managing APIs, as well as integrating with various applications and services on the BTP.

Protocols in the Context of SAP APIs

APIs can use various protocols for communication between the client and the server. Some commonly used protocols in the context of SAP APIs include

- **HTTP/HTTPS**: The Hypertext Transfer Protocol (HTTP) and its secure variant (HTTPS) are widely used for RESTful APIs, including SAP OData APIs.

- **SOAP**: Simple Object Access Protocol is an XML-based protocol for exchanging structured information in web services environments. It was used extensively in earlier versions of SAP Enterprise Services.

- **SAP RFC**: As previously introduced, Remote Function Call is a proprietary protocol used for communication between SAP systems and between SAP systems and external applications.

- **IDoc**: Intermediate Document is an SAP-proprietary protocol used for asynchronous data exchange between SAP systems and external systems. IDoc is often utilized in conjunction with the SAP Application Link Enabling (ALE) technology, which facilitates the integration and communication between distributed SAP systems and other, external applications. ALE ensures that IDocs are efficiently exchanged across different systems, enabling seamless and reliable data integration within complex IT landscapes

- **JCO**: Java Connector is a protocol used for Java-based applications to communicate with SAP systems using RFC or IDoc protocols. JCO is not a stand-alone protocol itself but rather a Java implementation of the previously described RFC and IDoc protocols. JCO allows Java applications to leverage these protocols to interact with SAP systems, such as by executing ABAP functions, reading and writing data, and handling asynchronous communication (in the case of IDocs).

The choice of protocol depends on a variety of factors, including the integration scenario's specific requirements, the version of the SAP system, performance considerations, and the programming language or platform used for developing the client application.

Statefulness vs. Statelessness

In the context of APIs, a *stateful API* maintains session information or state across multiple requests. This means that each request is aware of the context and data from previous requests within the same session or client interaction. Stateful APIs are commonly used in scenarios where the server needs to keep track of client-specific data or session state throughout the communication. Examples of stateful APIs in the SAP ecosystem include

- **SAP RFC**: RFC connections preserve the session state between the client and the SAP system, enabling the execution of multiple function calls in the same context.

- **Session-based web services**: Some SOAP-based web services in SAP may maintain session state through mechanisms like HTTP session or server-side session management.

- **Stateful OData services**: OData services generally are designed to be stateless, but there are scenarios where maintaining state across requests can be beneficial, such as for long-running operations or batch processing. Stateful APIs can provide benefits in terms of performance, security, and maintaining context across multiple requests. However, they also introduce additional complexity in managing session state, handling concurrent requests, and ensuring proper cleanup of resources.

A *stateless API* is designed in such a way that each request from the client to the server is treated as an independent transaction, with no information or context being stored on the server side between requests.

In a stateless API architecture, each request from the client must contain all the necessary information for the server to process the request without relying on any stored session data or server-side state. This approach is commonly used in RESTful APIs and is particularly well-suited for scalable and distributed systems.

Examples of stateless APIs in the SAP ecosystem include:

- **OData services**: SAP's OData services, which follow the RESTful architectural style, are designed to be stateless. Each request is self-contained and independent of any previous requests.

- **REST APIs**: Any REST APIs developed using SAP Business Technology Platform Integration Suite or other frameworks are typically stateless, adhering to the principles of REST architecture.

- **Microservices**: As SAP embraces a more cloud-native and microservices-based architecture, many of the APIs exposed by these microservices are designed to be stateless for better scalability and resilience. Stateless APIs offer several advantages, including improved scalability, easier load balancing, and better fault tolerance. However, they may require additional effort to handle scenarios where maintaining state or context across multiple requests is necessary, such as in long-running transactions or multistep processes.

Can SAP ERP Be the API Layer?

If you are familiar with SAP or have experience with SAP implementations, you likely know that SAP enables Create, Read, Update, and Delete (CRUD) operations on a variety of business objects, such as sales orders, purchase orders, materials, business partners, and so on through a straightforward API layer and an application layer system known as SAP Gateway, as illustrated in Figure 2-3.

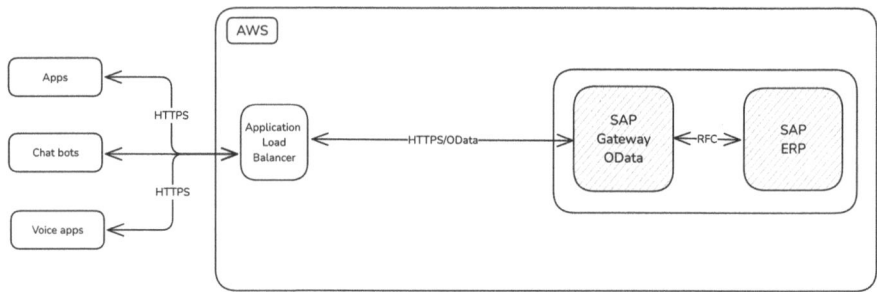

Figure 2-3. *SAP ERP as the API layer*

As you can see, SAP ERP can be your API layer. Just add an application load balancer that is responsible for balancing the load and distributing it to the target application, like SAP Gateway, and start consuming the APIs from other applications. But using SAP ERP as your API layer has three inherent challenges: it is limited to the SAP ecosystem; it is not serverless, so it doesn't scale well; and it is a potential single point of failure (SPOF).

A different approach (in our opinion, a better approach) is to continue using SAP Gateway to expose this API, as it offers benefits such as standardized API development and management within the SAP ecosystem, support for exposing SAP business objects as OData APIs, and its role as a hub for accessing other SAP ERP applications. Then utilize Amazon API Gateway as the front layer to provide features like API versioning, throttling, caching, and security. The diagram in Figure 2-4 illustrates this process. Developers can create, publish, maintain, monitor, and secure APIs at any scale with Amazon API Gateway, a fully managed service from AWS. It acts as a front door for applications to access data, business logic, or functionality from back-end services, such as SAP OData endpoints, or from AWS services such as Amazon DynamoDB and Amazon Kinesis.

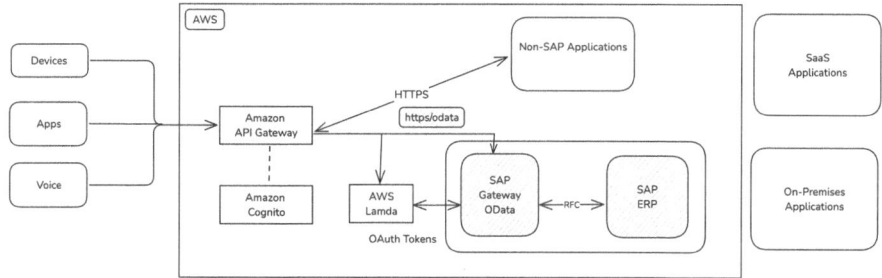

Figure 2-4. *Amazon API Gateway as the API layer*

With Amazon API Gateway, you get all the caching capabilities of globally available artifacts that are available in edge locations. You can meter the APIs and clearly segregate your APIs based on who can access what. You can exchange Security Assertion Markup Language (SAML) tokens with a back-end application. Amazon Cognito is a service that enables you to manage user authentication, authorization, and identity in your web and mobile apps. It allows users to securely sign in directly or through third-party identity providers, and provides seamless access to AWS services. If you set up Cognito in AWS and permit users to authenticate through it, they can access all authorized AWS services. However, with the use of a custom authorizer in API Gateway, the application can leverage the user's authentication in Cognito to authenticate the user in SAP as well. That's very important, especially with indirect licensing requirements. You don't want to use a service user to make all those updates, because it's a compliance issue, too. By using Amazon API Gateway, you can access non-SAP applications and consume them as APIs, in addition to your on-premises and SaaS applications.

With this architecture in place, you can securely expose your SAP business objects and functionality as APIs for your mobile app or web applications built outside of your SAP ecosystem. You can also expose SAP functionality with voice apps and chat bots powered by purpose-built services like Amazon Lex. All of these things become easy because each of them is just an API call away in most use cases.

Enterprises are also seeking to use machine learning to derive value from their mission-critical data to solve business problems and generate new revenue sources. For enterprises that have implemented SAP as their ERP solution, their teams are really deep in building and extending SAP applications, and they are consistently seeking ways to take SAP data and processes to the ML stack to derive value from them. By exposing SAP data and processes as APIs through Amazon API Gateway, data scientists and developers can easily integrate with AWS's ML services and capabilities, which are often just an API call away.

Can SAP API Management Be the API Layer?

While the strategy presented in the previous section leverages AWS services, enterprises with a strong commitment to the SAP ecosystem may consider SAP API Management as an alternative. It's important to note that SAP API Management is part of the SAP Integration Suite, which is available as a service on the SAP Business Technology Platform (BTP). SAP BTP itself is powered by AWS and utilizes many AWS services under the hood. SAP API Management provides a dedicated API management solution tailored for SAP environments, offering tight integration with other SAP products and services as illustrated in Figure 2-5 below:

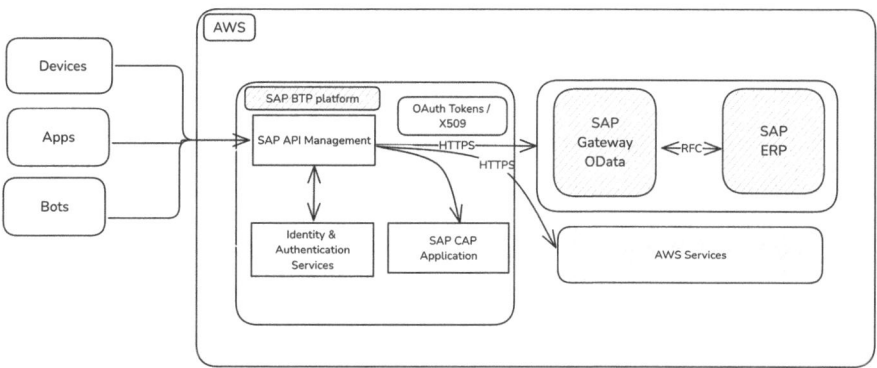

Figure 2-5. *SAP API Management as the API layer*

When formulating an API strategy for integrating SAP systems with cloud-based services or external applications, organizations often face a choice between leveraging cloud-native API management solutions like Amazon API Gateway or SAP API Management. Several key considerations should drive the decision:

- **Existing SAP landscape**: If your organization has a significant investment in SAP technologies and relies heavily on SAP systems, SAP API Management may be a more suitable choice, as it provides seamless integration with other SAP products and services. However, if your technology landscape is more diverse, with a mix of SAP and non-SAP systems, Amazon API Gateway may offer better flexibility and integration capabilities with a wide range of AWS services and third-party applications.

- **Cloud strategy alignment**: If your organization has adopted a cloud-first strategy and is already leveraging AWS services, Amazon API Gateway aligns well with this approach, providing native integration with other AWS services and a managed API gateway solution.

85

On the other hand, if your organization has a strong commitment to the SAP ecosystem and favors SAP's cloud offerings, SAP API Management could be a more suitable option because it integrates seamlessly with other SAP and cloud services.

- **API management requirements**: Both solutions offer robust API management capabilities, such as API creation, publishing, security, monitoring, and analytics. Evaluate the specific features and requirements of your API management needs, such as support for API versioning, rate limiting, caching, and integration with existing identity and access management solutions. Notably, SAP Integration Suite provides native adapters, integration content, and reference integration patterns that are specifically tailored for SAP environments, all of which are natively available in SAP API Management. This can be a significant advantage for organizations deeply embedded in the SAP ecosystem, as it streamlines the integration process and enhances compatibility with other SAP services and applications

- **Scalability and performance**: Amazon API Gateway is designed to be highly scalable and performant, leveraging the underlying AWS infrastructure and auto-scaling capabilities. Since SAP BTP is powered by AWS and utilizes AWS services under the hood, the scalability and performance capabilities of SAP API Management should be comparable to those of Amazon API Gateway. However, it's important to note that the specific implementation details, configuration, and integration with other SAP components may introduce some potential limitations or variations in the scalability and performance characteristics of SAP API Management. Therefore, it's still advisable to consider your specific use case, workload patterns, and integration requirements when evaluating the scalability and performance aspects of either solution.

- **Security**: Both Amazon API Gateway and SAP API Management provide robust security features and mechanisms to protect your APIs and ensure secure access. Amazon API Gateway integrates with AWS Identity and Access Management (IAM) for authentication and authorization, and it supports API keys, Amazon Cognito user pools, and AWS Lambda authorizers for custom authentication logic. It also offers features like SSL/TLS encryption, AWS WAF (Web Application Firewall) integration, and request/response data validation. SAP API Management leverages SAP's security frameworks and can integrate with various authentication and authorization mechanisms, such as SAML, OAuth, API keys, and SAP's identity and access management solutions. It also supports SSL/TLS encryption and provides features like rate limiting, IP filtering, and threat protection.

- **Cost and pricing model**: Evaluate both solutions' pricing models and estimate the overall cost based on your usage patterns, traffic volumes, and resource requirements. Amazon API Gateway typically follows a pay-as-you-go model based on API calls and data transfers, while SAP API Management may have different pricing tiers and licensing models.

- **Teams with different skillsets:** The skillsets of your IT and development teams are crucial in determining which API management solution will be most effective. If your teams are more familiar with AWS services and cloud-native development, Amazon API Gateway may be easier to adopt and manage. Conversely, if your teams have extensive experience with SAP technologies and frameworks, SAP API Management could be a more seamless fit, reducing the learning curve and enhancing productivity.

It's important to evaluate the trade-offs between SAP BTP API Management and Amazon API Gateway based on factors such as the existing technology stack, cloud strategy, API management requirements, security, scalability needs, developer experience, and cost considerations. Given that SAP BTP is based on AWS, it's crucial to carefully assess any potential overlaps or redundancies when using both solutions.

In some cases, a hybrid approach combining both solutions may be appropriate, with SAP API Management handling SAP-specific APIs and Amazon API Gateway acting as the front layer for non-SAP APIs and integration with AWS services. This approach could leverage the strengths of both platforms while minimizing potential redundancies or inefficiencies.

Ultimately, the API strategy should align with the organization's overall goals, technology landscape, and future roadmap, ensuring a secure, scalable, and flexible approach to exposing and consuming APIs across the enterprise. The fact that SAP BTP is powered by AWS services may provide synergies and integration opportunities that should be carefully evaluated in the context of the organization's specific requirements.

Decoupling with Microservices

APIs are just the front door for applications, but how can we decouple tightly coupled applications?

APIs provide a great connected experience for your customers, partners, and employees to interact with your business applications, including SAP ERP. While APIs provide a great connected experience, will they solve the problem of moving away from building tightly coupled extensions on top of your SAP ERP implementation?

The answer is no. Enterprises are seeking to build application extensions on top of their SAP ERP by coupling them loosely using a microservices-based architecture style and building those extensions outside of their ERP core. APIs become the front door for communication and integration between these decoupled services. Microservices, combined with an API-driven approach, pave the way for creating modular, scalable, and adaptable application extensions that can evolve independently from traditional monolithic systems.

Microservices and APIs are two approaches to modular software design. Modular programming aims to design smaller software components that interact with each other to perform complex functions. This is more efficient than designing software as one large code base for all functions. Microservices are an architectural approach that composes software into small, independent, highly specialized services. Each microservice solves a single problem or performs a specific task.

Before we jump into modern cloud-native application development for decoupling tightly coupled applications, let's look at how software architecture has evolved over the years.

The Evolution of Software Architecture

The software industry has witnessed a significant shift in architectural patterns, moving away from monolithic, tightly coupled applications toward a more modular and decentralized approach. The microservices architecture, which decomposes large applications into smaller, independent services, has gained widespread adoption. Today's focus has expanded to encompass serverless and event-driven architectures, which further enhance scalability, cost-efficiency, and responsiveness to real-time events. Figure 2-6 illustrates this shift.

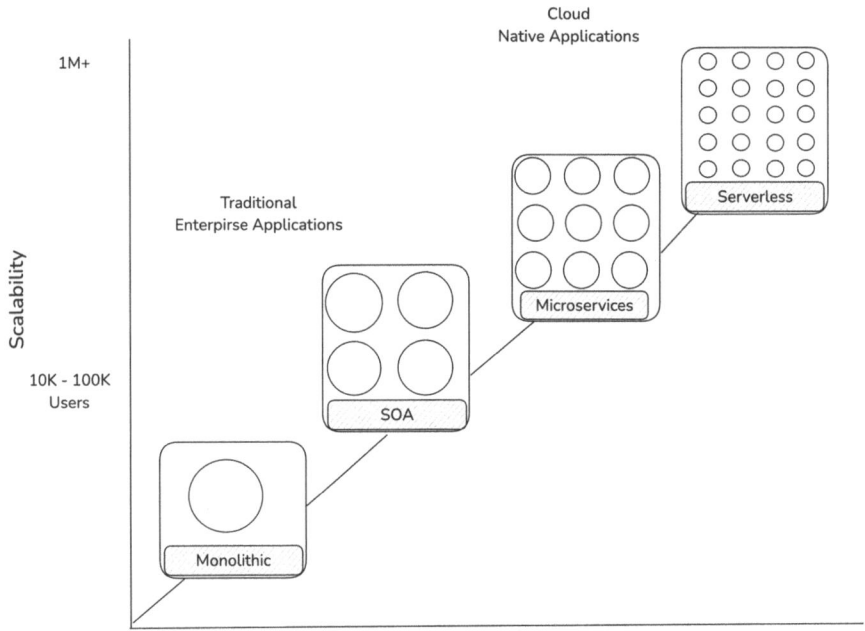

Figure 2-6. *Evolution of software architecture*

Meanwhile, the complexity and requirements of modern applications have evolved. Modern applications are cloud-based, necessitating the processing of vast amounts of data that often exceed the capabilities of traditional relational databases. Additionally, these applications demand global and instantaneous access, pushing the boundaries of traditional architectures.

Monolithic Architecture

Monolithic, in the context of software architecture, refers to a traditional, unified style for application design. A monolithic application is built as a single, indivisible unit, with all of its components, services, and modules combined into a single code base.

In a monolithic architecture, the application is self-contained and tightly coupled, meaning that all components are interdependent and cannot be easily separated or deployed independently. Any update or modification to one part of the application requires the entire monolithic code base to be rebuilt and redeployed.

Traditionally, large enterprise applications such as SAP ERP have been monolithic in origin. However, as applications grew in complexity and scale, the monolithic approach started to face challenges, such as:

- **Scalability**: It becomes difficult to scale specific parts of the application independently, as the entire monolith needs to be scaled together.

- **Flexibility**: Adding new features or making changes requires modifying the entire code base, increasing the risk of introducing bugs and making the application more complex.

- **Deployments**: Deploying updates or new features requires redeploying the entire application, leading to potential downtime and increased risk.

- **Technology lock-in**: The entire application is built using a specific technology stack, making it challenging to adopt new technologies or programming languages.

- **Resilience**: Achieving resilience in a monolithic architecture is challenging due to its "all or nothing" philosophy. If one component fails, it can bring down the entire application, making it difficult to isolate faults and maintain high availability.

It's worth reiterating that *monolithic*, in this context, means composed all in one piece, designed to be self-contained. The components of the program are interconnected and interdependent, rather than loosely coupled. Figure 2-7 depicts how SAP ERP software, despite being monolithic, has evolved over the years.

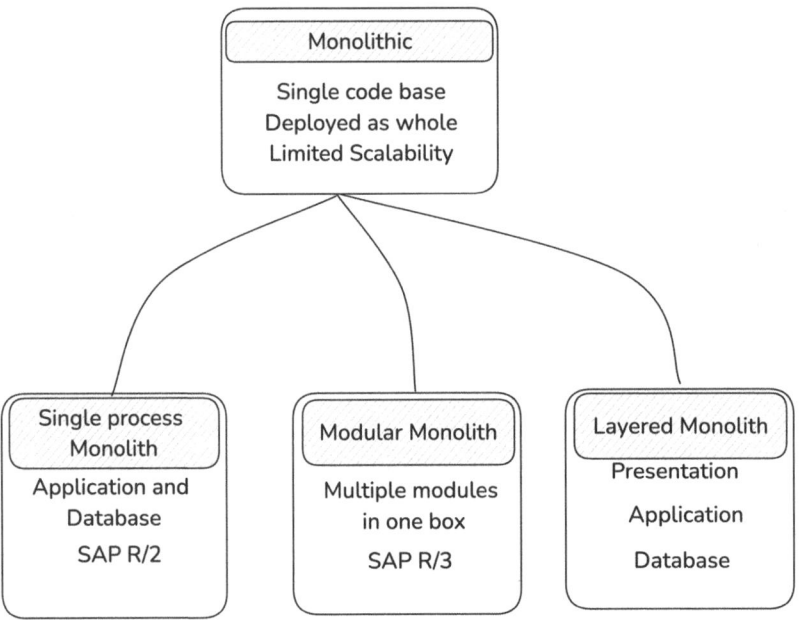

Figure 2-7. *Types of monolithic architecture patterns in the SAP ERP environment*

Service-Oriented Architecture

During the early to mid-2000s, service-oriented architecture (SOA) emerged as a response to the limitations of monolithic architectures. As enterprises began to develop more complex and distributed software systems, there was a growing need for greater flexibility, scalability, and reusability in application design.

Decoupling the application into smaller, independent components or services, each capable of independent development, deployment, and maintenance, was the core concept behind SOA. Significant advantages of this approach included enhanced flexibility to adapt to changing business needs and technological advancements, improved scalability through independent scaling of specific components, and increased fault tolerance due to the isolation of issues to individual services. Figure 2-8 illustrates this core concept.

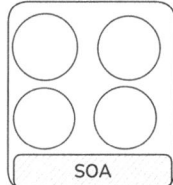

SOA is a Collection of services

These services communicate each other over a communication protocol

some means of connecting the services to each other is needed like a service broker.

SOAP webservices are the best examples

Figure 2-8. *Service-oriented architecture (SOA) architecture style*

The rise of web services and the widespread adoption of XML as a standard for data exchange further fueled the popularity of SOA during this period. While the hype around SOA has subsided, many of its principles and concepts have endured and continue to influence modern software architecture and development practices, shaping the way applications are designed and built today.

SAP was no exception, and during the mid-2000s, SAP introduced the concept of SAP NetWeaver as a business process platform, along with the Enterprise Service Repository and composable applications. The goal was to maintain a stable ERP core while adding innovation through enhancement packages (Enterprise SOA by Evolution) and, at the same time, enabling the development of side-by-side extensions through composable applications (Enterprise SOA by Design).

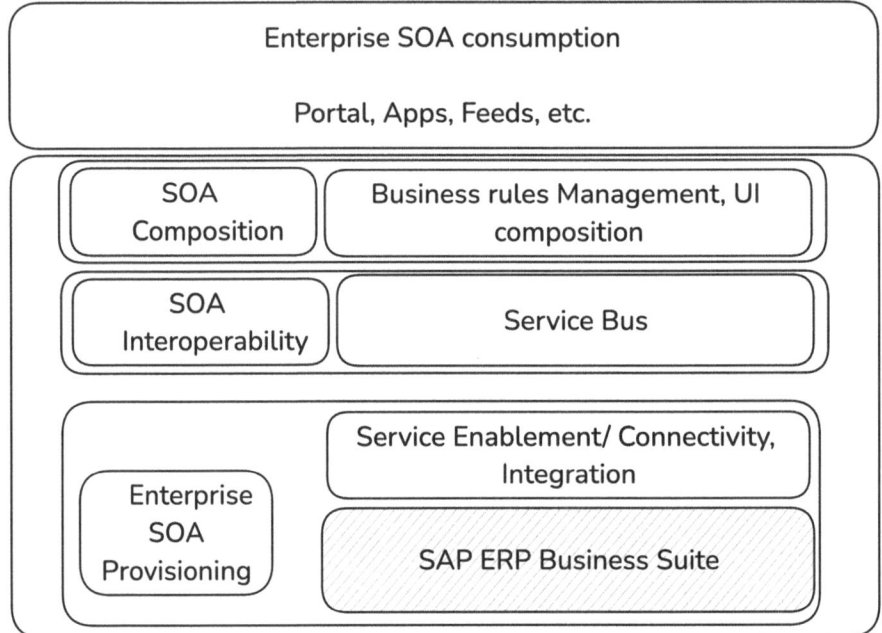

Figure 2-9. *SAP and SOA*

Service-oriented architecture was a step in the right direction, garnering significant hype and promising prospects. However, like many technologies, SOA became an overused term with varying interpretations among different stakeholders. Fundamentally, SOA entailed dividing an application into multiple services, typically via HTTP, and classifying these services into various types such as subsystems or tiers as illustrated in Figure 2-9 above.

Microservices

Microservices architecture and SOA are related concepts, but there are key distinctions between the two. SOA involves decomposing applications into multiple services but often relies on large central brokers, orchestrators, and enterprise service bus (ESB) implementations. The microservices architecture, on the other hand, prioritizes the deployment and scaling of small, independent services, as depicted in Figure 2-10. As a popular saying goes, "The microservices architecture is SOA done right."

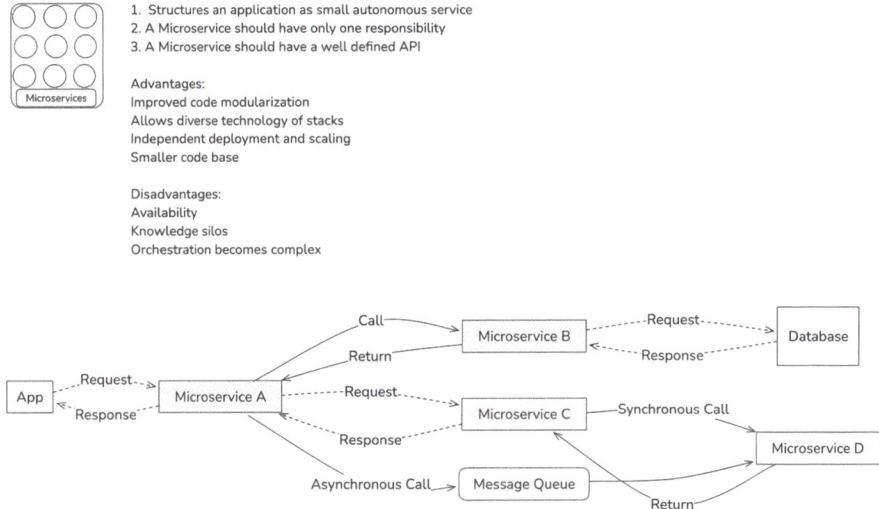

Figure 2-10. *Microservices architecture style*

One of the main ideas behind the microservices architecture is that each microservice owns its own data and its own domain logic. This is also one of the main design flaws for enterprise applications, as architects must decide whether to request data from other microservices or duplicate the data, introducing the idea of "eventual consistency," which is typically not a good idea in large enterprise applications.

As an example, suppose that you want to let your supply chain operation users know the state of a delivery based on where the carrier is so that they can better serve your customers by managing any exceptions along the supply chain. Trying to build the delivery status tracking functionality directly within the core SAP ERP system would likely result in a monolithic and potentially nonscalable solution. Decoupling this process and implementing it as a series of microservices is a more effective approach. Let's see how you could leverage AWS for this approach.

The key microservices involved in this delivery tracking scenario are the delivery request processing service, the carrier integration service, and the notification service.

- The delivery request processing service is responsible for receiving the delivery information from the SAP system. It receives delivery information updates from the SAP system, stores that information in Amazon Dynamo DB, and triggers the carrier integration service.

- The carrier integration service is responsible for integrating with various carrier APIs to fetch the latest delivery status updates by tracking carrier movements. It translates the carrier-specific status updates into a common format understood by the notification service and provides a consistent interface to retrieve the latest delivery status.

- The notification service is responsible for fetching the latest delivery status by interacting with the carrier integration service and publishing the status to the supply chain operation team.

Figure 2-11 illustrates the deployment of these key microservices as AWS Lambda functions, fronted by Amazon API Gateway. AWS Lambda is an event-driven, serverless compute service that allows you to run your code in response to various events or triggers without the need to manage any underlying infrastructure.

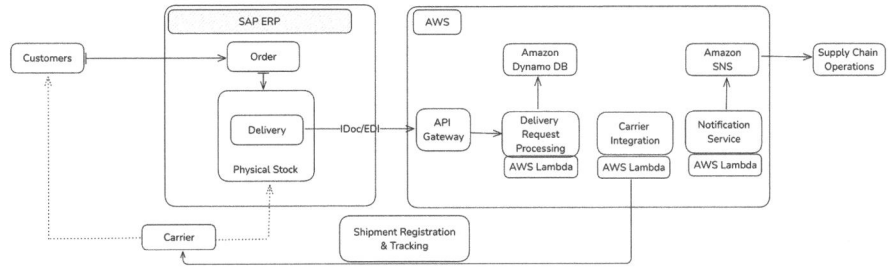

Figure 2-11. *Microservices architecture pattern for the "delivery status tracking" example use case*

The benefits of using AWS Lambda for the microservices in this architecture include scalability, serverless operation, cost optimization, reduced maintenance, and seamless integration with other AWS services. AWS Lambda automatically scales the compute resources based on the incoming traffic and events, eliminating the need to manage any underlying infrastructure. This allows you to focus on the business logic of the microservices rather than worrying about the underlying compute resources. Additionally, the pay-per-use pricing model of AWS Lambda can lead to significant cost savings, as you only pay for the resources consumed by the microservices. With AWS Lambda, you also don't need to worry about OS updates, security patches, or scaling the underlying infrastructure, as AWS handles all of these maintenance tasks.

The integration between AWS Lambda and other AWS services, such as Amazon API Gateway and Amazon DynamoDB, enables a cohesive and efficient microservice architecture, supporting delivery tracking functionality while seamlessly integrating with the existing SAP core ERP system.

SAP, like many other software vendors, has embraced the concept of microservices and embarked on a gradual journey to adapt its software to this architecture pattern. Figure 2-12 illustrates how SAP first "simplified"

SAP S/4HANA by eliminating any unnecessary elements from the original ERP core. Subsequently, an ongoing iterative process involves decoupling modules from the core and releasing them as new native cloud applications or, alternatively, delivering smaller functions as services on the SAP BTP platform.

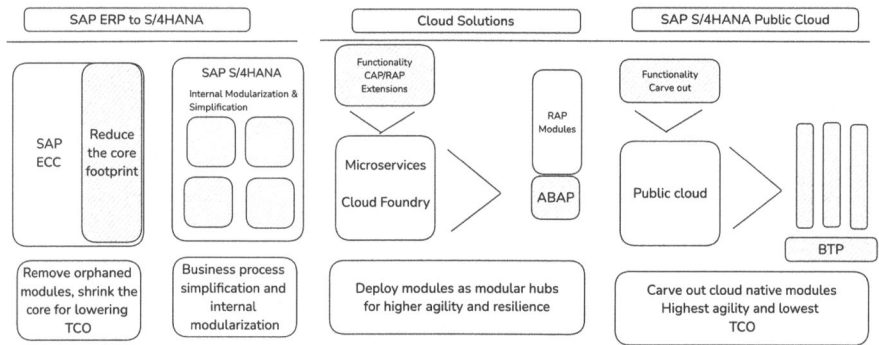

Figure 2-12. *SAP's transformative journey to microservices*

Serverless

The concept of serverless computing was the next step in the evolution of software architecture. Microservices and serverless computing serve different purposes and are not mutually exclusive.

While microservices provide benefits in terms of scalability and flexibility, they still require developers to manage and scale the infrastructure.

As Illustrated in Figure 2-13, Serverless computing completely removes the need for infrastructure management, allowing developers to focus solely on writing code. With serverless functions, cloud providers manage the infrastructure and automatically allocate computing resources as needed in response to events or requests. Reusable code, written by developers, executes in stateless containers and incurs charges only during their execution.

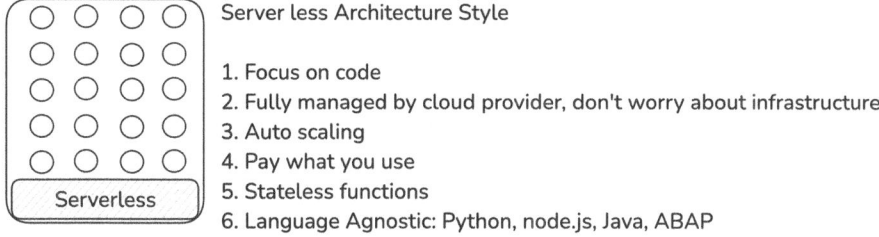

Figure 2-13. *Serverless architecture style*

Serverless computing means faster and easier development and deployment for you. This increases the possibility for enterprises to run a number of experiments faster than ever before, helping to innovate and stay ahead of the game. Figure 2-14 further highlights the key benefits of the serverless architecture style compared to traditional on-premises or virtual machine-based approaches.

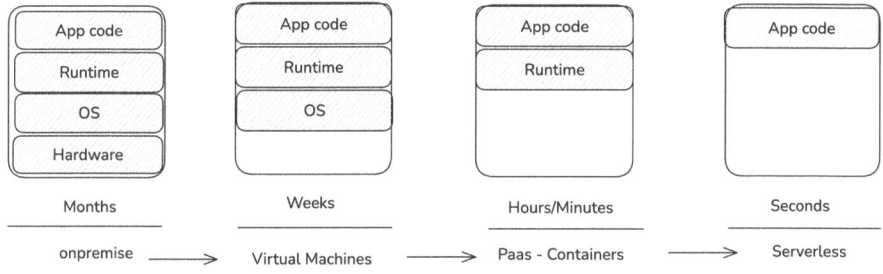

Figure 2-14. *Advantages of serverless architecture*

This serverless approach enables a diverse range of use cases that can complement the microservices architecture, including building web APIs, processing file uploads, creating event-driven workflows, responding to database changes, running scheduled tasks, and implementing reliable message queue systems—all without the need to manage the underlying infrastructure.

Coupling: The Integration Magic Word

When we talk about building loosely coupled or tightly coupled applications in the modern architecture world, the concept of coupling is a crucial consideration. Coupling isn't binary. It's important to understand that coupling is not a simple binary state of "coupled" or "decoupled." Instead, it exists on a spectrum (see Figure 2-15), with varying degrees of coupling between different components or services in the system. Some level of coupling is often necessary, but the goal is to minimize it to the greatest extent possible.

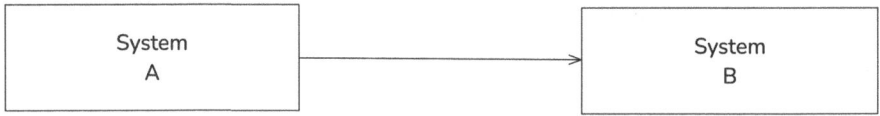

Coupling is the mesasure of independent variability between connected systems

Figure 2-15. *What is coupling?*

Coupling isn't one-dimensional. Coupling is a multifaceted concept, and it's not limited to a single dimension. There are different types of coupling, such as *technology coupling*, like the integration that can be built only in JAVA; *data format coupling*, which supports a specific data format like JSON or CSV; and *temporal coupling*, which allows either synchronous or asynchronous communication. Each type has its own implications and trade-offs. Effectively managing coupling in an integration architecture requires considering these various forms of coupling and their impact on the overall system design.

When you build distributed system architectures, decoupling comes with a cost, both at the design stage and during runtime. At the design stage, you'll need to invest time and effort into defining clear messaging contracts, establishing messaging channels, and implementing the necessary infrastructure to support asynchronous communication. At runtime, the overhead of event dispatching, transformation, and delivery can introduce latency and complexity that must be accounted for.

In an integration architecture's balancing act, the goal is to strike the right balance between coupling and decoupling. Excessive coupling can lead to rigidity and reduced flexibility, while excessive decoupling can introduce unnecessary complexity and overhead. Your system's specific requirements and constraints will determine the optimal level of coupling, which you should carefully evaluate throughout the design and implementation process. By understanding the nuances of coupling in an integration architecture, you can make informed decisions that lead to a more resilient, scalable, and maintainable system.

Integration Patterns

As we discussed in the previous chapter, coupling is crucial when building modern distributed applications. The goal is to achieve the right balance, whether tight or loose, while maintaining connections between components. In this section, we'll explore integration patterns representing architectural principles for both tightly and loosely coupled systems. These patterns provide guidance on designing the appropriate level of coupling based on the specific requirements and constraints of your distributed application architecture.

Synchronous (Request-Response) Model

The synchronous (request-response) model is a traditional architectural style where components communicate in a synchronous manner. As shown in Figure 2-16, in this model, the interaction between components follows a request-response pattern. One component, known as the client, sends a request to another component, called the server. The server then responds directly to the client's request.

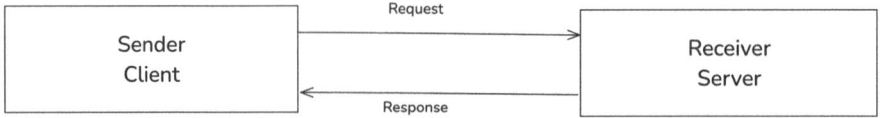

Figure 2-16. *Synchronous integration pattern*

The key characteristic of this model is that the client waits for the server's response before continuing with its own processing. This is often referred to as a *blocking call*. The advantages of this pattern are that it offers very low latency communication, as it involves a quick call without a lot of distributed components. Additionally, it allows for a fast fail, which is always beneficial during development. However, in fully distributed systems, failures can be difficult to detect early on, a disadvantage compared to tightly coupled architectures.

The synchronous model has some notable disadvantages, such as slow responses and becoming a single point of failure, particularly when it comes to receiver failure and throttling. If the server component fails or becomes unavailable, the client request will also fail, leading to a potential service disruption. The client is dependent on the server's availability and responsiveness, and the failure of the receiver (the server), as shown in Figure 2-17, can disrupt the entire system's functionality.

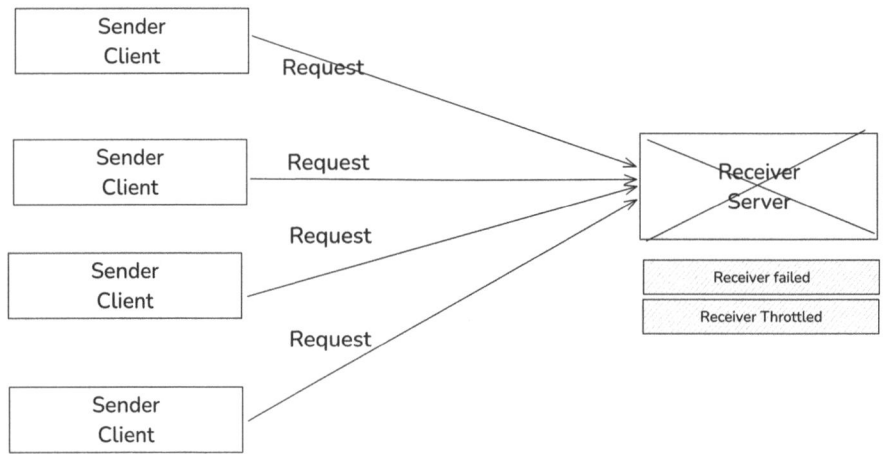

Figure 2-17. *Disadvantages of synchronous integration*

In today's world, *slow is the new broken.* The synchronous model can be susceptible to throttling, where the system becomes overloaded and no longer can handle the incoming requests in a timely manner. When the server receives more requests than it can process, it will start queuing or dropping requests, leading to increased latency and potential timeouts for the clients. This throttling effect can cascade through the system, as the overloaded server may block other clients from making progress, creating a domino effect.

The best approach is to go with an asynchronous pattern. For example, in an SAP environment, asynchronous integration can be used for processes like data synchronization with external systems, where immediate feedback isn't necessary. An SAP S/4HANA system can send data updates to an external CRM system and continue processing other tasks while the data is synchronized in the background. This avoids overloading the system and allows it to remain responsive, even under heavy load.

However, sometimes you have to go with synchronous integration, particularly when real-time validation is critical. For instance, when processing a sales order in SAP S/4HANA, the system must synchronously check inventory availability and credit limits before confirming the order. Immediate feedback is essential in this case to ensure the order can be fulfilled and the transaction can proceed without delay.

While asynchronous integration offers advantages like improved scalability and resilience, it also comes with its own set of challenges, such as increased complexity and delayed processing. Therefore, it's important to evaluate the specific requirements of each scenario and choose the integration pattern that best meets the needs of your system. The following sections describe three ways that you can achieve asynchronous integration: queue, router, and message bus.

Asynchronous Point-to-Point Model (Queue)

The asynchronous point-to-point model uses message queues to facilitate decoupled communication between components. In this approach, the producer publishes messages to a queue, and the consumer retrieves those messages when it is ready to process them, as depicted in Figure 2-18.

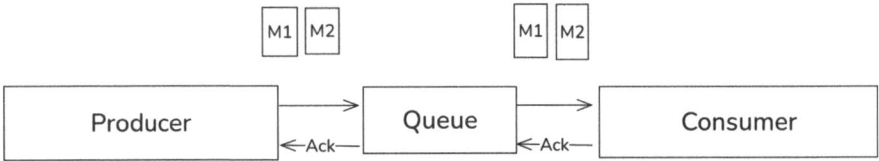

Figure 2-18. *Asynchronous point-to-point model queue pattern integration*

This asynchronous communication decouples the producer and consumer, as they no longer need to be available at the same time.

Message queues act as intermediaries, storing the messages in a durable and reliable manner. This loose coupling allows greater flexibility, as you can add, remove, or modify components without affecting the overall system. The asynchronous nature and use of message queues also improve scalability and resilience.

In the asynchronous point-to-point model, message queues provide the ability to implement robust retry mechanisms and dead-letter queues, as depicted in Figure 2-19. When a consumer fails to process a message, the message can be automatically retried, either immediately or after a configured delay. This ensures that messages are not permanently lost due to temporary issues or failures.

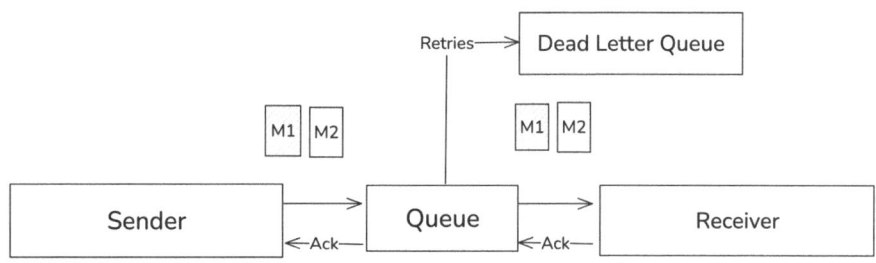

Figure 2-19. *Retries are possible with a dead-letter queue*

Another key characteristic of the asynchronous point-to-point model is that only one consumer can successfully consume each message from the queue. This ensures that messages are processed exactly once, without the risk of duplicate processing or race conditions. You can also easily scale consumers up or down to handle increased workloads, and the queue can absorb temporary spikes in message production. As shown in Figure 2-20, if a consumer fails, other available consumers can process the messages from the queue, ensuring a more resilient system.

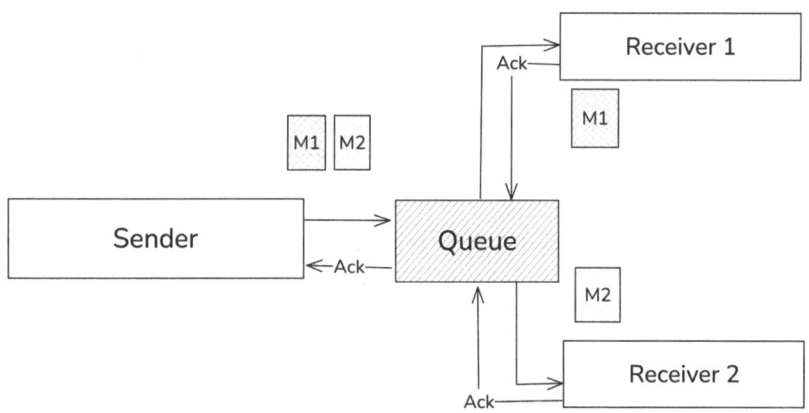

Figure 2-20. *Only one receiver can consume each message*

While the asynchronous point-to-point model with message queues offers several benefits, such as improved scalability and resilience, it also presents some potential drawbacks that you should consider. One key disadvantage is the challenge of response correlation. In the asynchronous approach, the request and response are decoupled through the message queue, making it difficult to track the life cycle of a specific request or to correlate the response with the original request, especially in more complex, multistep business processes. Maintaining the necessary context and state across these asynchronous interactions requires additional application-level logic and coordination, which can increase the complexity of your system.

Another disadvantage is the issue of backlog recovery time. When your system experiences a high volume of messages or a temporary consumer failure, a backlog of messages can accumulate in the queue. Recovering from such a backlog can be time-consuming, as consumers need to work through the accumulated messages, potentially causing delays in processing new messages. This backlog recovery time can impact the overall responsiveness and timeliness of your system, especially in time-sensitive use cases. These drawbacks highlight the trade-offs involved in choosing the asynchronous point-to-point model over a synchronous request-response approach.

Amazon Simple Queue Service (SQS) is an excellent choice for implementing the asynchronous point-to-point model. As a fully managed message queuing service provided by AWS, Amazon SQS offers a reliable and scalable way to decouple the components of your distributed application. As shown in Figure 2-21, Amazon SQS provides managed message queues that allow producers to publish messages and consumers to retrieve and process those messages asynchronously. The service ensures message durability and reliability, with built-in retry mechanisms and dead-letter queues to handle processing failures.

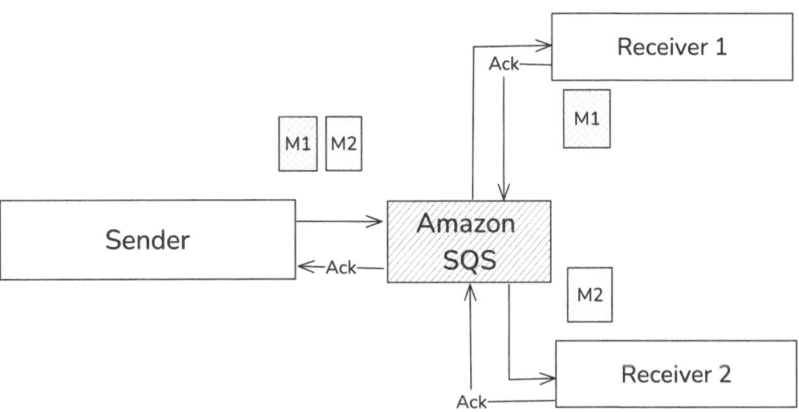

Figure 2-21. *Implementing asynchronous integration with Amazon SQS*

One of the key benefits of using Amazon SQS is its scalability and high availability. The service automatically scales to handle fluctuations in message traffic, ensuring your application can manage increased workloads without provisioning or managing additional infrastructure. Amazon SQS also maintains high availability, keeping your message queues accessible and reliable even during regional outages or service disruptions.

Asynchronous Point-to-Point Model (Router)

Asynchronous Point-to-Point Pattern (Router) can be useful when building integrations that require routing messages to specific receivers based on the type of message. However, as described in Figure 2-22, the complexity of maintaining routing logic with this pattern often increases over time and may not scale well, as the sender has to maintain the routing logic, which can increase location coupling and reduce flexibility.

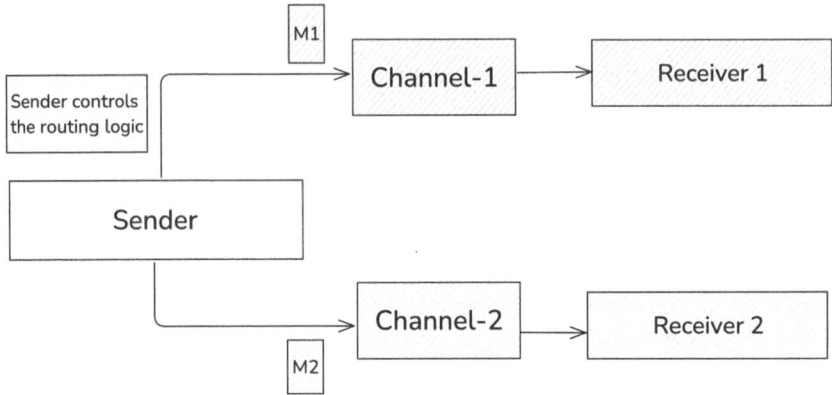

Figure 2-22. *Asynchronous point-to-point model router pattern integration*

Asynchronous Point-to-Point Model (Message Bus)

The asynchronous point-to-point model can also leverage a message bus to facilitate decoupled communication between components. In this approach, shown in Figure 2-23, the message bus acts as a central communication channel, allowing producers to publish messages and consumers to subscribe to and process those messages.

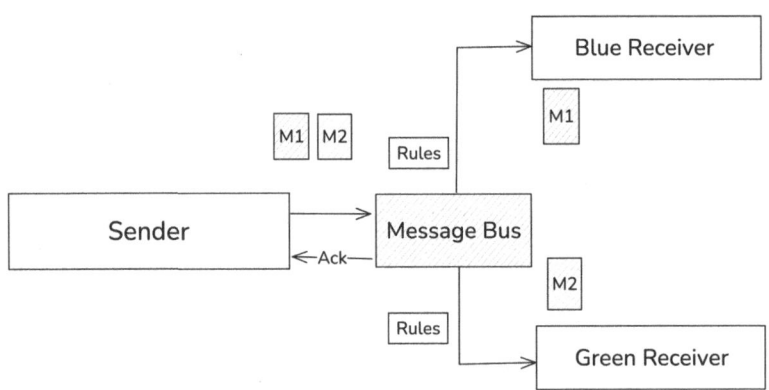

Figure 2-23. *Asynchronous point-to-point model message bus pattern integration*

The message bus provides a shared, event-driven infrastructure that enables the various components of your application to interact with each other in an asynchronous and loosely coupled manner. Producers simply publish messages to the bus without needing to know the specific consumers that will process those messages. Conversely, consumers subscribe to the message bus, receiving and processing the messages that are relevant to their functionality.

This architectural style promotes a high degree of flexibility, as you can easily add, remove, or modify components without affecting the overall system. The message bus handles the message delivery, routing messages to the appropriate consumers based on their subscriptions. Additionally, the bus can provide advanced features like message transformation, content-based filtering, and message enrichment to meet the specific requirements of your application.

The asynchronous nature of the message bus model also enhances the scalability and resilience of your system. Consumers can scale independently to handle increased workloads, and the bus can distribute messages across multiple consumer instances to balance the load. If a consumer fails, the bus can continue to deliver messages to the remaining healthy consumers, minimizing the impact of the failure.

By adopting the asynchronous point-to-point model with a message bus, you can enjoy the benefits of decoupled communication, improved flexibility, and enhanced scalability and fault tolerance. This architectural approach empowers you to build modern, distributed applications that can adapt to evolving requirements and handle diverse message-processing needs.

Amazon EventBridge is an excellent choice for implementing the asynchronous point-to-point model with a message bus pattern. Amazon EventBridge is a powerful serverless event bus service that enables you to build highly scalable, fault-tolerant, and event-driven applications, enabling highly decoupled and scalable event-driven architectures. As illustrated in Figure 2-24, the service seamlessly ingests events from a wide range of sources, including AWS services, SaaS applications, and your own custom applications, and then intelligently routes those events to the appropriate targets based on predefined rules and event schemas.

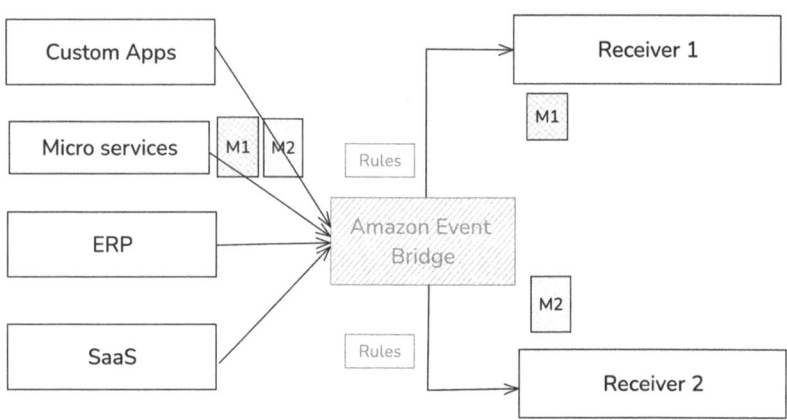

Figure 2-24. *Implementing an asynchronous integration point-to-point model (message bus pattern) with Amazon EventBridge*

Amazon EventBridge's capabilities extend beyond just event routing, as it also allows you to transform and enrich events before delivery, ensuring they meet the specific requirements of your consumers. The service's support for event replay and dead-letter queues further enhances the reliability of your event-driven system, allowing you to investigate and address processing failures.

By integrating with other AWS serverless services, Amazon EventBridge empowers you to construct fully event-driven architectures, which we are going to discuss next. In these event-driven architectures, AWS Lambda functions, Amazon Simple Notification Service (SNS) topics, and Amazon SQS queues can be triggered in response to incoming events. This serverless integration, combined with robust security features, makes EventBridge an excellent choice for implementing the asynchronous point-to-point model with a message bus pattern.

Event-Driven Architecture

In modern architecture, everything is asynchronous, driven by events, and orchestrated by an event broker, which is a service that mediates the communication of event messages between producers and consumers. As depicted in Figure 2-25, *event-driven architecture (EDA)* is an architectural style that promotes loosely coupled systems. In an event-driven architecture, the different components of the system communicate asynchronously by producing, detecting, and reacting to events.

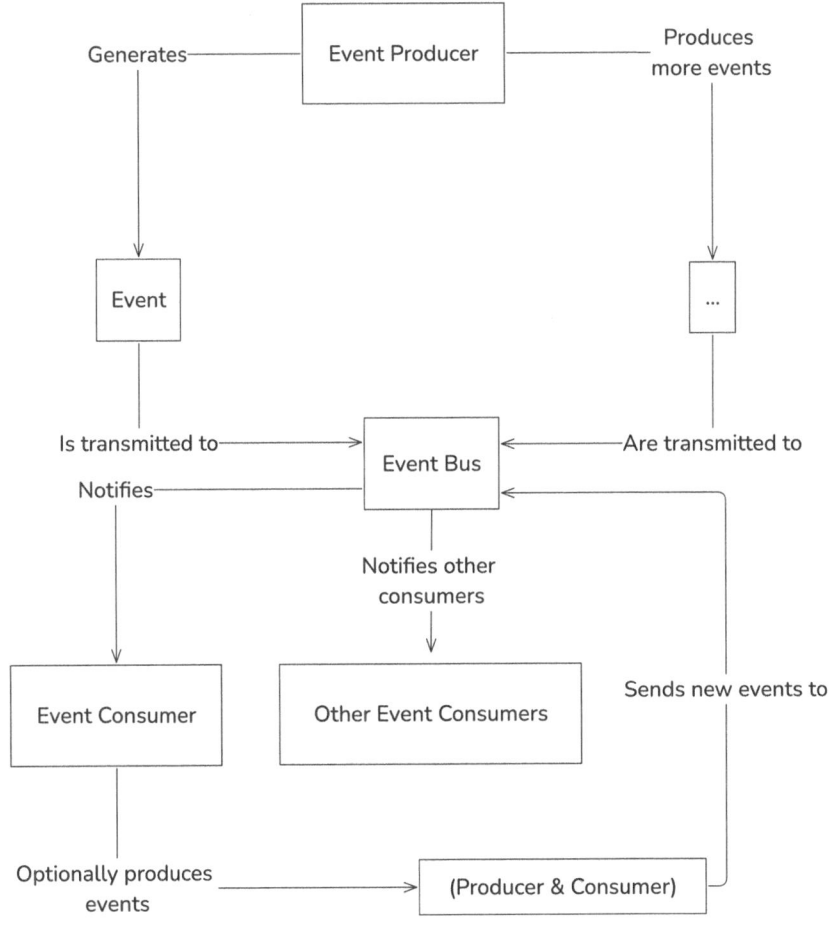

Figure 2-25. *How event-driven architecture works*

What Is an Event?

An *event* is a significant change in state, either within a system or in the external environment, that is of interest to one or more components of that system. Events are immutable, meaning they can't be changed.

The key properties of an event are

- **Event type**: The type or category of the event, which defines the nature of the change in state. Examples could include "order placed", "payment received", or "user signed up."

- **Event data**: The payload or information associated with the event, which provides the relevant details about the change in state. This could include things like order details, payment amount, user profile information, and so forth.

- **Event timestamp**: The time at which the event occurred, which is important for understanding the sequence of events and for any time-sensitive processing.

- **Event source**: The origin or producer of the event, which identifies the component or service that generated the event. This is useful for tracing the event back to its source.

- **Event context**: Additional metadata or attributes that provide relevant context about the event, such as the user, device, or location associated with the event.

These properties of an event are essential for helping the various components of an event-driven system understand the nature of the change, the relevant data, and the context in which the event occurred. This information allows the system to appropriately process, react to, and potentially transform the events as they flow through the architecture.

The effective definition and handling of events are crucial for building robust, scalable, and maintainable event-driven applications.

In the context of event-driven architectures, you can choose between two main types of events: sparse events and full-state events (see Figure 2-26). *Sparse events* represent incremental changes or specific occurrences within your system, capturing a focused change in state, such as "order created" or "user profile updated". These small, targeted events promote a decoupled, reactive, and scalable architecture, as you can process individual state changes independently. In contrast, *full-state events* represent the complete state of an entity or object at a given point in time, providing consumers with the full set of data that defines the current state. The choice between sparse events and full-state events depends on the specific requirements of your system, with many architectures utilizing a combination of both approaches to meet diverse needs.

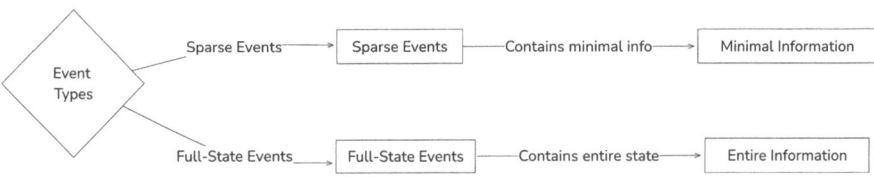

Figure 2-26. *Difference between sparse events and full-state events*

Event-driven architecture is not new, and we didn't invent it, but we are really doubling down on it. We think it's the best way to build decoupled system architectures. As illustrated in the Figure 2-27 below, event-driven architecture involves the use of state changes in a system, also known as events, to trigger actions in other parts of the system. These events are published through event brokers, such as message queues or event buses, which act as the central hub. The event brokers allow different components to publish and subscribe to events without needing to know about each other directly. Although event-driven architecture has been around for a while, nowadays, with event brokering services like SAP BTP Event Mesh and AWS services like Amazon EventBridge, implementing loosely coupled systems has become much easier.

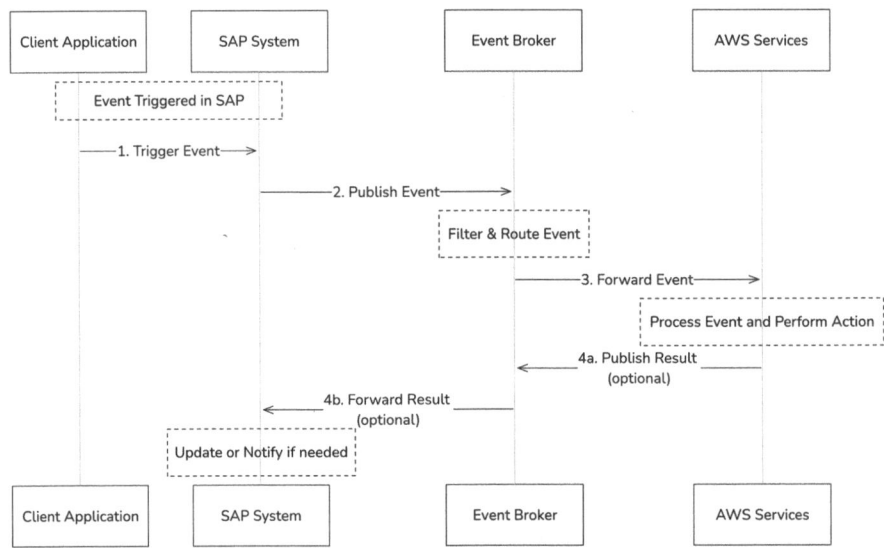

Figure 2-27. *How event-driven architecture works with SAP and AWS services*

Event-Driven Architecture with SAP S/4HANA, SAP BTP, and AWS Services

Figure 2-28 illustrates a scenario in which events from SAP are orchestrated with AWS services. SAP S/4HANA's Enterprise Event Enablement feature empowers the system to broadcast over 220 different business events through an event broker such as SAP BTP Event Mesh. This capability enables real-time notification of state changes to remote applications. For instance, when a business transaction like a delivery receipt against a purchase order gets posted in SAP S/4HANA, it triggers a corresponding event.

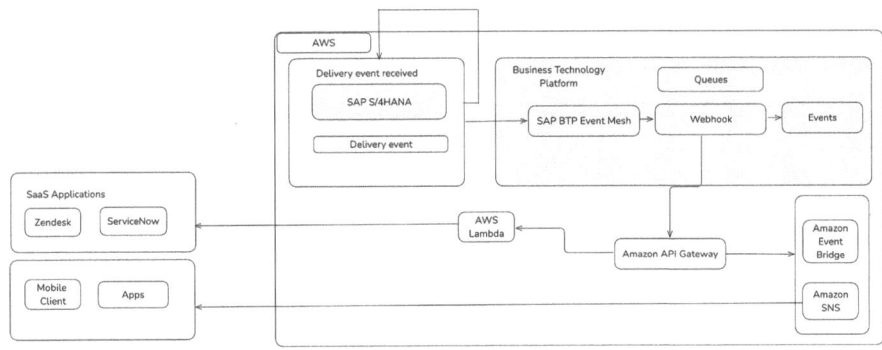

Figure 2-28. *How event-driven architecture works with SAP*
S/4HANA, SAP BTP, and AWS services

The triggered event publishes a message to a designated topic in
the BTP Event Mesh service using a WebSocket connection. BTP Event
Mesh acts as an event broker, storing the published messages in a queue
while simultaneously distributing them to subscribed applications. This
decoupled architecture allows for asynchronous communication between
SAP S/4HANA and other systems.

The BTP Event Mesh service seamlessly integrates with various AWS
services, such as Amazon API Gateway. This integration enables the
propagation of events to diverse target applications at scale. Advance
Message Queuing Protocol (AMQP) clients can also directly consume
messages from the BTP Event Mesh queue by subscribing to the relevant
topics, providing flexibility in event consumption.

Leveraging serverless services like Amazon SNS or Amazon
EventBridge, the events originating from SAP S/4HANA can be propagated
to various target applications, including mobile clients. This architecture
ensures scalable and asynchronous communication with multiple
consumers, enabling real-time notifications and event-driven workflows.

Serverless functions and event-driven architectures are closely related concepts that are often used together to build modern applications. As mentioned earlier in the chapter, serverless functions are a type of cloud computing service where the cloud provider manages the infrastructure and automatically scales resources as needed.

Not all event-driven architectures are serverless, but all serverless architectures are event-driven.

Event-driven architectures, on the other hand, are software design patterns where the flow of data and processing is determined by events rather than by a centralized control structure. In an event-driven architecture, different components of the system communicate with each other by publishing and subscribing to events. When an event occurs, the system triggers the appropriate response, which could include running a serverless function.

Therefore, serverless functions are often used as a key component of event-driven architectures. SAP is also embracing this architecture trend and enabling customers to develop side-by-side extensions on the SAP Business Technology Platform to implement serverless and event-driven applications.

Summary

In this chapter, we explored the fundamentals of modern architecture, delving into the transformative potential of leveraging cloud services such as those offered by AWS to modernize and extend SAP applications. We discussed the evolution of architectural patterns, from monolithic to service-oriented, and ultimately embracing microservices, APIs, serverless computing, and event-driven architectures. In the next chapter, we will delve into the world of advanced analytics and machine learning, illustrating how enterprises can leverage AWS services to derive valuable insights from their SAP data, enabling data-driven decision-making and unlocking new opportunities for innovation.

CHAPTER 3

Modern Data Strategy

Data is a strategic asset—data is the new oil for enterprises.

In Chapter 2, we explored the fundamentals of modern architecture, delving into the intricacies of microservices and APIs. These architectural paradigms have revolutionized how we design and deploy applications, offering unparalleled flexibility, scalability, and resilience. Microservices enable us to break down complex applications into smaller, manageable services that can be developed, deployed, and scaled independently. APIs, on the other hand, facilitate seamless communication between these microservices, ensuring interoperability and integration across diverse systems.

As we transition from the structural aspects of modern architecture to the strategic realm of data management, it is imperative to understand how a well-defined data strategy can unlock the full potential of your SAP environment on AWS. While modern architecture provides the foundation, a comprehensive data strategy ensures that the data flowing through this architecture is effectively managed, integrated, and leveraged for maximum impact.

Modern data strategy is not merely about managing data; it is about leveraging data as a strategic asset to drive innovation, operational efficiency, and competitive advantage. In the context of SAP on AWS, a robust data strategy encompasses the seamless integration of data across various SAP modules, siloed applications, and AWS services, ensuring data availability, integrity, and security while enabling advanced analytics and insights. This involves not only storing and processing data efficiently but also ensuring that data is accessible to the right stakeholders at the right time, in the right format.

This chapter will guide you through the key components of a modern data strategy tailored for SAP on AWS. We will explore best practices for data archival, data integration, and data analytics, highlighting how AWS services can be leveraged to enhance your SAP data landscape. By understanding the principles of data governance, data storage, and data processing, you will be able to create a data strategy that supports your business objectives and technological requirements. We will begin by defining data strategy objectives and exploring data storage and management, focusing on selecting the right storage solutions, implementing data lifecycle management, and developing cost-effective data archiving strategies. Next, we will delve into the specifics of data integration and interoperability, examining how to connect SAP data with AWS services seamlessly, ensuring alignment with your business goals. We will then explore establishing a framework for data governance and implementing mesh-based architectures.

Furthermore, we will discuss advanced analytics and insights, showcasing how AWS analytics services can be used to derive meaningful insights from your SAP data. Finally, we will address the critical aspects of security and data protection, ensuring that your data strategy not only enhances performance but also safeguards sensitive information.

The following topics will be covered in this chapter:

- Data archiving

- Data lakes

- Data virtualization and data federation

- End-to-end enterprise analytics

- Data Mesh

By the end of this chapter, you will have a comprehensive understanding of how to formulate and implement a data strategy that aligns with your business goals and technological landscape, ensuring that your data is not just managed but harnessed for strategic growth. This holistic approach will empower your organization to transform data into actionable insights, driving innovation and maintaining a competitive edge in the ever-evolving business environment.

Data Archiving

Data is growing at an unprecedented rate—there's a lot more data than people think.

Data is ubiquitous and is being exponentially generated in enterprise. However, the majority of this data is dark data. The Gartner glossary defines dark data as the information assets that organizations collect, process, and store during regular business activities, but which they generally fail to use for other purposes. Those other purposes could include analytics, business relationships, and direct monetization. Within the lifespan of a typical data platform like SAP, which averages around 15 years, the amount of data stored and analyzed grows by an order of magnitude every five years. This means that starting from terabytes, customers can expect to reach petabytes of data over the lifetime of their data platforms. Managing this exponential growth of data generated in systems like SAP effectively requires robust archiving strategies.

Data archiving is an important methodology for reducing data volumes in the system, complying with legal data retention requirements, and improving the performance in the system. Data archiving is a secure and reliable process through which data from closed business processes— meaning data that is no longer needed in online business—is written from the database to an archive, and then deleted from the database.

As illustrated in Figure 3-1, data archiving is essential for several reasons. Firstly, it significantly reduces costs by moving inactive data to lower-cost storage solutions, which helps manage overall data expenses more effectively. This is particularly important when comparing the high costs of in-memory storage, which, while offering unparalleled speed and performance for real-time analytics, is not cost-effective for storing large volumes of infrequently accessed data. By contrast, low-cost storage options provide an economical alternative for archival data, ensuring substantial cost savings.

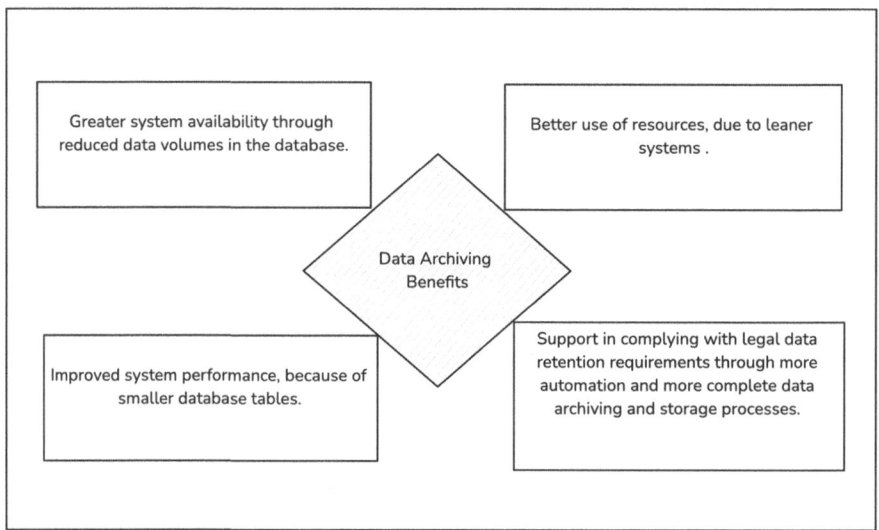

Figure 3-1. *Data archiving benefits*

Secondly, archiving older data enhances the performance of active SAP systems, as these systems no longer need to process extensive historical data, thereby optimizing their efficiency. Thirdly, data archiving ensures compliance with legal and regulatory requirements, which often mandate the retention of data for several years.

Lastly, during migration processes, archiving reduces the volume of data that needs to be moved, thereby lowering SAP HANA licensing costs, migration expenses, and simplifying the process. SAP HANA licensing is typically based on the amount of data stored in-memory, so by archiving older or less relevant data, organizations can significantly reduce their licensing fees. This approach ensures that only relevant and current data is migrated to the new system, improving its performance. Overall, a strategic approach to data archiving balances the high costs of in-memory storage with the benefits of low-cost archival solutions, optimizing both performance and cost-efficiency.

Types of Data to Archive

Identifying the types of data suitable for archiving is crucial in the context of SAP. Generally, the types of data to be archived include the following:

- Historical data that no longer is actively used but must be retained for regulatory or business reasons

- Transaction data, such as old transaction records not needed for daily operations

- Outdated master data that is infrequently accessed

- Audit logs, required for compliance but rarely accessed

- Various documents, including invoices, reports, and contracts, that need to be retained for legal or historical purposes

- Data from legacy systems that must be retained for reference or legal purposes but is not needed for daily operations

By systematically identifying and archiving these types of data, organizations can ensure efficient data management, compliance with regulatory requirements, and optimal performance of their SAP systems.

Data Management Evolution: Archiving, SAP ILM, and Data Aging in SAP HANA

In the evolving landscape of enterprise data management, SAP has introduced several strategies and tools to handle the growing volumes of data efficiently. These include data archiving, SAP Information Lifecycle Management (ILM), and Data Aging in SAP HANA. Each of these strategies, depicted in Figure 3-2, plays a crucial role in managing data effectively, reducing costs, and ensuring compliance with legal and regulatory requirements.

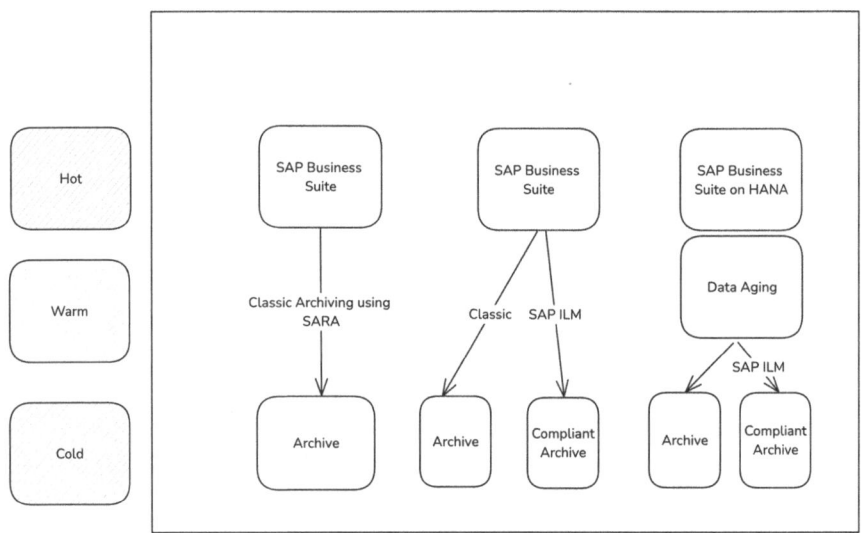

Figure 3-2. *Data management evolution: Archiving, SAP ILM, and Data Aging in SAP S/4HANA*

Classic Archiving

This remains a fundamental practice for managing data growth. By systematically moving inactive data to low-cost storage, organizations can maintain system performance and reduce costs while ensuring compliance with regulatory requirements.

SAP Information Lifecycle Management

SAP ILM extends the capabilities of traditional data archiving by providing advanced tools for managing the complete life cycle of data, from creation to deletion. SAP ILM is particularly useful for automating data retention management, enabling the automated retention and deletion of data based on predefined rules. This automation ensures that data is stored only as long as necessary, reducing storage costs and maintaining compliance with data retention policies. Additionally, SAP ILM supports compliance with regulations such as the European Union General Data Protection Regulation (GDPR) by facilitating the blocking and deletion of personal data, ensuring that data management practices adhere to stringent legal requirements. Furthermore, SAP ILM aids in the decommissioning of legacy systems by allowing organizations to retire old systems while retaining access to necessary historical data. This capability helps streamline IT infrastructure and reduces the costs associated with maintaining obsolete systems.

Data Aging

Data Aging is a feature in SAP HANA designed to manage the memory footprint of the database by organizing data based on its "temperature"— how frequently it is accessed. Frequently accessed data, known as *hot data*, remains in the main memory to ensure optimal performance. *Warm data*, which is accessed less frequently, is kept on the SAP HANA data disk rather than in-memory. This approach helps reduce memory usage while still allowing relatively quick access to important data. Unlike data tiering, Data Aging does not involve moving data to external storage solutions; it keeps all data within the SAP HANA environment, optimizing the system's resource usage.

In addition to Data Aging, SAP HANA provides Native Storage Extension (NSE) and SAP BW Extension Nodes to further optimize data management. NSE allows large volumes of data to be stored on disk instead of in-memory, which helps reduce costs while maintaining efficient access to data that doesn't need the high performance of in-memory storage. BW Extension Nodes function similarly in SAP BW/4HANA environments, enabling the segregation of data across different storage tiers within the SAP HANA environment. These features are particularly useful for managing large datasets that don't require the speed of in-memory processing but still need to be accessible for business processes.

Data tiering is a broader data management strategy that involves moving data to different storage layers based on access frequency and performance needs. In an SAP environment, data tiering typically moves data from high-cost, high-performance storage (like in-memory) to more cost-effective external storage solutions. Hot data stays in memory, while warm and cold data are moved to less expensive storage tiers, such as SSDs, HDDs, or even cloud-based storage solutions. This approach reduces the cost of storage by placing older, less frequently accessed data in slower, cheaper storage, while keeping the most critical data in the fastest, most expensive storage for quick access.

What to Consider Before Archiving SAP Data

When planning to archive SAP data, it's essential to consider several key factors to ensure an effective and efficient process. As depicted in Figure 3-3, these include the volume of data and its impact on system performance, the decision of whether to build a custom solution or buy a commercial solution, the total cost of ownership (TCO), and the importance of maintaining data context.

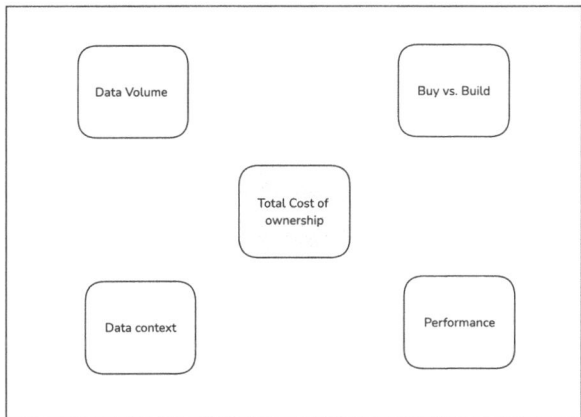

Figure 3-3. *Key factors to consider when archiving SAP data*

Data Volume and Performance

The amount of data to be archived is a critical factor. Large volumes of data require robust archiving solutions that can handle significant data loads without compromising system performance. Assessing the volume helps in selecting the right storage solutions and determining the frequency of archiving processes.

Archiving should improve the performance of the SAP system by reducing the amount of active data. It's essential to monitor system performance before and after archiving to ensure that the expected performance improvements are realized. The goal is to offload inactive data while ensuring that the active dataset is optimized for fast retrieval and processing.

Build vs. Buy

The following list describes the benefits and drawbacks to consider when deciding whether to build or buy a solution for archiving SAP data:

- **Build**: Developing an in-house archiving solution can be tailored to specific business needs and integrated closely with existing systems. However, it requires significant investment in terms of time, resources, and expertise. The maintenance and updating of the solution are ongoing responsibilities.

- **Buy**: Purchasing a commercial archiving solution can be quicker to deploy and often includes support and updates from the vendor. These solutions are designed to handle standard archiving needs and can be more cost-effective in the long run, especially if they come with advanced features like compliance management and integration with cloud storage. However, integrating the commercial archiving solution with your existing SAP landscape, data sources, and other enterprise systems may require significant time, effort, and technical expertise, which can add unexpected costs and challenges.

Total Cost of Ownership

When calculating the TCO of archiving SAP data, consider the following:

- **Initial costs**: This includes the cost of purchasing software, hardware, and the initial setup. For in-house solutions, this also covers development costs.

- **Operational costs**: These are the ongoing costs of running the archiving solution, including storage costs, system maintenance, and support.

- **Cost savings**: Effective data archiving reduces the load on primary databases, potentially lowering the costs associated with high-performance storage and improving system efficiency, which can translate to operational savings. It's important to compare the TCO of in-house solutions versus commercial solutions to determine the most cost-effective option.

Maintaining Data Context

Maintaining the context of archived data is crucial for ensuring that the data remains accessible and meaningful for business operations, audits, and compliance purposes. This involves preserving all relevant metadata so that archived data can be easily searched and retrieved. It also includes maintaining the integrity and authenticity of the data to ensure it can be reliably used for reporting and compliance. Furthermore, compliance with legal and regulatory requirements is essential, which means keeping necessary data for mandated periods and ensuring it can be securely accessed and audited as needed. This comprehensive approach ensures that archived data continues to serve its purpose effectively without compromising on accessibility, reliability, or legal compliance.

SAP Data Archiving Options with AWS Services

Leveraging AWS services such as Amazon Simple Storage Service (S3) and AWS Storage Gateway for SAP data archiving provides scalable, cost-effective, and secure storage solutions (see Figure 3-4).

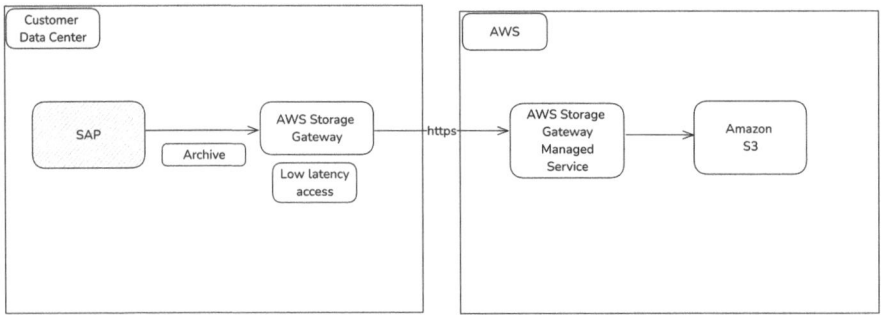

Figure 3-4. *Leveraging Amazon S3 using AWS Storage Gateway for archiving*

Amazon S3 provides secure, durable, highly scalable object storage at a very low cost. You can store and retrieve any amount of data, at any time, from anywhere on the Web through a simple web service interface. Designed to provide 99.999999999% durability and 99.99% availability over a given year, Amazon S3 is ideal for SAP system backups and other mission-critical primary data storage needs.

AWS Storage Gateway addresses the hybrid cloud storage challenges enterprises face. AWS Storage Gateway allows you to connect to and use key cloud storage services such as Amazon S3. AWS Storage Gateway easily deploys on premises, as a virtual machine (AWS supports all the major hypervisors), or via a preconfigured hardware appliance.

By providing a local cache, AWS Storage Gateway enables low-latency access to frequently accessed data. AWS Storage Gateway supports access via standard storage protocols (NFS, SMB, iSCSI, and iSCSI VTL) for on-prem applications to access data stored in Amazon S3. So, no changes are required to existing applications that reside in your landscape.

You can also use Amazon Elastic File System (Amazon EFS) and Amazon FSx for NetApp ONTAP to store your archive file in a highly available, scalable, and durable manner (see Figure 3-5). Amazon EFS and FSx for ONTAP can be mounted as your archive file system and you can archive your data from SAP to this file system through SAP transaction code SARA.

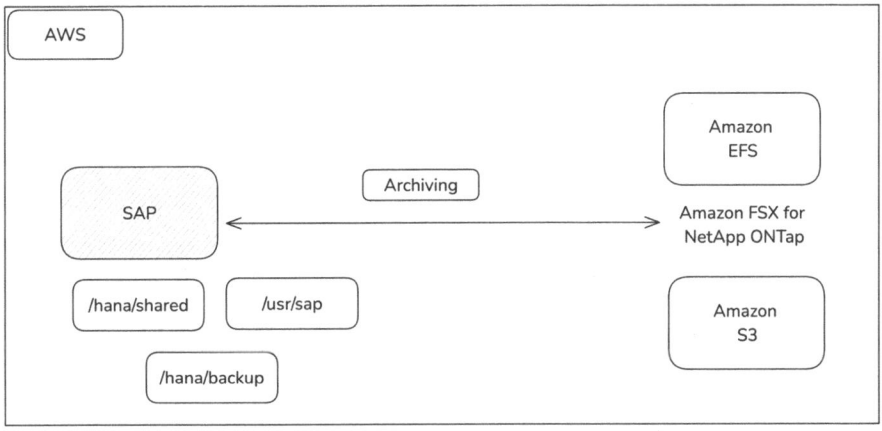

Figure 3-5. *Leveraging Amazon EFS for cold storage*

Amazon EFS is ideal for applications requiring shared file storage across multiple instances and supports dynamic scaling. Amazon FSx for NetApp ONTAP offers the storage capabilities of NetApp's ONTAP file system with AWS's scalability and flexibility. Amazon FSx for ONTAP supports shared storage and advanced data management features, making it suitable for SAP data archiving.

Archiving SAP data using AWS services offers a scalable, cost-effective, and secure solution for managing data growth, improving system performance, and ensuring compliance. By leveraging AWS services like Amazon S3, Amazon EFS, and Amazon FSx, organizations can efficiently manage their SAP data archiving needs, ensuring that data remains accessible, secure, and cost-effective.

While data archiving focuses on the long-term storage of data for compliance and historical reference, now let's transition to the topic of data lake, which is a more dynamic approach to data management that not only stores vast amounts of data but also enables real-time analytics and insights.

129

Data Lakes: A Modern Approach to Data Management

Data is everywhere.

Data is more diverse than ever and is continuously generated, increasing exponentially. Every five years, the amount of data grows tenfold and often needs to be preserved indefinitely. Starting with terabytes, this can quickly grow to petabytes over the platform's lifetime. Traditional data warehouses, while effective for certain analytics challenges, are not cost-effective at the scales most organizations now face. While archiving focuses on the efficient and compliant storage of inactive or less frequently accessed data, organizations also need an operational strategy to access continuously generated data from mission-critical applications and derive value from it. As shown in Figure 3-6, this exponential growth of data is a phenomenon that cannot be ignored.

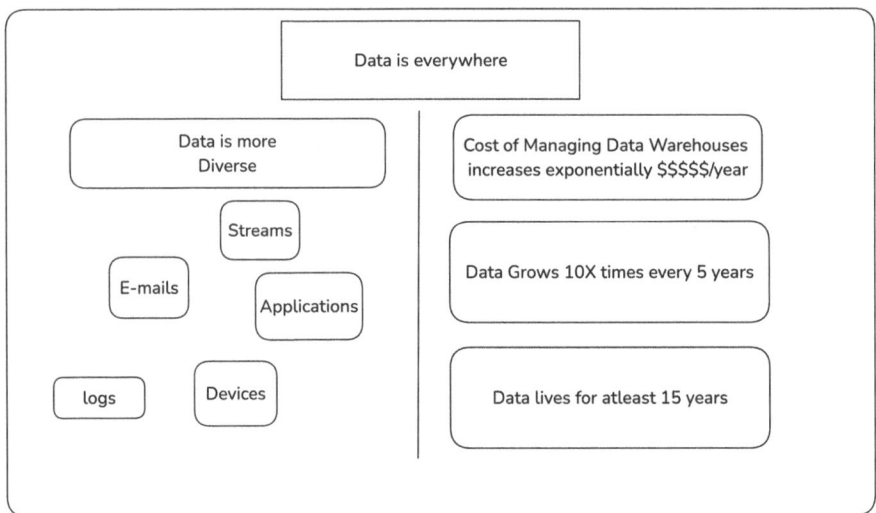

Figure 3-6. *Data diversity and complexity*

Today's enterprises are dealing with more than just structured data. Now, customers need to effectively capture, store, curate, and analyze semi-structured and unstructured data. This includes a variety of sources such as e-mails, chat logs, social networking feeds, network logs, application logs, clickstreams, system and development logs, and more.

Everyone needs access to this data, and the personas accessing it are vastly different from what they were a few years ago. Organizations have data scientists who require specialized tools and platforms to access, process, and analyze vast datasets for insights and predictive analytics. Tech-savvy business users demand flexible tools to slice and dice information, enabling detailed analysis to inform business decisions. Executives need operational reporting systems that provide high-level summaries and dashboards for strategic decision-making.

Data is generated across multiple systems; it's no longer just a single ERP application that generates data. Organizations now have SaaS applications, social media feeds, IoT devices, and many more sources contributing to data generation. This data includes structured and unstructured formats, making it necessary to bring all the data into one single area, catalog it, and ensure users can access it easily. This is the whole concept of the *data lake*: a centralized repository that allows for efficient storage, cataloging, and access to diverse data types, supporting a wide range of analytical and operational needs. The illustration in Figure 3-7 depicts this convergence of traditional data silos into a unified data lake, which is a key strategy for organizations to effectively manage and derive value from the large volume and diversity of data.

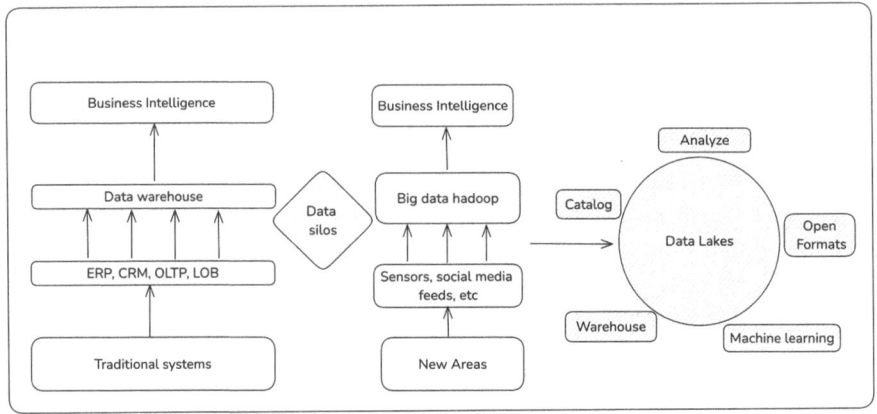

Figure 3-7. *Convergence of traditional data silos into data lakes*

Data lakes serve as a modern approach to handling this diverse and growing volume of data. Unlike traditional data archiving, which focuses on the storage of inactive data, data lakes provide a centralized repository for storing all structured, semi-structured, and unstructured data at any scale. This enables organizations to run different types of analytics—from dashboards and visualizations to big data processing, real-time analytics, and machine learning—to guide better decisions.

Let's take a look at how to build scalable data lakes on AWS.

How to Build Data Lakes on AWS

Data lakes on AWS are designed to handle large volumes of diverse data sources, providing a scalable, cost-effective, and comprehensive solution for data storage and analysis. By utilizing AWS services such as Amazon S3, Amazon Athena, AWS Glue, Amazon Redshift, and Redshift Spectrum, organizations can build a robust data lake infrastructure.

Amazon S3 serves as the backbone of your data lake, offering scalable and durable object storage. It can handle massive amounts of data, including structured, semi-structured, and unstructured data. With various storage classes like S3 Standard, S3 Standard-IA, and S3 Glacier, you can optimize costs based on data access patterns, ensuring a cost-effective storage solution.

Amazon Athena allows you to run SQL queries directly on data stored in Amazon S3. Being serverless, Amazon Athena eliminates the need for managing infrastructure, making data analysis straightforward and cost-effective. You only pay for the queries you run, which makes it a flexible and economical option for interactive data querying.

AWS Glue is a managed extract, transform, and load (ETL) service that automates data preparation and transformation tasks. Amazon Glue can discover, catalog, and transform data, making it easier to move data into the data lake and prepare it for analysis. The AWS Glue Data Catalog acts as a central repository for metadata, storing information about data stored in Amazon S3 and enabling easy data discovery and management.

Amazon Redshift is a fully managed data warehouse service that provides fast query performance using SQL-based tools. With Amazon Redshift Spectrum, you can run queries against exabytes of data in Amazon S3 without having to load it into Amazon Redshift, facilitating seamless integration between your data warehouse and data lake.

Steps to Build a Scalable Data Lake on AWS

Building a data lake on AWS involves several steps, each leveraging different AWS services to handle data ingestion, storage, cataloging, processing, and security, as depicted in Figure 3-8 and described in the following list:

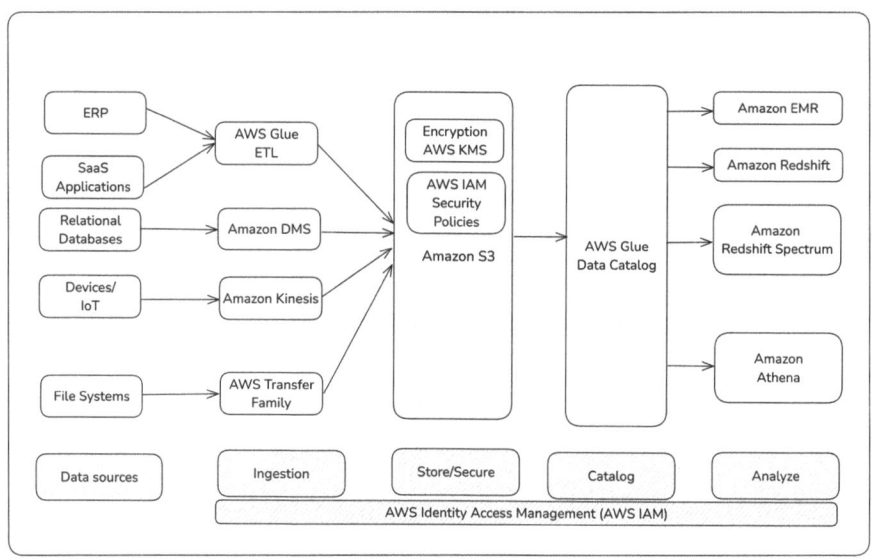

Figure 3-8. *Steps to build data lakes on AWS*

- **Data ingestion**: Ingesting data from various sources is the first step. AWS provides several services for this purpose. AWS Database Migration Service (AWS DMS) helps migrate data from on-premises databases, other cloud providers, or different AWS regions. For real-time data streaming, Amazon Kinesis collects, processes, and analyzes data from sources such as IoT devices, application logs, and social media. AWS Glue ETL automates data preparation and transformation tasks by discovering, cataloging, and transforming data for analysis. For file transfers, AWS Transfer Family securely scales your recurring business-to-business file transfers to AWS storage services using SFTP, FTPS, FTP, and AS2 protocols.

- **Data storage**: Amazon S3 is the primary storage layer for your data lake, offering scalable, durable, and cost-effective object storage. It supports various storage classes like S3 Standard for frequently accessed data, S3 Standard-IA for infrequently accessed data, and S3 Glacier. Utilizing the Amazon S3 versioning and lifecycle policies helps manage data over time and optimize costs based on access frequency.

- **Data cataloging**: AWS Glue Data Catalog acts as a central repository for metadata, storing information about data stored in Amazon S3. This cataloging makes data easily discoverable and manageable, enabling efficient data management and analytics. By integrating with other AWS analytics services, the Glue Data Catalog facilitates seamless data discovery and management.

- **Data processing and analytics**: AWS offers various services for processing and analyzing data stored in the data lake. AWS Glue is used for ETL processes to clean and transform data. Amazon Athena provides an ad hoc querying service that allows you to analyze data directly on Amazon S3. Amazon EMR (formerly called Amazon Elastic MapReduce) is a managed Hadoop framework that processes large amounts of data using open source tools such as Apache Spark, Hadoop, and Hive. For more complex and large-scale data warehousing and querying needs, Amazon Redshift and Redshift Spectrum are utilized.

- **Data security and compliance**: Ensuring data security and compliance is crucial in a data lake environment. AWS Identity and Access Management (AWS IAM) manages user access and permissions securely across AWS services. AWS Key Management Service (AWS KMS) manages encryption keys for securing data at rest and in transit. Amazon Macie uses machine learning to automatically discover, classify, and protect sensitive data stored in Amazon S3. AWS CloudTrail tracks user activity and API usage, ensuring compliance and providing audit capabilities.

Building a scalable data lake on AWS involves integrating various services to handle data ingestion, storage, cataloging, processing, and security. By leveraging services such as Amazon S3, AWS Glue, Amazon Athena, and others, organizations can create a robust data lake environment that supports diverse data analytics needs and drives strategic decision-making.

Integrating SAP Data into an AWS Data Lake

Let's dive into how you can effectively bring SAP data into the AWS data lake, starting with Amazon S3 as a landing zone. From this point onward, the focus will be on transferring SAP data to Amazon S3, as once your data is in Amazon S3, you can leverage the extensive range of AWS services to perform various operations.

Before diving into the various architectures and extraction patterns for integrating SAP data into an AWS data lake, it's crucial to keep a few key considerations in mind. These considerations will help guide your decisions and ensure that the patterns you choose align with your business needs and technical requirements. Your choice of extraction patterns may vary depending on what data you are dealing with and why you need it.

Volume, Velocity, Variety (VVV Approach)

First and foremost, it's essential to consider the three Vs of data: volume, velocity, and variety. Figure 3-9 illustrates this "VVV approach."

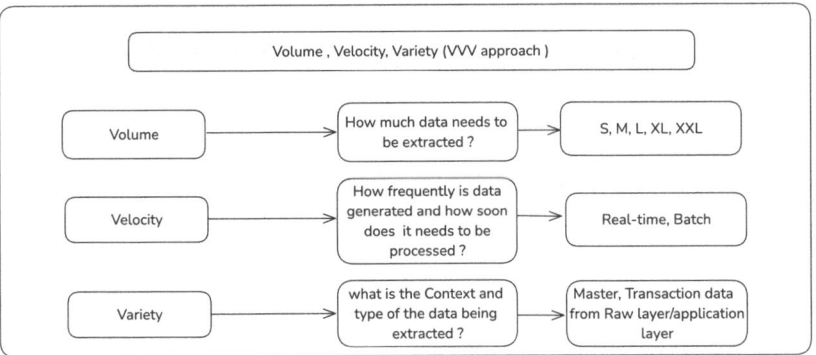

Figure 3-9. *Data extraction consideration using VVV approach*

- **Volume**: Assess the volume of data and performance requirements. This involves understanding how much data needs to be extracted. The volume of data refers to the sheer amount of data generated by SAP systems. SAP environments can produce vast amounts of data from various modules, including financial transactions, inventory management, and other business processes. Large volumes of data might necessitate more robust and scalable extraction tools and architectures to maintain performance.

- **Velocity**: Velocity refers to the speed at which data is generated, processed, and made available for analysis. In an SAP environment, real-time data processing is often critical for decision-making and operational efficiency. For real-time data replication ensuring that data is continuously synchronized between SAP systems and your data lake, use tools that employ a "push" strategy. For less time-sensitive data, a "pull" strategy can be employed using ETL services, which can be effective for periodic data extraction and transformation.

In a push strategy, data is actively sent from the source systems to the target data lake. Source systems "push" the data to the destination at regular intervals or in real time as changes occur. In a pull strategy, the target system (data lake or data warehouse) requests data from the source systems. The target system "pulls" the data either based on its schedule or on-demand.

- **Variety**: Consider the context of the data being extracted. Different types of data (e.g., transactional, master, log data) may require different handling and transformation processes. This also involves distinguishing between raw data and application layer data. Each type of data serves different purposes and has unique characteristics and requirements:

  - **Raw data**: Data that is collected directly from source SAP systems at the database layer without any business rules applied. This data is unprocessed and retains its original format.

  - **Application layer data**: Data that has been processed, aggregated, or transformed by application logic to serve specific business needs. This data is typically derived from raw data but has been tailored to meet the requirements of end users and business applications.

When extracting SAP data, it is also crucial to consider the licensing requirements associated with the tools and systems used. As shown in Figure 3-10, SAP systems and third-party tools often come with specific licensing terms that dictate how data can be accessed, processed, and stored.

Ensure you have the appropriate licenses to access and extract data from your SAP systems. This includes understanding any restrictions or limits imposed by your SAP licensing agreements. If using third-party ETL tools or data integration platforms, verify that their use complies with their licensing terms. This includes any costs associated with data volume, user access, and specific functionalities.

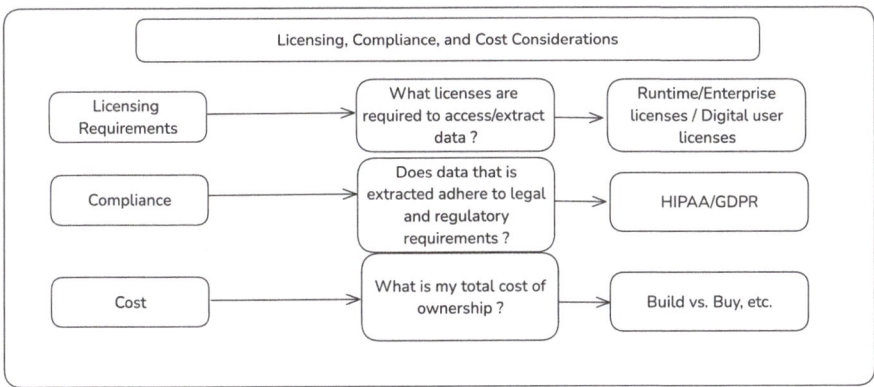

Figure 3-10. *Licensing, compliance, and cost considerations*

It is also critical to ensure that your data extraction process adheres to legal and regulatory requirements. This might involve data encryption, user access controls, and audit trails to maintain compliance with standards such as GDPR, HIPAA, or industry-specific regulations.

When integrating SAP data into an AWS data lake, the choice between building a custom solution or buying a commercial product depends on several factors. Building a custom solution offers complete customization and flexibility, allowing for tailored integration that fits specific business needs. However, it requires significant time, resources, and ongoing maintenance, which can be costly and complex. This approach is ideal for highly specialized processes and organizations with strong in-house development capabilities.

On the other hand, buying a commercial solution provides faster deployment, reliable vendor support, and predictable costs. These solutions are typically less customizable but are well suited for standard business processes and organizations looking for quick, reliable integration without the burden of extensive development and maintenance. The decision hinges on business requirements, budget, time constraints, resource availability, and the need for scalability and flexibility. By carefully evaluating these factors, organizations can choose the approach that best aligns with their strategic goals and technical capabilities.

SAP Data Extraction Patterns

With the decision to build or buy your data integration solution in place, the next crucial step is to explore various SAP data extraction patterns. These patterns determine how data will be moved from SAP systems into your AWS data lake. Each pattern has its own set of use cases, advantages, and challenges, and understanding these will help you select the best approach for your specific needs.

In this section, we will dive into the details of different SAP data extraction patterns such as direct database extraction, real-time data replication, ETL processes, API-based extraction, and hybrid approaches. Each pattern addresses different requirements in terms of data volume, velocity, and variety, ensuring that you have a comprehensive strategy for managing your SAP data effectively. By examining these patterns, you can align your data extraction processes with your overall data strategy, ensuring seamless integration and optimal performance in your data lake.

Setting the Stage: SAP Solutions

When discussing SAP data extraction and integration into AWS data lakes, it's essential to understand the context of SAP solutions. As illustrated in Figure 3-11, Our focus will be on ERP solutions such as SAP ERP Central Component (SAP ECC) and SAP S/4HANA, as well as warehousing solutions like SAP Business Warehouse (SAP BW). Regardless of whether you are running your SAP ERP/BW applications on any databases or are transitioning to S/4HANA, the principles discussed here apply to both scenarios.

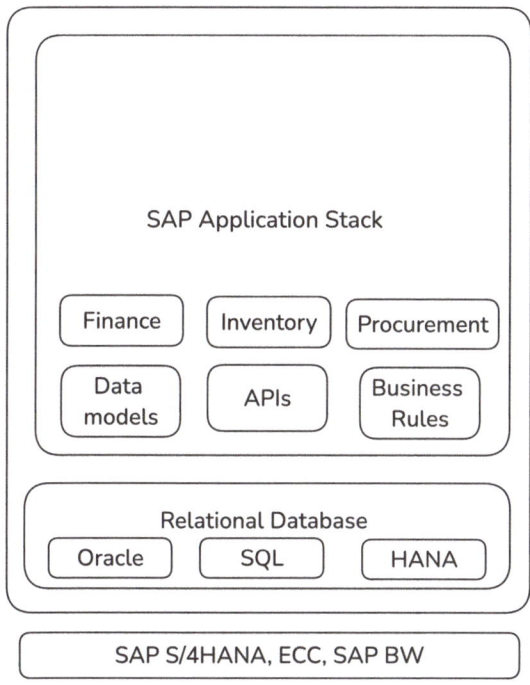

Figure 3-11. *SAP application stack*

At the core of SAP systems, whether you are using ECC or S/4HANA, is a relational database. Even with SAP S/4HANA, which leverages the in-memory computing capabilities of the HANA database, the underlying structure remains relational. The vast majority of SAP applications are built on the Advanced Business Application Programming (ABAP) stack. This means that the business process logic and data management are primarily executed through ABAP code. While HANA offers advanced capabilities like code push-down for performance optimization, the core business logic typically resides in the ABAP layer.

Database Extraction Pattern

One common pattern for extracting SAP data,as illustrated in Figure 3-12, is to go at the database level. This approach involves pulling data directly from the SAP database, often utilizing third-party adapters that extract data from change logs rather than connecting directly to the database as an application. These adapters can read transactional file logs written by the database, capturing changes in near real time without putting excessive load on the database itself.

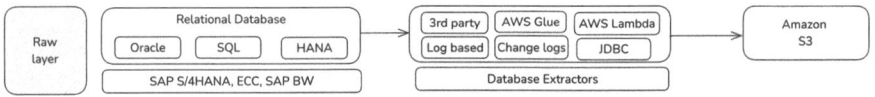

Figure 3-12. *Database-level extraction*

Using Third-Party Adapters

Many customers have successfully implemented third-party adapters to streamline their data extraction processes. These adapters connect to the database's transactional file logs, capturing inserts, updates, and deletions efficiently. This method ensures that the data extraction process is nonintrusive and minimizes the performance impact on the SAP systems.

Using AWS Services

The following AWS services also can be used to extract SAP data:

- **AWS Glue**: As previously introduced, AWS Glue is a fully managed ETL service that can be configured to extract data from SAP databases. By writing custom code, you can use AWS Glue to connect to your SAP database using JDBC drivers. Glue scripts can be written in Python or Scala, allowing for flexibility in processing and transforming the extracted data before loading it into Amazon S3 or other destinations.

- **AWS Lambda**: AWS Lambda functions can be used to write custom extraction scripts that connect to the SAP database. By using JDBC drivers or SAP-provided packages for HANA (available in Python, Node.js, or Java), AWS Lambda functions can directly pull data from the database. This serverless approach ensures scalability and cost-efficiency, as you only pay for the compute time consumed by the function.

Using SAP Database Client Libraries

For SAP HANA databases, SAP provides client libraries that facilitate direct data extraction. These libraries support various programming languages, including Python, Node.js, and Java. Depending on your preferred development environment, you can leverage these libraries to connect to the HANA database and extract data efficiently.

- **Python**: The hdbcli client allows Python applications to connect to SAP HANA and perform data operations.

- **Node.js**: The `@sap/hana-client` library enables Node.js applications to interact with SAP HANA databases.

- **Java**: The SAP HANA JDBC driver allows Java applications to connect to SAP HANA and execute SQL queries for data extraction.

Direct database extraction is a powerful method for integrating SAP data into an AWS data lake, but several factors must be considered. Data from SAP systems is often highly normalized and raw, lacking functional context, which necessitates complex joins and stitching in the target environment. This pattern is ideal for large-scale replication scenarios, capturing changes efficiently via transactional log files. However, it requires significant enterprise licenses for both SAP systems and third-party extraction tools, potentially increasing costs. The process can impose a performance load on source systems, so efficient querying and scheduling during off-peak hours are essential. Security and compliance are critical, requiring encryption, access controls, and adherence to regulations such as GDPR. By addressing these considerations, organizations can ensure a smooth and efficient data extraction process that maintains system performance and data integrity.

Application-Level Data Extraction

Moving up one level from direct database extraction, the application stack offers another layer for data extraction that retains the business context inherent in SAP's packaged applications, as shown in Figure 3-13. In SAP ERP applications, business logic largely resides in the ABAP layer. Even with the code push-down capabilities of the SAP HANA database, the ABAP stack remains a critical entry point for accessing business context through APIs.

Extracting data at the application level ensures that the business logic and relationships within the data are preserved, providing a more meaningful and context-rich dataset.

144

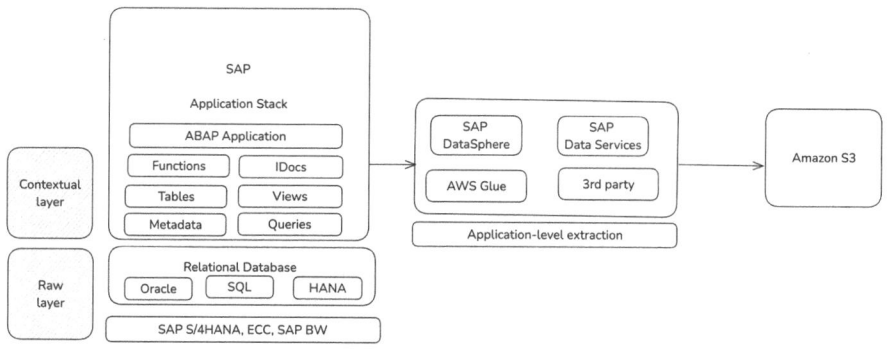

Figure 3-13. *Application-level extraction*

The ABAP stack provides direct access to SAP tables, views, and predefined queries. These can be used to extract data in a structured manner, preserving the business logic. SAP provides various frameworks like IDOC (Intermediate Document, described in Chapter 2), which facilitate data exchange between SAP systems and external systems.

SAP exposes business processes and objects as functions, which are reusable components that encapsulate specific business logic and can be used to extract data while ensuring that business rules are applied. Business Application Programming Interfaces (BAPIs) are standardized programming interfaces that allow external applications to interact with SAP systems. They provide a controlled way to extract data while maintaining the integrity and context of the data.

Application-level orchestration tools, such as SAP Data Services and SAP Datasphere, leverage integration frameworks within the ABAP stack to extract data from SAP applications. These tools help extract and store data in Amazon S3 through default connectors.

SAP Data Services and SAP Data Intelligence are comprehensive data integration and ETL tools that enhance data quality and management across an enterprise. SAP Data Intelligence, a containerized solution powered by the SAP Business Technology Platform, provides robust functionalities for data integration, data quality, and cleansing

SAP Datasphere is a cloud-based data management solution designed to simplify data management and integration. It extends the capabilities of traditional data warehouses and data lakes by providing a unified data experience.

These tools can natively connect with SAP applications using Remote Function Call (RFC SDK) libraries, pulling data from remote function modules, tables, views, and queries. Furthermore, SAP Data Services can install custom ABAP code in the target SAP application to push data from the application, enhancing performance in certain scenarios.

SAP applications support HTTP access to function modules, allowing AWS Glue or AWS Lambda to access these modules using HTTP. SAP has also published the PyRFC library for Python, which can be utilized in AWS Glue or AWS Lambda to integrate natively using RFC SDK. Additionally, SAP IDocs can be integrated with S3 using an HTTP push pattern

Key Considerations for Application-Level Extraction

You should consider the following aspects of application-level extraction:

- **Business context preservation**: Application-level extraction retains the business context, simplifying the extraction process by pulling all related data and its associations mapped through function modules. For example, extracting all sales order data for a particular territory can be done with all related data intact, reducing the need for additional business logic mapping outside SAP.

- **Change data capture (CDC)**: Not all SAP function modules or frameworks support CDC capabilities by default, limiting the ability to capture incremental changes efficiently.

- **Cost and development effort**: Utilizing AWS native services like AWS Glue or AWS Lambda can reduce the TCO by eliminating the need for third-party applications. However, this may require increased custom development efforts to establish HTTP or RFC integrations with SAP applications.

- **Performance considerations**: Application-level extraction may have potential performance limitations compared to database-level extraction due to the additional load on SAP application servers. Pulling data using function modules and other frameworks can add significant processing overhead to the SAP system.

By considering these factors, organizations can optimize their SAP data extraction processes, balancing the need for context-rich data with performance and cost considerations.

Semantic Data Extraction Layer

As you move up the stack in SAP's architecture, the approach to data extraction becomes increasingly sophisticated, offering more preserved business context, semantic data context, and metadata. In this context, "semantic" refers to the meaning, relationships, and associated metadata of the data within its business environment. Semantic data extraction ensures that the extracted data maintains its business logic, relationships, inherent meaning, and important metadata, making it more useful and context-rich for analytics and integration. This higher-level extraction enables organizations to derive more meaningful insights and ensures that data is not just raw but also aligned with the underlying business processes and objectives.

147

The main reason for moving up the stack is to leverage the predefined business logic and interrelationships already established in SAP systems. Higher-level extraction methods ensure that the extracted data maintains its semantic richness, providing more meaningful insights and reducing the need for complex transformations. As shown in Figure 3-14 below, SAP offers various tools and frameworks to facilitate this higher-level extraction.

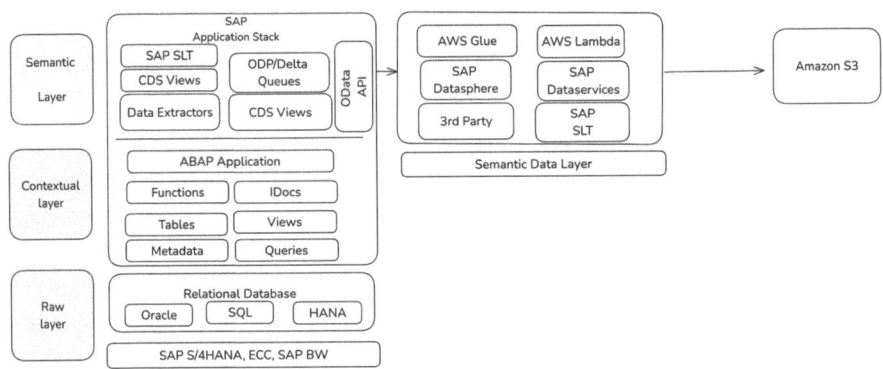

Figure 3-14. *Sematic data layer for ingesting data into AWS data lakes*

For instance, SAP DataSources and Extractors are designed to pull data from SAP ERP systems into SAP BW. These tools retain business logic during extraction by providing predefined data structures and relationships, ensuring that the extracted data is contextually complete. Core Data Services (CDS) views in HANA and S/4HANA encapsulate complex business data and provide semantically rich data models. These views simplify data access by embedding business logic into the data model, ensuring that the extracted data maintains its intended relationships and meanings.

SAP Landscape Transformation (SLT) facilitates real-time data replication by writing triggers on the underlying database to pull data as changes occur, supporting various SAP and non-SAP targets. For replication targets, SAP SLT supports by default SAP HANA, SAP BW, SAP Data Services, and a set of non-SAP databases. For replicating data to targets that are not supported by SAP yet, customers can implement their own customizations using the Replicating Data Using SAP LT replication server SDK.

Operational Data Provisioning (ODP) is another framework designed by SAP for replicating data using a provider and subscriber model. It supports both full data extraction and change data capture (CDC) via operational delta queues. Various SAP data sources, including DataSources, CDS views, and SAP SLT, can act as providers for ODP. Exposing ODP data sources as OData services allows for easy integration with external tools like SAP Data Sphere, AWS Glue, AWS Lambda or third party ETL tools. This provides a flexible and efficient way to pull data from SAP systems.

Open Data Protocol (OData) services play a significant role in this context by enabling REST-based integration with SAP data. OData services provide a standardized way to access and manipulate data using HTTP protocols. By exposing ODP data sources as OData services, organizations can leverage a web-based API to interact with SAP data seamlessly. This means that external tools, including AWS Glue and AWS Lambda, can query and extract data from SAP systems over HTTP, simplifying the integration process and reducing the need for custom connectors.

Using ODP with AWS services further enhances data extraction capabilities. AWS Glue and AWS Lambda can pull data from ODP-exposed OData services, offering a cost-effective solution for data extraction without the need for third-party applications. While this approach reduces the TCO, it may require additional custom development to establish HTTP or RFC integrations with SAP applications. Additionally, SLT can push data to SAP-supported targets, and AWS Glue can pull this data into Amazon S3, ensuring real-time data availability for analytics.

There are several considerations to keep in mind when using these higher-level extraction methods. Business context preservation is crucial, as it ensures that the extracted data retains all table relationships, customizations, and package configurations, reducing the need for additional transformation efforts. Change data capture is supported via operational delta queues in ODP, allowing for efficient incremental data capture. Using AWS native services reduces the need for third-party applications but may increase custom development efforts for HTTP-based integrations. Performance considerations are also important, as application-level extraction may introduce performance limitations compared to database-level extraction due to the additional load on SAP application servers.

By leveraging higher-level data extraction sources such as SAP data Extractors, CDS views, SLT, and ODP, organizations can ensure that the extracted data retains its business context and interrelationships. This approach simplifies the data extraction process, provides seamless CDC mechanisms, enhances data integrity, and reduces the need for extensive post-extraction transformations. Integrating these methods with AWS services like Glue and Lambda provides a robust, cost-effective solution for comprehensive data management and analytics.

Having explored the capabilities of data lakes and data ingestion patterns, next we'll explore the equally important data virtualization and federation techniques, which provide seamless access and integration of disparate data sources without the need for physical consolidation.

Data Virtualization and Data Federation

Traditional data extraction methods, while effective for certain scenarios, can be complemented or even replaced by data virtualization and data federation to address specific needs.

Data Virtualization

Data virtualization is a technique that creates a unified view of data from multiple sources without physically moving or copying the data. Data virtualization uses a middleware layer that connects to the source systems and provides a virtual schema that can be queried by the users or applications. Data virtualization enables real-time access to the latest data, reduces data duplication and storage costs, and simplifies data management and governance.

SAP HANA Smart Data Access (SDA), a powerful feature available since HANA 1.0 SPS 6, enables you to perform data manipulation language (DML) statements on external data sources. You can create virtual tables in SAP HANA that point to tables in remote data sources, as depicted in Figure 3-15.

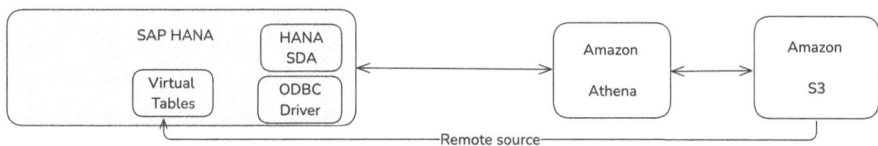

Figure 3-15. *Data virtualization using SAP HANA SDA*

Using the SAP HANA SDA feature and ODBC drivers from Amazon Athena, you can federate queries from SAP HANA using Athena. This allows you to combine data from SAP HANA with data available in an Amazon S3 data lake without needing to copy that data to SAP HANA first. Queries are executed by Amazon Athena, and the results are sent to SAP HANA.

Data Federation: Integrating Disparate Data Sources

Data federation is a software process that combines data from multiple sources by moving or caching partial data for query purposes. This allows the multiple databases to function as one, providing a unified view of the data. Data federation is part of the data virtualization framework. The concept of data virtualization grew with data federation but sprouted extra features, applications, and functions. Data virtualization, therefore, has a huge range of functions outside of data warehouse compilation. It includes metadata repositories, data abstraction, read and write access to source data systems, and advanced security.

Data federation is virtual database that combines heterogenous data sources and provides with a common access point.

Let's take a look at an example architecture as illustrated in Figure 3-16 below. If you use data lakes in Amazon S3 and use SAP HANA as your transactional data store, you may need to join the data in your data lake with SAP HANA (for example, to build a dashboard or create consolidated reporting).

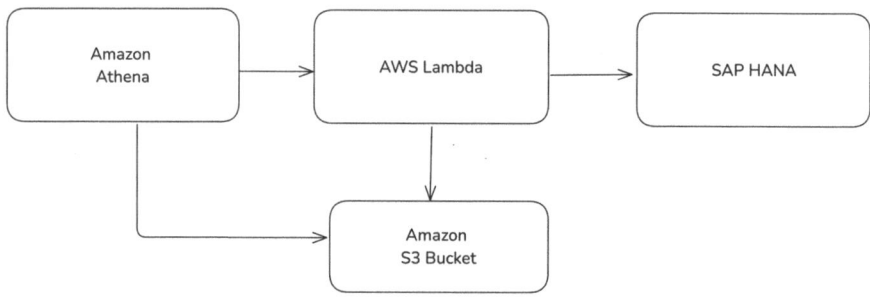

Figure 3-16. *Data federation using Amazon Athena*

In such use cases, Amazon Athena Federated Query allows you to seamlessly access the data from your SAP HANA database without building ETL pipelines to copy or unload the data to the Amazon S3 data lake or SAP HANA. This removes the overhead of creating additional ETL processes and shortens the development cycle.

As previously introduced, Amazon Athena is an interactive query service that makes it easy to analyze data in Amazon S3 using standard SQL. If you have data in sources other than Amazon S3, you can use Athena Federated Query to query the data in place or build pipelines to extract data from multiple data sources and store them in Amazon S3. With Athena Federated Query, you can run SQL queries across data stored in relational, nonrelational, object, and custom data sources.

Amazon Athena uses data source connectors that internally use AWS Lambda to run federated queries. Data source connectors are prebuilt and can be deployed from the Athena console or from the AWS Serverless Application Repository. When a federated query is run, Athena identifies the parts of the query that should be routed to the data source connector and runs them with AWS Lambda. The data source connector makes the connection to the source, runs the query, and returns the results to Athena. If the data doesn't fit into Lambda's runtime memory, it spills the data to Amazon S3 and is later accessed by Athena. Based on the user submitting the query, data source connectors can provide or restrict access to specific data elements.

Comparison of Data Federation and Data Virtualization

Both data federation and data virtualization offer powerful solutions for integrating and accessing data across multiple sources in an SAP environment. Understanding the key differences between these approaches can help organizations choose the right strategy based on their specific needs and goals. The following table summarizes the main distinctions between data federation and data virtualization:

Aspect	Data Federation	Data Virtualization
Definition	Combines data from multiple sources by moving or caching partial data for query purposes.	Provides real-time access to data across multiple sources without moving it.
Performance	Balances between performance and real-time access; suitable for complex, high-demand queries.	Offers real-time access, but may have performance trade-offs in very complex environments.

(continued)

Aspect	Data Federation	Data Virtualization
Data movement	Involves partial data movement or caching to optimize performance for specific queries.	No data movement; data remains at its original source.
Use case	Ideal for scenarios where performance optimization is critical and data sources are stable.	Best suited for agile, on-demand data access and analytics, particularly in dynamic environments.
Complexity of data sources	Can handle complex queries involving multiple data sources but may require configuration to optimize performance.	Simplifies access to multiple data sources by providing a unified view without complex configurations.
Business needs	Suitable for businesses needing both high performance and access to large, complex datasets.	Ideal for organizations requiring quick, flexible access to diverse data sources for analysis.

Building on the discussion of data virtualization and federation, which streamline access to diverse data sources, we now move to end-to-end enterprise analytics, which leverages these integrated datasets to deliver comprehensive insights and drive informed decision-making across the entire data pipeline.

End-to-End Enterprise Analytics

In the previous sections, we discussed various data extraction methods and the importance of data federation and virtualization. These techniques are foundational for integrating data from multiple sources, ensuring that the data retains its business context, and enabling seamless access for analysis. However, to truly harness the power of this integrated data, we need to shift our focus to end-to-end enterprise analytics (see Figure 3-17). This transition involves not only integrating and managing data but also leveraging advanced analytics and machine learning capabilities to drive business insights and innovation.

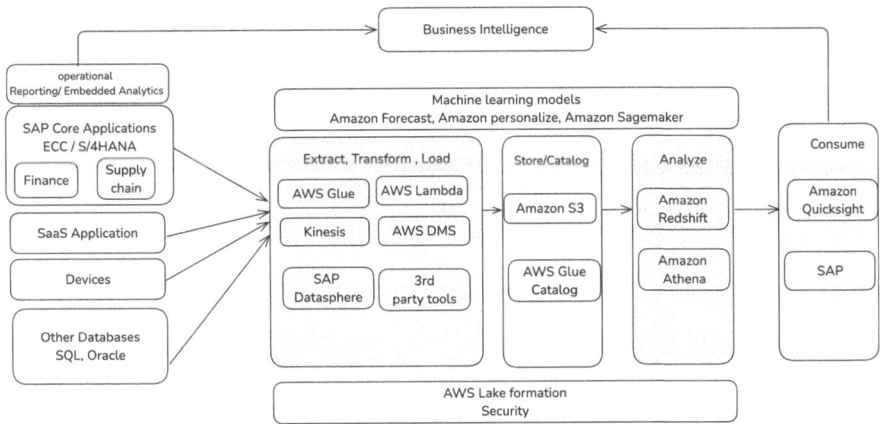

Figure 3-17. *End-to-end enterprise analytics with SAP data lakes on AWS*

At the end of the day, what do enterprises want to do? They don't just want to pull data out of SAP and put it on Amazon S3—that's merely the starting point. SAP ECC or S/4HANA already provides operational reporting out of the box. Especially with S/4HANA, embedded analytics efficiently handle operational reporting, enabling customers to perform such reporting directly within SAP.

However, enterprises are also investing in non-SAP applications like SaaS applications, IoT, and stand-alone databases. This diversification necessitates bringing all of this data into a common landing layer. To achieve this, you can utilize purpose-built ETL services like AWS Glue, AWS Lambda, AWS Database Migration Service (AWS DMS), or third-party tools to pull data from SAP and other ecosystems and place it into Amazon S3.

Streamlining with AWS Lake Formation

AWS Lake Formation simplifies the setup and management of your data lake. It automates the tasks of securing, creating, and partitioning data buckets, making it easier to organize and manage your data. With AWS Lake Formation, you can streamline data partitioning, apply security policies, and set up access controls to ensure that only authorized users can access specific datasets.

Applying Machine Learning

Once your data is in Amazon S3, you can leverage advanced machine learning to derive deeper insights. AWS offers a suite of out-of-the-box ML services. For example, Amazon Forecast can be used for demand forecasting, and Amazon Personalize helps build recommendation engines. Moreover, you can bring your custom ML models and utilize Amazon SageMaker to train and deploy them at scale. SageMaker provides a comprehensive environment for developing, training, and deploying ML models, ensuring you can tailor the models to your specific business needs.

Performing Analytics

Storing data in Amazon S3 allows you to perform a variety of analytics using AWS services. Amazon Redshift, a fully managed data warehouse, enables complex querying and reporting. Amazon Athena lets you run SQL queries directly on data stored in Amazon S3 without needing to move the data, making it a powerful tool for ad hoc querying.

User Access

Providing users with seamless access to data is crucial. Through APIs, users can interact with the data without needing deep technical knowledge. SAP S/4HANA can federate queries, allowing you to query data stored in Amazon Redshift without moving it into HANA. This approach ensures that data remains centralized, reducing duplication and improving efficiency. Visualization tools like Amazon QuickSight can be used to create dashboards and reports, providing users with intuitive ways to interact with and understand the data.

Achieving End-to-End Enterprise Analytics

By integrating data from various sources into a centralized data lake on Amazon S3, you establish a robust foundation for enterprise analytics. Amazon S3's cost-effectiveness, global availability, and high durability make it an ideal choice for storing both structured and unstructured data. Building on this foundation with advanced analytics and machine learning capabilities allows organizations to transform data into actionable insights.

We discussed how enterprises can create scalable data lakes on AWS by combining data from multiple sources to derive value from their data. In the previous sections, we highlighted the importance of data extraction, federation, and virtualization as foundational steps for integrating data from various sources. This integration enables organizations to harness the power of this data through advanced analytics and machine learning, ultimately driving business insights and innovation. However, to further enhance your data management capabilities, it's essential to explore the concept of Data Mesh.

Data Mesh: A Paradigm Shift in Data Architecture

Data Mesh is an *architecture* principle. Data Mesh is a paradigm shift in data architecture that decentralizes data ownership and management to domain-specific teams, treating data as a product. This contrasts with traditional centralized data lakes, where all data is ingested into a single repository. While data lakes provide scalable storage and access to diverse data sources, they often struggle with governance, quality, and agility due to their centralized nature.

Data Mesh addresses the limitations of monolithic data architectures by promoting a federated approach where data is treated as a product and owned by cross-functional teams. This paradigm emphasizes the decentralization of data ownership, governance, and management to domain-specific teams, enabling greater scalability, agility, and responsibility.

Importance of Data Mesh from a Data Lake Perspective

Traditional data lakes centralize data storage and processing, which can lead to bottlenecks, governance challenges, and difficulties in scaling as data volumes grow. Data Mesh addresses these issues by decentralizing data ownership and enabling domain teams to manage their data independently. This approach not only distributes the workload but also aligns data management closer to the domain experts, who understand the data best.

In a Data Mesh architecture, as shown in Figure 3-18 below, data is organized around business domains, such as finance, marketing, supply chain, and so on. Each domain owns its data pipelines, ensuring that data quality, governance, and access controls are maintained within the domain. This structure fosters greater agility and responsiveness, as domain teams can quickly adapt to changes and innovate without being hampered by centralized processes.

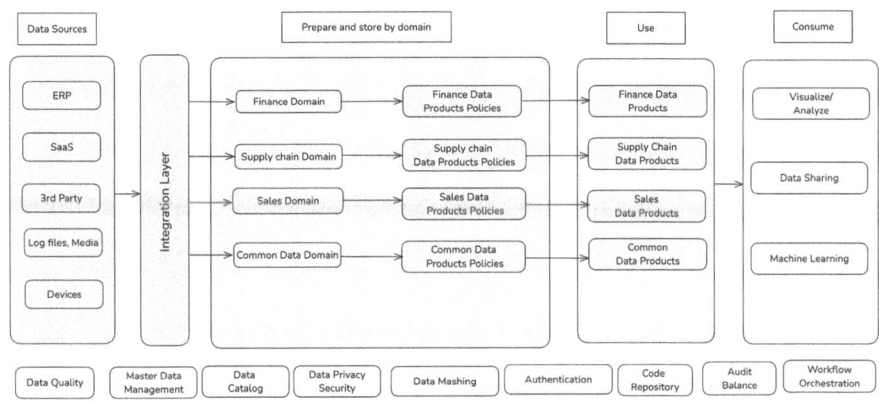

Figure 3-18. *Domain-centric data architecture*

Achieving Data Mesh Architecture for SAP Data Using AWS Services

Enterprises can leverage AWS services to implement a Data Mesh architecture effectively. By integrating SAP data with AWS's robust suite of data services, organizations can create a decentralized, domain-centric data infrastructure that promotes data ownership, governance, and scalability.

The primary AWS service used to achieve Data Mesh architecture is Amazon DataZone, which provides a comprehensive environment for managing data across an organization. It enables the creation of data zones, which are logical groupings of data resources based on domains or projects. Each data zone can have its own policies, access controls, and governance frameworks, allowing for decentralized management of data.

Enterprises leveraging SAP ERP can use Amazon DataZone to segment data by business domains. For example, the finance domain can have its own data domain where financial data from SAP systems is ingested,

159

processed, and stored. Similarly, the supply chain domain can have its own data zone for supply chain data. This segregation ensures that domain teams have full control over their data, aligning with the principles of Data Mesh.

Enhancing Data Governance with Data Mesh Architecture

Data access control is challenging, is managed differently across organizations, and often requires manual approvals, which can be a time-consuming process and hard to keep up to date, resulting in personas not having access to the data they need.

The following are the key benefits of Amazon DataZone:

- **Govern data access across organizational boundaries**: Without relying on individual credentials, Amazon DataZone ensures that the right user accesses the right data for the right purpose, adhering to your organization's security regulations. It also helps provide transparency on data asset usage and approve data subscriptions with a governed workflow. The usage auditing capabilities allow for the monitoring of data assets across various projects.

- **Connect data and people through shared data and tools to drive business insights**: Amazon DataZone helps increase your business team's efficiency by helping organizations collaborate seamlessly across teams and providing self-service access to data and analytics tools. Using Amazon DataZone features like business terms search, enterprises can share and access cataloged data, making data accessible to all the configured users to learn more about the data they want to use with the business glossary.

- **Automate data discovery and cataloging with ML**: Enterprises can now reduce the time needed to manually enter data attributes into the business data catalog and minimize the introduction of errors with ML-powered cataloging in Amazon DataZone. More and richer data in the data catalog improves the search experience, too. This reduces the time that enterprises spend searching for and using data from weeks to days.

These benefits make data accessible to everyone in the organization, streamlining the process of finding and using data.

Transitioning to a Data Mesh architecture offers significant advantages for organizations seeking to enhance their data lakes. By decentralizing data ownership and management, Data Mesh promotes scalability, agility, and responsibility. For SAP customers, integrating SAP data with AWS services provides a powerful combination that supports the principles of Data Mesh.

AWS services like DataZone, Lake Formation, Glue, and SageMaker enable the creation of domain-centric data layers, robust data governance, and advanced analytics capabilities. By adopting a Data Mesh architecture, organizations can achieve end-to-end enterprise analytics, driving innovation and informed decision-making.

In conclusion, Data Mesh represents a paradigm shift in how organizations manage and utilize their data. By decentralizing data ownership and leveraging the power of AWS and SAP, businesses can build a flexible, scalable, and robust data infrastructure that meets the demands of modern data-driven enterprises. For more information on other data-leveraging mechanisms, including Amazon Bedrock Knowledge Bases, Amazon Q, and Amazon Q for QuickSight, refer to Chapter 6.

Summary

In this chapter, we explored essential components of a modern data strategy, including data archiving, data lakes, data virtualization and federation, and end-to-end enterprise analytics. We discussed how effective data management—through archiving and leveraging data lakes—can optimize system performance and enhance flexibility. We also examined the roles of data virtualization and federation in integrating diverse data sources and enabling real-time access. Finally, we highlighted the importance of a cohesive analytics ecosystem for transforming data into actionable insights. By implementing these strategies, your organization can align data management with business goals, driving innovation and maintaining a competitive edge.

CHAPTER 4

Extending SAP Business Processes

SAP business process modernization transforms inefficiencies into growth, driving efficiency, agility, and innovation at every step.

In a rapidly evolving business landscape, where technology-driven unicorn startups are increasingly replacing established S&P 500 companies, the imperative for transformation is clear. A striking statistic from Innosight's Corporate Longevity Forecast (`https://www.innosight.com/wp-content/uploads/2021/05/Innosight_2021-Corporate-Longevity-Forecast.pdf`) indicates that average company lifespan of the S&P 500 companies are expected to come down to 15-20 years this decade, underscoring the critical need for established businesses to innovate or face the risk of obsolescence.

This chapter focuses on the necessity of business process innovation in this high-stakes environment, particularly through the lens of SAP and AWS technologies. It's no longer just about making small incremental improvements to an existing business process; with all the cloud technologies available, redefining the business process in its entirety brings more value to the business. In this chapter, we aim to guide you on a transformative journey, demonstrating how the integration of SAP with AWS services can lead businesses not just to adapt but to fundamentally

B. Baruah et al., *Evolve from Infrastructure to Innovation with SAP on AWS*, https://doi.org/10.1007/979-8-8688-0890-6_4

reinvent themselves. This journey is about moving from simply competing in the current market to completely reshaping business models and strategies to thrive in the face of rapidly emerging technological disruptors.

This chapter presents real-world use cases of successful innovation initiatives undertaken by leading organizations. By studying these practical examples, you can gain confidence in pursuing uncharted territories, kindle new ideas, and identify best practices to increase the likelihood of achieving your desired outcomes. These use cases serve as a testament to the power of innovation and a roadmap for navigating potential challenges along the way.

The following topics will be covered in this chapter:

- Strategic innovation approach

- Leveraging the right technology stack

- Joint Reference Architectures from SAP and AWS

- Extending SAP business processes with the AWS SDK for SAP ABAP

- Integration of AWS B2B Data Interchange into SAP business processes

Strategic Innovation Approach

As much as you are ambitious to change the status quo and deliver new capabilities to your business, you don't want to burn yourself in that process either—for example, committing to your business that you will deliver a new machine learning–based prediction capability for financial forecasting and realizing later that the problem is more complex than you initially thought. That is overpromising and underdelivering, which erodes the trust and your brand value in your organization. By using the evaluation framework shown in Figure 4-1, you can make a meaningful commitment.

164

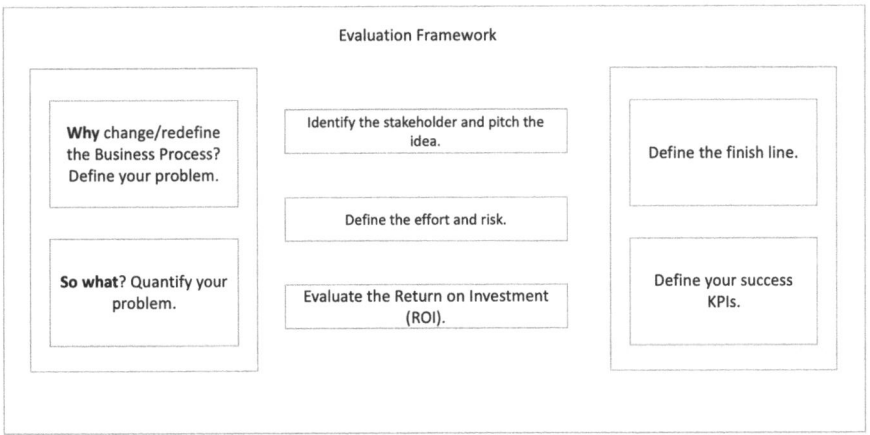

Figure 4-1. *Evaluation framework*

Use this framework in the early stage of your exploration to evaluate if you have a use case that sets you up for success. You will gather data points for all the boxes in this framework. With that, you can assess your vision's worth.

How to Use the Evaluation Framework

As a real-life example of how to use this evaluation framework, suppose you are using a third-party solution that integrates with your SAP system and uses optical character recognition (OCR)-based technology to process your handwritten vendor invoices and post them into the SAP system. Now, you are going through the documentation about the AWS SDK for SAP ABAP that can call Amazon service called Amazon Textract. You are reading its capabilities and you believe that this solution could do a better job than the current product with easier integration because of the AWS SDK for SAP ABAP. You also find out that the AWS service would be cheaper than the existing solution. You are excited and want to propose this idea to your leadership. Let's use the evaluation framework shown in Figure 4-1 to evaluate your idea.

The first step is to define the current state and target state. Starting with the current state, which involves two steps, you first write your problem statement-"The why". You describe that your current OCR system is costly, needs multiple teams to manage and takes time to onboard a new vendor because the system needs to be trained for each new vendor invoice with structural changes.

Next, you quantify the problem statement. The objective is to answer the question "So what?" regarding your problem. Suppose you are spending $100,000 for the license of your OCR solution and $25,000 for the operational cost per year. Onboarding a new vendor's invoice processing currently takes approximately two weeks, and every format change takes a week or so. Diving deeper, you determine that you are processing 15,000 pages of documents per month. Next, using the AWS Pricing Calculator (`https://calculator.aws/#/`), you calculate the cost of using Amazon Textract to perform this processing would be $270 per month, or $3,240 per year, resulting in ~98% cost savings compared to $125K. Now this is a compelling reason.

So, you now can define your target state:

- We will simplify the architecture by reducing the $25,000 operational cost every year.

- We will reduce the license cost by 98% (~$98K) per year.

- For new vendor onboarding, we are aiming for two weeks (no change here, considering the ABAP development efforts to handle each new invoice format).

So, this exercise demonstrates that a compelling reason for switching to Amazon Textract is cost savings. Over a five-year period, it's a significant savings ($608K). But, it is a compelling reason only if the solution meets the success criteria of processing all the documents with 95% accuracy with no human intervention. You would have to clearly define these KPIs.

But, wait a minute! You still need to assess the efforts and risks of making the change before you pitch your great idea! You need to identify what could go wrong. In this case, risks could include a skills gap, feature gap, and resistance to change. The goal is to avoid finding a roadblock midway into your project, such as realizing that Amazon Textract can't replace a particular technical capability that is unique to your current solution.

So, you need to measure the risks and the time investment. To address this risk, you decide to run a mini proof of concept (POC) for two weeks. You select all the types of invoices in your business and build text extraction applications in the lower environments. Your investment for this POC is two weeks of developer time and less than $50 for Amazon Textract (assuming your volume will be fractional compared to production use case). Against $608K savings over five years, spending $50 + two weeks of developer effort is a compelling pitch to convince your leadership to permit this POC with the data you have. By using this strategy, you have not yet committed, but you have achieved a safe experimentation platform.

Next, you identify the stakeholders and pitch the idea. We discuss about this in detail in Chapter 8, section: "How To Get Business Buy-In?". After obtained permission to execute your POC, you do the nitty-gritty work to implement the AWS SDK for SAP ABAP, configure Textract for multiple invoice types, and write ABAP code to implement the solution. Once your POC is successful, you have all the data points you need between your current state and target state box in the evaluation framework. You have efforts, known issues, and risks. Now you are already set for success! So, this exercise gives you a clear picture of the value that you will bring to your business with your vision.

With this section, you have learned how to identify a successful business process innovation use case. You know how to use the evaluation framework to measure multiple dimensions of your vision and set a success path. So, you picked your battle wisely! Now, you need an arsenal to fight it. That's what the next section is about. It introduces an array of technologies to plan your strategy.

Leveraging the Right Technology Stack

There are 200+ AWS services and 80+ SAP Business Technology Platform (BTP) services. The purpose of this section is not to list and describe them all. That would be like displaying all the tools in front of you with their manual. That is not the intent. Rather, we will explain the strategy to pick the right technology stack for your organization's needs.

It's simple! You have only two options: through SAP BTP services, and through AWS native services via the AWS SDK for SAP ABAP. Although the choices are simple, determining which technology stack is the best path is more challenging due to the rapid technological advancements that offer numerous equally viable options. To help you make sure you are choosing the most optimal path, the following subsections explain both options— using SAP BTP and using AWS native services. We will conclude with a methodology to choose the right solution.

SAP Business Technology Platform

SAP BTP has 80+ services with various purposes. As an analogy, consider this as a fully stacked pantry given to a chef. What that chef can make with the 90+ ingredients is limitless. It's up to the chef's imagination and the customer's demand. Similarly, consider SAP BTP as the pantry and you are the chef. You can mix and match the technologies and deliver new solutions to your business.

SAP Discovery Center

To explain each SAP BTP service to customers, SAP offers not only extensive online documentation but also a service discovery portal, SAP Discovery Center, which offers the following three sections to help customers discover what they need:

- **Services**: A customer can learn the purpose of a service, for example use case, the underlying cloud provider on which they are available, and pricing details.

- **Missions**: Provides model end-to-end working solutions with artifacts, with step-by-step guides to implement them. This resource enables customers to jumpstart their project faster.

- **Reference Architectures**: Provides architecture guidance that is vetted by SAP and AWS architects.

We encourage you to visit and explore the SAP Discovery Center to gain an overall understanding of the crucial SAP BTP services.

Critical Architectural Considerations

Now, let's shift gears and identify a few strategic points that will influence how you architect your solution. We will begin with performance and security considerations.

The first strategic point to consider is that many SAP "legacy" products are already moving into SAP BTP or would be, in the near future. SAP BTP services run on cloud service providers (CSPs) such as AWS, Microsoft Azure, Google Cloud Platform (GCP), and SAP's own cloud environment. AWS is the only CSP that runs all the SAP BTP services that are made available to CSPs. Why does this matter? You would want to host your SAP system on the same CSP which hosts all the BTP services, that are made available to the CSPs, because that way all your SAP systems and BTP services are on the same network, benefiting you with low- latency, improved performance and enhanced security. You can use the AWS PrivateLink to achieve private connection between SAP BTP services and your SAP systems (including S/4HANA) on AWS. This is a crucial decision point when you are selecting a CSP to host your SAP S/4HANA on RISE.

The second strategic point of note is that with the new SAP S/4HANA release, SAP is promoting a strategy called *clean core*, the backbone of which is SAP BTP. *Clean* means up-to-date, unmodified, documented, consistent, efficient, and cloud compliant; *core* refers to the foundation of the SAP ecosystem—the standard SAP functionalities, processes, data, integrations, and operations in the S/4HANA system.

Why is clean core important? Because it allows accelerated version upgrade cycles to the latest releases of S/4HANA, thereby allowing you to access the latest innovation rolled out by SAP to gain competitive edge, thereby enhancing security and performance, data simplification, and lower TCO. Custom code cannot be avoided completely, because in certain cases, modifying the standard code makes the customer's business process unique, giving them a unique edge. So, clean core is a strategy that balances the value created by custom code against the liability that comes with it. With a clean core approach, you focus on extending your SAP S/4HANA system via extensions built on SAP BTP.

A strategic point that you are familiar with from previous chapters is that you can accelerate innovation and reform business processes by integrating your SAP systems with AWS services. By utilizing SAP BTP, including SAP Open Connectors, part of the SAP Integration Suite, you have access to prebuilt connectors for AWS services like Amazon Simple Storage Service (Amazon S3) and Amazon Simple Queue Service (SQS). This integration offers a broad range of applications, from analytics to machine learning, enhancing your business's flexibility and scalability. By leveraging these tools, you can modernize your SAP landscape and streamline operations, tapping into the constructive collaboration between SAP and AWS technologies for your digital transformation.

SAP Integration Suite is also a great candidate if you are aiming to integrate SAP SaaS applications. For example, if you want to build a data lake out of the SAP Concur system in AWS using Amazon S3, a combination of SAP Integration Suite and SAP BTP Datasphere is a great way to do that.

AWS Technology Stack

AWS's technology stack, featuring over 200 services provides advanced capabilities across various domains, including object storage, data analytics, AI/ML, IoT etc. AWS probably has a service for every imaginable business problem. But how to integrate those with a SAP System? Thats where the AWS SDK for SAP ABAP comes into picture. The AWS SDK for SAP ABAP makes it easy for ABAP developers to extend transform SAP-based business processes by harnessing the power of AWS services using the SAP ABAP language. With the AWS SDK for SAP ABAP you can reduce architectural complexity by eliminating the need to manually create complex integrations between SAP and AWS services, in both self-managed and RISE with SAP deployments. You can maintain a strong security posture by encrypting all payloads with HTTPS and enforcing SAP-level permission.

AWS SDK for SAP ABAP

As an SAP customer, the AWS SDK for SAP ABAP offers you a seamless way to integrate 200+ AWS services into your SAP environment. This tool is designed to be familiar to SAP developers, using native ABAP constructs to simplify the integration process. What might have taken thousands of lines of code for a complex integration could be done with simple four-line code in the AWS SDK for SAP ABAP. An ABAP developer can tap into the full potential of AWS services without leaving their comfort zone. This also opens the opportunity to experiment with the latest services. For example, within two weeks of the release of Amazon Bedrock, a generative AI service, the AWS SDK for SAP ABAP offered the feature to be able to connect to Amazon Bedrock.

On a high level, the AWS SDK for SAP ABAP is easy to implement and secure. The AWS SDK for SAP ABAP documentation explains the process in great detail, so we are distilling it to address the key points.

171

First, you have to import transports to implement the AWS SDK for SAP ABAP. Those transports are exclusive. For example, if you wish to use Amazon Textract and Amazon Bedrock for your application, you need to import the two transports corresponding to these two services along with the base transports. This is to ensure that you are implementing only the transport that you need.

The second aspect is security. The AWS SDK for SAP ABAP gives you the ability to control granular authentication and access at the SAP and AWS layer, and it's so integrated that you can control it via SAP's roles and profiles, too. The AWS SDK for SAP ABAP is designed to be flexible, accommodating different hosting scenarios and simplifying the process of connecting SAP systems with AWS services. This integration enables you to leverage AWS's powerful cloud capabilities within your existing SAP applications, enhancing functionality and efficiency.

The AWS SDK for SAP ABAP is your go-to service for infusing AWS cloud capabilities into your day-to-day transactions. Because of its seamless integration, with few lines of code, the ABAP team does not need to wait for an integration team to set up the channels. The ABAP developers have full control of calling the AWS services and handling the output within the same ABAP program they are working. This reduces the complexity multifold and opens a new playground for them.

Methodology for Choosing a Technology Stack

Determining which technology stack to use - SAP BTP services or AWS native services might be overwhelming at times. The right choice often depends on your specific use case, and there is no one-size-fits-all answer. However, a few guiding principles can help streamline the decision-making process.

If your goal is to integrate an SAP SaaS product with your SAP S/4HANA system, SAP BTP services are typically the preferred option. On the other hand, if you're looking to enhance or simplify SAP business processes within your SAP S/4HANA system using functionality available through AWS, then AWS SDK for SAP ABAP is likely the best fit. In some scenarios, both SAP BTP and AWS native services can address the same needs, providing flexibility in how you approach solutions. For example, if it a use case for SAP data extraction both SAP Datasphere and AWS services like AWS Glue or Amazon AppFlow can be the answer. So, they all can work together or independently in the way that you need. By focusing on your success criteria, you can choose the right formula for you. The general flow described next can help you.

The first step is to evaluate the capabilities to achieve what you need in each service you intend to use. Pick the best fit for your problem statement. For example, if you need a complex IDoc mapping integration that is available only in the SAP Integration Suite, you would choose that over AWS services.

Once you have identified your options, as a second step, evaluate the cost of each (all the SAP BTP and AWS services have public pricing, which makes this process transparent). Determine the trade-offs and pick a solution that works for you. To further guide you with this, AWS and SAP have released Joint Reference Architectures, described in the next section.

In various chapters throughout this book, you will see various architecture patterns using SAP BTP services and AWS services for different use cases. Some of these architecture patterns, which SAP and AWS has come up with, via collaboration between the two companies, are what we are going to discuss next. The following section will familiarize you with the process of choosing among the variety of SAP BTP and AWS services, depending on the use case.

Joint Reference Architectures from SAP and AWS

In this section, you are going to learn about the Joint Reference Architectures (JRAs) from SAP and AWS and how to use them as a guide to infuse innovation into SAP business processes using SAP BTP and AWS services. This section includes various use cases to help demonstrate how to leverage a Joint Reference Architecture.

As previously mentioned, there are 200+ AWS services and 90+ SAP BTP services. As the numbers suggest, the combinations are countless. How do you select a perfect blend to achieve your business outcomes? For that purpose, experts from SAP and AWS worked together to produce Joint Reference Architectures (`https://community.sap.com/t5/technology-blogs-by-sap/sap-and-aws-joint-reference-architectures-to-maximize-utilization-and/ba-p/13549809`), a collection of solutions developed and tested by the architects at SAP and AWS based on customer use cases they have seen. This is a good starting point for you to identify some common use cases that can deliver value to your business.

The JRAs serve as a comprehensive guide for businesses aiming to maximize their SAP investments through integration with SAP BTP services, powered by AWS and AWS native services. Each JRA acts as a roadmap, offering insights into how these technologies can be blended to address various business scenarios. The JRAs share principles and service capabilities to help businesses achieve their goals efficiently and cost-effectively, making them a valuable resource for organizations looking to innovate and transform their operations.

The SAP and AWS JRAs are divided into three major pillars:

- Data-to-value architecture

- Integration and app development

- Platform foundation

Let's look at each of these categories, along with use cases to help clarify them.

Data-to-Value Architecture

Data-to-value architecture explains how to use SAP BTP services like Datasphere to extract, store, or federate access to the data stored in AWS data lakes or warehouses like Amazon Redshift. By using this architecture, customers can combine SAP and AWS cloud-native apps data and deliver new insights to their business. This reduces the barrier caused by data silos.

Example Use Case: Federation of SAP and Non SAP Data

Problem

Businesses often struggle to make informed decisions due to data being siloed and inaccessible. For example, SAP data might be stored in SAP Datasphere while non-SAP data is stored in Amazon S3 or Amazon Redshift. The challenge is to harness the vast amounts of data generated by various business processes and turn it into actionable insights. While trying to achieve this objective, technical and security challenges often are encountered due to different technical platforms that become roadblocks.

Solution

SAP Datasphere in SAP BTP offers promising features to solve the problem just described. With its federated data access across various technology platforms, SAP Datasphere can seamlessly integrate the data that is stored in the AWS cloud data lake with SAP data. In addition, it provides a data modeling and analytics feature, which makes it a single-stop solution to solve complex data problems.

Architecture

Understanding the architecture requires a high-level knowledge about SAP Datasphere, Amazon S3 data lake, Amazon Redshift, and Amazon Athena, all of which were introduced in Chapter 1. The architecture diagram presented in Figure 4-2 illustrates the flow of data between various components in a data integration and analytics platform.

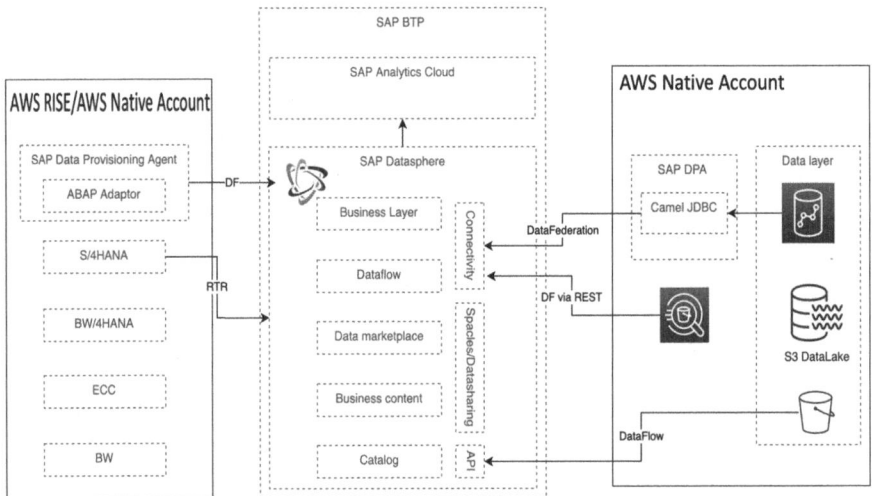

Figure 4-2. *Data-to-value architecture example*

This Joint Reference Architecture is explained in detail in a two part SAP Community blog series:

- Federating Queries from SAP Datasphere to Amazon Athena to Derive Insights: `https://community.sap.com/t5/technology-blogs-by-sap/federating-queries-from-sap-datasphere-to-amazon-athena-to-derive-insights/ba-p/13475132`

- Federating Queries in HANA Cloud from Amazon
  Athena using Athena API Adapter to Derive Insights:
  https://community.sap.com/t5/technology-blogs-
  by-sap/federating-queries-in-hana-cloud-from-
  amazon-athena-using-athena-api/ba-p/13476091

As depicted in Figure 4-2, SAP Datasphere plays a central role in aggregating and modeling data from various sources. On the left side of the diagram are the traditional SAP applications, which may operate under RISE with SAP or a customer's perpetual license within the AWS environment. The SAP Data Provisioning Agent (SAP DPA) is equipped with an ABAP adaptor to facilitate data extraction from ABAP foundation-based systems such as SAP S/4HANA and BW/4HANA, as well as NetWeaver-based systems like SAP ECC or SAP BW. This data is then transfered to SAP Datasphere within SAP BTP.

Consider a scenario where an organization has constructed a data lake using AWS cloud-native services, incorporating a cloud-native data warehouse like Amazon Redshift and storage in Amazon S3 buckets. Should the business need to amalgamate data from these disparate sources for analytical purposes to inform business decisions, SAP Datasphere is well equipped to handle this task. It can connect to Amazon Redshift using CamelJDBC or perform queries against data stored in Amazon S3 using AWS Athena. Additionally, SAP Datasphere can initiate data flows from Amazon S3 through API calls to retrieve the necessary information.

Once SAP Datasphere has gathered data from both traditional SAP applications and cloud-native data lakes and warehouses, it can proceed to model the data within its business layer. Subsequently, this modeled data can be forwarded to SAP Analytics Cloud to facilitate reporting. This process effectively bridges data silos and leverages the power of multiple data sources, enabling businesses to make informed decisions without the need to physically move data, thanks to the federated query capabilities.

Let's look at this architecture diagram in detail. We have divided the overall architecture into three components: data ingestion, SAP Datasphere, and data federation.

Data Ingestion

The data ingestion process starts from the left side of the diagram, where data sources like SAP S/4HANA, BW/4HANA, ECC, and BW feed data into the SAP Datasphere component via facilitators like SAP Data Provisioning Agent and ABAP Adaptor.

Usually, SAP Datasphere acts as a central data integration or warehousing layer, receiving data from various sources. But, apart from data warehousing capabilities, SAP Datasphere can also act as only an extraction service. Based on a customer's needs, it can extract the SAP data from S/4HANA and send it to a data lake on Amazon S3. Typically, customers who want to have a central (SAP and non-SAP) data lake on Amazon S3 would opt for this option. This is not what is depicted in this architecture diagram, but this option is worth mentioning.

SAP Datasphere

Within SAP Datasphere, there are several subcomponents:

- **Business layer**: This layer handles data transformation, cleansing, and business logic operations.

- **Dataflow**: This component manages the flow and movement of data within SAP Datasphere. (Bringing data from Amazon S3 was discussed earlier.)

- **Data marketplace**: This acts as a central repository or catalog for the integrated data.

- **Business content**: This component stores and manages business-specific data and content.

- **Catalog**: This component maintains a catalog or metadata repository of the available data sources and assets.

Data Federation

By using the architecture that federates or queries data live from its source systems, you can leverage the value of combining datasets from various sources like Amazon Redshift, SAP S/4 HANA, and so forth. SAP Datasphere data federation architecture helps combine the data from sources outside of SAP (e.g., Amazon S3) and SAP business data on the fly to make richer business semantic models:

- With built-in connectors in SAP Datasphere for Amazon Athena, data can now be federated live through virtual models in SAP Datasphere, for use in Live Analytical dashboard on SAP Analytics Cloud.

- With built-in connectors for Amazon Redshift in SAP Datasphere, Amazon Redshift data can be queried live in SAP Datasphere to combine it with SAP data for analytics use cases.

Result

The integration leads to more accurate insights, identification of new business opportunities, and a data-driven culture that supports strategic decision-making.

Integration and App Development

The integration and app development pillar of the Joint Reference Architecture provide a backbone for building applications either by extending the existing ones or creating new ones. These applications can effortlessly access data from non-SAP sources hosted on AWS, such as Amazon Aurora, or from custom-built native AWS business applications.

Example Use Case: Event-Driven User Notification Process

Businesses often struggle to efficiently integrate critical event notifications from various SAP and non-SAP business processes. This challenge can lead to delays in response times and hinder the overall effectiveness of business operations.

Problem

Every time a business partner is created in the SAP S/4HANA system, a notification needs to be sent to multiple stakeholders via e-mail, text, or another method.

Solution

The proposed solution involves event-driven architecture leveraging SAP Event Mesh and Amazon Simple Notification Service (SNS) to seamlessly deliver notifications for events in SAP S/4HANA via SAP BTP. (Chapter 2 discusses event-driven architecture in detail.) This integration enables real-time alerts, ensuring that key business processes are triggered promptly in response to events from diverse sources.

Architecture

The architecture depicted in Figure 4-3 illustrates the integration between AWS and SAP BTP, focusing on how events from SAP S/4HANA can be used to trigger notifications via AWS Simple Notification Service (SNS).

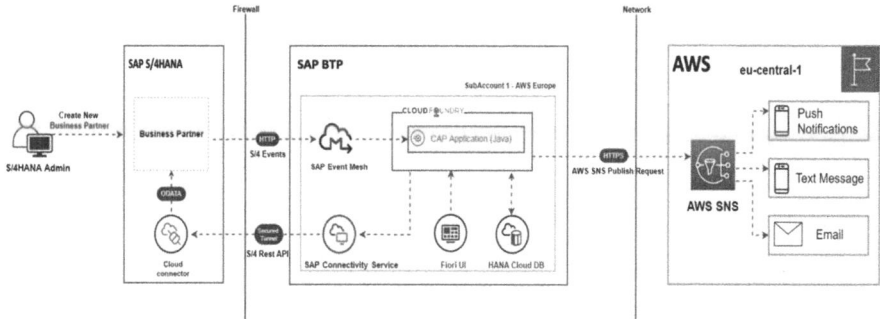

Figure 4-3. *Event-driven architecture pattern for user notification*

This Joint Reference Architecture is explained in detail in this SAP Community blog:

> Receive Notifications from Amazon Simple Notification Service for SAP S/4HANA BTP Extension App: `https://community.sap. com/t5/technology-blogs-by-sap/receive- notifications-from-amazon-simple- notification-service-for-sap-s/ba-p/13538475`

Here's the flow of the architecture based on the diagram in Figure 4-3:

1. **S/4HANA administrator**: An admin creates a new business partner in the SAP S/4HANA system.

2. **Business partner**: This new data triggers events within the SAP S/4HANA system.

3. **S/4 events**: These events are then sent through the SAP Cloud Connector, which securely connects on-premises systems to cloud applications.

4. **SAP Connectivity Service**: The events are passed to the SAP Connectivity Service within SAP BTP.

5. **SAP Event Mesh**: The SAP Event Mesh receives these events and enables event-driven integration.

6. **CAP application (Java)**: A Java-based SAP Cloud Application Programming Model (CAP) application processes the events. The recommend approach to integrate SAP CAP in Java with AWS SNS is to use the AWS SDK for Java. This SDK provides a native set of APIs that enables your Java applications to interact seamlessly with AWS services, including Amazon SNS. By leveraging the AWS SDK for Java, you can efficiently publish messages to SNS topics, manage subscriptions, and handle other SNS-related tasks natively within your CAP-based Java applications.

7. **AWS SNS**: The processed events are sent as a publish request to AWS SNS in the eu-central-1 region.

8. **Notifications**: AWS SNS then delivers the notifications through various channels such as push notifications, text messages, or e-mails.

Result

By implementing this solution, businesses can achieve a more intelligent and interconnected system, leading to improved efficiency in business processes. The integration allows for the automation of event-driven business processes, reducing the reliance on manual operations and enabling a more agile response to changing conditions. Notifications from Amazon SNS ensure that stakeholders are promptly informed, facilitating swift action and decision-making within the SAP ecosystem.

Platform Foundation

The platform foundation pillar of the Joint Reference Architecture emphasizes delivering a highly available platform to ensure optimal performance and an exceptional customer experience for applications.

Example Use Case: High Availability for SAP Build Work Zone

Businesses require continuous operation of their critical applications, with zero tolerance for downtime. Certain SAP BTP services are part of these critical applications. This use case presents the example of SAP Build Work Zone (formerly Fiori Launchpad), part of SAP's portfolio of low-code solutions that enables users to create and customize business applications with minimal coding. It allows users to build, extend, and enhance SAP applications through a user-friendly interface, facilitating the rapid development of enterprise applications. If you have multiple development projects in critical phases, you want to ensure that SAP Build Work Zone has high availability.

Problem

For services like SAP Build Work Zone, standard edition, which serves as a central point of entry for end users to access various applications, high availability and responsiveness are crucial. The limitation of SAP BTP's subaccount to a single host in a single region poses a challenge for achieving this level of business continuity.

Solution

To enhance the availability of the SAP Build Work Zone service and prepare for potential failures, the solution involves using Amazon Route 53. This service acts as a highly available and scalable DNS web service that can manage traffic policies and health checks, redirecting users to the correct available target in case of an outage. By cloning the SAP Build Work Zone service in a different SAP BTP subaccount hosted in another region, traffic can be redirected seamlessly to the right destination.

Architecture

The architecture flow illustrated in Figure 4-4 is designed to ensure business continuity for the SAP Build Work Zone, a standard edition service, by leveraging Amazon Route 53.

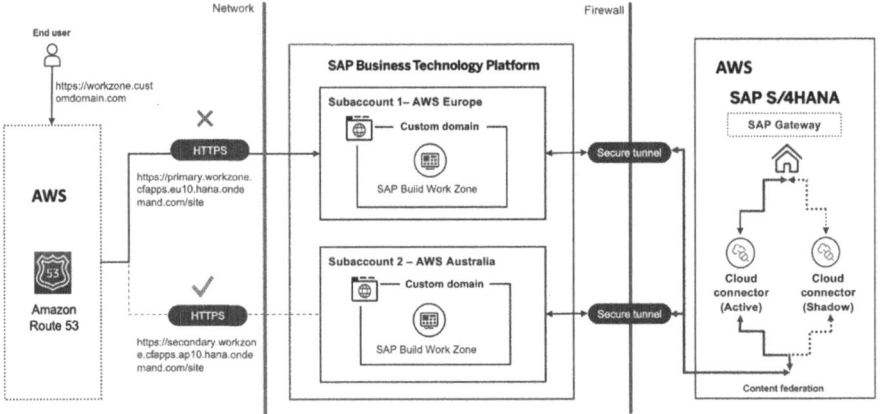

Figure 4-4. *Architecture pattern for implementing high availability for SAP BTP Services*

Let's dive deep into the architecture. To keep it simple, we have divided the discussion into multiple sections.

End-User Access

The end user attempts to access the SAP Build Work Zone service through a custom domain URL (e.g., `https://workzone.customdomain.com`).

Amazon Route 53

The user's request is received by Amazon Route 53, which is configured to manage the DNS for the custom domain. Route 53 is responsible for directing traffic and handling failover in case of an outage.

SAP Business Technology Platform

Two SAP Build Work Zone instances are provisioned in two separate subaccounts, one in the AWS Europe region (Subaccount 1) and the other in the AWS Australia region (Subaccount 2).

- Each subaccount contains an SAP Build Work Zone instance, and both are configured to serve the end users.

- If the primary SAP Build Work Zone instance hosted in the AWS Europe region (e.g., `https://primary.workzone.cfapps.eu10.hana.ondemand.com/site`) is healthy and available, Amazon Route 53 routes the user's request to this primary instance.

- If the primary instance is not available, Amazon Route 53 automatically redirects the traffic to the secondary SAP Build Work Zone instance hosted in the AWS Australia region (e.g., `https://secondary.workzone.cfapps.ap10.hana.ondemand.com/site`).

- **Secure tunnel**: A secure tunnel is established between the SAP Build Work Zone instances and the SAP S/4HANA system, ensuring secure data transmission.

SAP S/4HANA on AWS

The SAP S/4HANA system is hosted within AWS and is connected to the SAP Build Work Zone instances through SAP Gateway. The Cloud Connector (Active and Shadow) ensures secure connectivity and content federation between the SAP S/4HANA system and the SAP Build Work Zone instances.

By using this architecture, in the event of a regional outage or failure of the primary instance, the traffic is automatically rerouted to the secondary instance, minimizing downtime and ensuring continuous access for the end users.

Result

Implementing this architecture ensures the high availability and reliability of the SAP Build Work Zone service. It addresses the need for business continuity by providing a failover strategy that minimizes downtime and maintains user access to critical services. This approach is beneficial for various SAP BTP services based on business use cases, such as employee self-service portals and field portals for service industry employees, where consistent access to business information is essential.

As you have seen in this section, JRAs combine the best capabilities of SAP BTP services with the best capabilities of AWS services to help deliver desired customer innovation and value. JRAs can help close gaps in functionality and processes and deliver last mile of customer innovation with rapid time to value. Now, we will see how customers can extend SAP business processes with the AWS SDK for ABAP.

Extending SAP Business Processes with the AWS SDK for ABAP

The AWS SDK for SAP ABAP makes it easy for ABAP developers to extend and transform SAP-based business processes by harnessing the power of

AWS services seamlessly using the SAP ABAP language. The AWS SDK for SAP ABAP fills a long-standing gap between the ABAP builder community and AWS services. It accelerates business process innovation by giving ABAP developers access to all 200+ AWS services. It reduces architectural complexity by eliminating the need to manually create complex integrations between SAP and AWS services, in both self-managed and RISE with SAP deployments. At the same time, it maintains a strong security posture by encrypting all payloads with HTTPS and enforcing SAP-level permissions.

The following three use cases demonstrate how business processes can be modernized by using the AWS SDK for SAP ABAP.

Use Case: Modernize Document Processing in SAP

A leading online automobile sales company in the United States receives up to 1,000 checks a day from its customers for invoices that it has issued as part of the billing process. The company was processing these checks by manually entering them into SAP, a time-consuming process. The manual process included processing of the checks, doing a customer match in SAP based on the MICR account number and bank code, and finally posting these in SAP as part of accounts receivable (AR) to clear the invoices. This required two or three clerks to process around 1,000 checks per day.

Problem

Manually processing checks was time-consuming and error-prone, leading to a medium error rate. It took an average of three days for customers to get a confirmation that a check was accepted or not. The company wanted to improve this process to achieve two outcomes: give customers a faster confirmation on checks, and reduce the manual efforts of the associates who were processing the checks.

Solution

By integrating Amazon S3 and Amazon Textract with the SAP check processing application, via the AWS SDK for SAP ABAP, the auto sales company solved its problem. The solution minimized the customer confirmation time from three days to minutes for 80% of the checks, because the solution was able to process 80% of the checks received automatically, close the AR, and post it to GL. In 20% of cases, there were anomalies requiring manual intervention, for reasons outside the control of the solution, such as inaccurate customer master data in SAP, overpayments, underpayments, and so forth.

Architecture

Figure 4-5 shows the high-level architecture of the solution built using the AWS SDK for SAP ABAP.

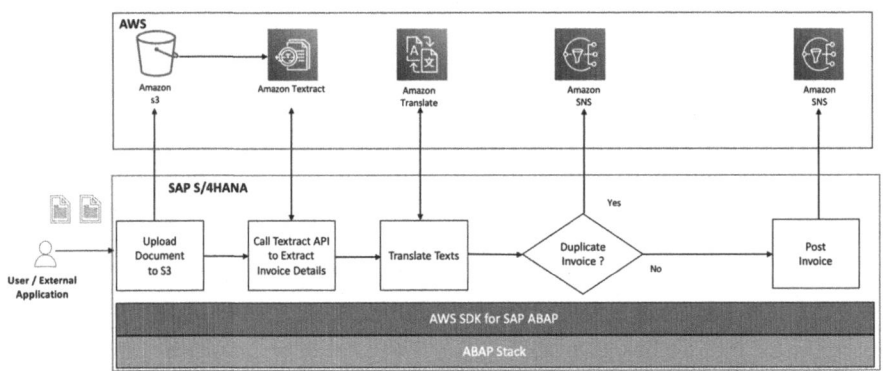

Figure 4-5. *Architecture pattern for document processing using Amazon Textract via the AWS SDK for ABAP*

From the perspective of an end user (clerk), the integration between SAP and AWS technologies happens seamlessly in the background, leveraging the AWS SDK for SAP ABAP. Here's how the process works:

1. **Upload document to S3**: The clerk logs into the SAP system and initiates the document processing workflow. They upload the document (e.g., an invoice) to the SAP system, which then transparently uploads the document to an Amazon S3 bucket using the AWS SDK for SAP ABAP.

2. **Call Textract API to extract invoice details**: Once the document is in the S3 bucket, the SAP system calls the Amazon Textract API through the AWS SDK for SAP ABAP. Textract uses machine learning to intelligently extract relevant details from the document, such as MICR details in a check, especially the account number and routing number, which are key to linking the check to a customer.

3. **Confidence score evaluation**: The extracted data from Textract is returned to the SAP system along with a confidence score indicating the reliability of the extraction. The SAP system evaluates this confidence score against a predefined threshold (e.g., 95%).

4. **Successful extraction**: If the confidence score exceeds the threshold, the SAP system considers the data extraction successful. It sends a successful communication notification through Amazon Simple Notification Service (SNS), facilitated by the AWS SDK for SAP ABAP. In the customer master data of SAP, the routing number and account number are stored. Once these details are extracted from the check, a search is made in SAP to identify who the customer is and what open AR exists for that customer. Once this is identified, the check is matched to the correct open AR, cleared, and posted to the GL.

5. **Manual approval for low confidence**: If the confidence score is below the threshold, the SAP system triggers a manual approval process. It sends a failure communication through Amazon SNS, again using the AWS SDK for SAP ABAP. This notification prompts a human user to review and correct the extracted data within the SAP system.

Result

Automating the check processing procedure via the new AWS SDK for SAP ABAP has resulted in several benefits for the auto sales company, including scalability, improved accuracy, and reduced manual effort. With the new automated process, the company can process 15,000 to 20,000 checks per month, reducing manual effort and errors while improving accuracy.

The new automated process also enables the clerks to focus on more value-added tasks and exceptions, rather than spending most of their time on manual data entry and processing. With the new intuitive front-end screen that accepts a PDF input file of scanned checks and the use of Textract, clerks can electronically process information from the checks quickly and accurately, matching them with the appropriate invoices.

Automating the check processing procedure via the AWS SDK for SAP ABAP resulted in the following benefits:

- **Faster processing time**: The new automated process is much faster than the previous manual process. With the ability to process thousands of checks per month, the auto sale company can quickly clear invoices and receive payments, which improves cash flow and helps the company stay on top of its billing.

190

- **Accuracy:** 100% accuracy in terms of Textract being able to interpret the MICR numbers. Customer had tried other solutions in the past and was never able to achieve 100% accuracy.

- **Greater visibility and control**: With the new automated process, the auto sale company has greater visibility and control over its billing process. The system provides real-time updates and alerts, allowing the company to identify and resolve issues quickly. This level of visibility and control enables the company to make better-informed decisions and ensures that its billing process is efficient and effective.

- **Improved customer satisfaction**: The new automated process also improves customer satisfaction by providing faster and more accurate invoice processing. Customers can now expect their payments to be processed quickly and accurately, which reduces the likelihood of payment disputes and improves overall customer satisfaction.

- **Cost savings**: By automating the check processing procedure, the auto sale company has also achieved cost savings. The company no longer needs to employ as many clerks to manually process checks, which reduces labor costs and increases efficiency.

Throughout this process, the end user interacts solely with the SAP system's user interface, unaware of the underlying integration with AWS services. The AWS SDK for SAP ABAP acts as a bridge, enabling the SAP system to seamlessly leverage AWS services like Amazon S3, Textract, and Amazon SNS for intelligent document processing and data extraction. Without the AWS SDK for SAP ABAP, the auto sales company likely would not have optimized this business process, primarily for these reasons:

191

- **Security and authentication**: SAP and AWS
  have different security models. AWS, as a cloud
  service, employs sophisticated authentication and
  authorization procedures that are difficult to correctly
  code from SAP ABAP.

- **Data formatting**: Data formats are completely different
  between AWS and SAP—and also differ among systems
  and services. Without the AWS SDK for ABAP, for each
  data transmission from SAP to AWS, developers had to
  custom code the correct translation between the two
  messaging formats, writing custom code for Amazon S3
  and different custom code for Amazon SQS.

- **Connectivity complexity**: AWS has REST endpoints
  to which data should be sent. But programming
  this can be extremely challenging, especially for
  ABAP developers who had to work in an unfamiliar
  environment where properly coding for a single REST
  endpoint might require more than 1,000 lines of code.

- **Multiple point-to-point integration**: In the absence
  of dedicated integration, customers have to rely
  on middleware solutions to transfer data between
  SAP applications and AWS services. The developers
  would then create their own point-to-point solutions
  that required multiple network hops through the
  middleware

This integration via the AWS SDK for SAP ABAP allows the end user to
benefit from AWS's advanced cloud capabilities while working within the
familiar SAP environment, enhancing the overall efficiency and accuracy
of document processing tasks.

Use Case: Increasing Efficiency in Financial Consolidation by 90%

A global insurance company had a high volume of accounts receivable (AR) and accounts payable (AP) data spread across multiple data sources, formats, and quality levels. This diverse set of data needed to be transformed and posted into SAP General Ledger (GL) accounts. This was done manually.

Problem

This organization was relying on the manual effort of ten full-time employees every month to consolidate its finance data spread across multiple silos of applications into SAP GL. Because this process was manual, it was error-prone and impacted quality of reporting and forecasting.

Solution

By integrating SAP data with AWS AI/ML services, via the AWS SDK for SAP ABAP, the insurance company reduced its manual processing by 90%. The custom model that the customer built predicted the errors and corrected them while consolidating into SAP GL. The company improved its accuracy while also reducing manual efforts.

Architecture

Figure 4-6 shows the high-level architecture of the solution.

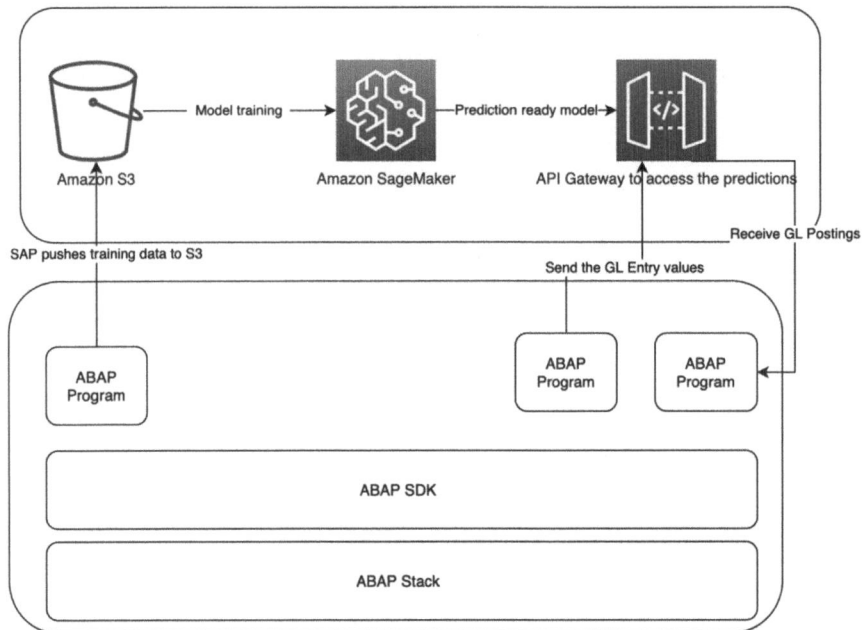

Figure 4-6. *Architecture pattern for predicting and correcting finance data*

The solution leverages machine learning to automate error correction and data processing for SAP systems by integrating with the AWS services Amazon S3, Amazon SageMaker, and Amazon API Gateway. The entire process can be divided into three stages: data preparation, model training, and model integration.

Data Preparation

The training data for machine learning is prepared as follows:

1. Past GL data and correction values are extracted from the SAP system and pushed into an Amazon S3 bucket.

2. This data stored in Amazon S3 becomes the training dataset for machine learning model training.

Model Training

Once the training data is available, the next stage involves training the model. Key points of training the ML mode are as follows:

1. Amazon SageMaker (AWS's ML service, introduced in Chapter 1) is used to create a custom model by training it on the data from Amazon S3.

2. The output of this model training process is exposed via an Amazon API Gateway endpoint, making the trained model accessible for predictions.

Model Integration

After the model is trained, the SAP system leverages this model to make predictions on new data. This is completed via the following steps:

1. The SAP system, through ABAP programs leveraging the AWS SDK for SAP ABAP, sends new GL entry values to the Amazon API Gateway endpoint.

2. The deployed model receives these values, makes predictions for error correction, and returns the predictions back to the SAP system.

3. The ABAP programs within SAP consume these predictions to automate error correction and data processing workflows.

Result

This integration allows the SAP system to take advantage of AWS ML capabilities like Amazon SageMaker for intelligent data processing. The solution also utilizes core AWS services like Amazon S3 for data storage and Amazon API Gateway for model deployment, all orchestrated through the AWS SDK for SAP ABAP.

Use Case: Improving Location Accuracy

A leading retailer had an SAP application to manage its delivery service. It did not have a robust address validation process, resulting in deliveries to incorrect addresses, increasing the operational cost.

Problem

Customer master addresses were inaccurate, which influenced supply deliveries and customer satisfaction. It was not possible to take corrective measures because of insufficient data.

Solution

Integrating Amazon Location Service with the SAP application, the retailer auto-corrected the wrongly entered addresses and increased the delivery address accuracy. This resulted in fewer deliveries to incorrect addresses and reduced operational cost.

Architecture

Figure 4-7 shows the high-level architecture of the solution.

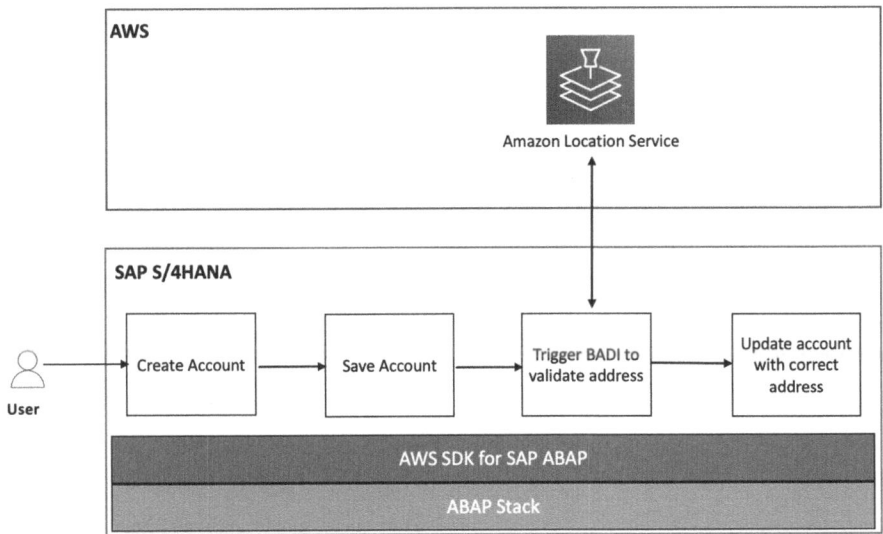

Figure 4-7. *Architecture pattern for implementing address correction with Amazon Location Service*

This solution leverages Amazon Location Service to validate and correct addresses within the SAP S/4HANA system. Here's a breakdown of the workflow:

1. **Create account**: The user provides account details, including the address information, through the SAP S/4HANA interface.

2. **Save account**: The account details, along with the entered address, are saved in the SAP S/4HANA system.

3. **Trigger BADI to validate address**: After saving the account, a Business Add-In (BADI) is triggered to initiate the address validation process. This BADI integrates with Amazon Location Service.

4. **Amazon Location Service**: Amazon Location Service receives the address data from the SAP system and performs address validation and correction using its geocoding capabilities.

5. **Update account with correct address**: The corrected address data is sent back to the SAP S/4HANA system, which then updates the account with the validated and corrected address information.

Result

This solution takes advantage of the AWS SDK for SAP ABAP, which enables the integration between the SAP S/4HANA system and Amazon Location Service. The ABAP stack is the runtime environment within SAP that facilitates this integration. By leveraging Amazon Location Service, this solution ensures accurate and standardized address data within the SAP system, improving data quality and enabling more efficient business processes.

Now that you have seen multiple example use cases wherein you may use the AWS SDK for SAP ABAP to modernize your SAP business processes, let's move on to another topic that is also a very important part of multiple business processes for many organizations: how AWS services can help with electronic data interchange in SAP.

Integration of AWS B2B Data Interchange into SAP Business Processes

Electronic data interchange (EDI) has revolutionized the way businesses exchange documents with their trading partners, as depicted in Figure 4-8. By transitioning from paper-based processes to electronic formats, companies have experienced significant benefits such as cost reduction, faster processing times, and decreased error rates.

Lot of people working on
Lot of paper docs, which takes
Lot of time

No people
No paper
Near real time

Figure 4-8. *How EDI changed the world*

The two primary methods of connecting with trading partners in the realm of EDI are direct connections and connections through value-added networks (VANs). Direct connections involve establishing a direct link between two trading partners' systems, allowing for the seamless exchange of EDI documents. On the other hand, VANs act as intermediaries, providing a centralized platform through which multiple trading partners can exchange EDI documents.

Regardless of the method chosen, there are two crucial functions involved in EDI:

- **Translation or mapping of data**: This process involves converting the data from the internal format used by one trading partner's system into a standardized EDI format that can be understood by the recipient's system. This ensures compatibility and seamless communication between different systems.

- **Data transfer using protocols**: Once the data has been translated into the appropriate format, it needs to be securely transferred between the sender and the recipient. Various protocols, such as AS1, AS2, AS3, AS4, Connect Direct, SFTP (Secure File Transfer Protocol), and others, are utilized for this purpose. These protocols govern the way data is packaged, transmitted, and received, ensuring reliability, security, and integrity throughout the exchange process.

By effectively managing these two functions, organizations can leverage EDI to streamline their business operations, enhance collaboration with trading partners, and achieve greater efficiency and accuracy in document exchange.

Typically, most of all organizations have 100 to 300 EDI integrations with their trading partners, which poses the following challenges:

- **Complex EDI configuration and maintenance**: As the number of trading partners increases, managing EDI configurations becomes more complex. Each partner may have different requirements, formats, and protocols, leading to a tangled web of configurations that are difficult to maintain and update.

- **High cost**: Implementing and maintaining EDI systems can be costly, especially for organizations with a large number of trading partners. Costs include software licenses, hardware infrastructure, data transmission fees (especially for high volumes of data), and personnel costs for IT teams dedicated to managing EDI systems. Poor data quality can exacerbate these costs by increasing the need for manual intervention and troubleshooting.

- **Difficulty to get visibility into errors**: Identifying and resolving errors in EDI transactions can be challenging without sufficient visibility into the process. Errors may occur due to issues with data formatting, transmission errors, or discrepancies between systems. Without adequate monitoring and error reporting mechanisms, it's difficult to pinpoint the root cause of errors and take corrective action promptly.

- **Inability to utilize data for insights**: Despite the high volume of transactions processed through EDI, organizations may struggle to leverage this data for actionable insights. Traditional EDI systems often lack robust analytics capabilities, making it challenging to extract meaningful insights from transactional data. This limits the ability to identify trends, optimize processes, or make data-driven decisions based on EDI data.

AWS B2B Data Interchange can help address these challenges. It is a fully managed service, enabling you to easily exchange EDI-based transactions with your business partners.

Core features of this service are

- **Trading partner management**: Organizations can simplify the process of trading partner management by creating and managing a roster of trading partners, organized by business name and e-mail address. They can save time by assigning reusable EDI trading profiles to each partner. By using this service, you can invite trading partners to begin transacting in a predeveloped customer vendor portal.

- **Inbound EDI transformation**: Flexibility can be enabled in inbound EDI processes by supporting various communication methods and leveraging reusable mapping templates across multiple transactions and partners. Additionally, by using this service, you can automatically transform X12 documents uploaded to an Amazon S3 bucket into XML or JSON formats.

- **Transaction visibility**: Visibility into EDI transactions is enhanced by easily identifying errors and anomalies, categorized by trading partner, date, and time. The trading partner onboarding process can be monitored with status updates, including stages such as invitation, onboarding initiation, and completion. Amazon CloudWatch Contributor Insights can be utilized to track top trading partners, document types, and ongoing usage trends.

- **Insightful analytics:** Enterprises can leverage automation to streamline business operations and mitigate risks associated with EDI processes. They can harness the power of AWS for EDI, accessing cutting-edge AI/ML capabilities.

Use Cases of AWS B2B Data Interchange specific to SAP

Two prevalent use cases of AWS B2B Data Interchange, specific to SAP, observed amongst customers are data lake creation and integrating EDI data into SAP systems.

Use Case: Creating a Data Lake

You can utilize high-volume EDI transactions to create a data lake, which then can be leveraged for various purposes such as the following:

- **Sending QA and compliance reports to regulatory agencies:** The data lake serves as a repository for EDI data, enabling organizations to extract relevant information for generating quality assurance (QA) and compliance reports. These reports demonstrate adherence to regulatory requirements and standards, facilitating seamless communication with regulatory agencies.

- **Monitoring warehouse and inventory levels**: By aggregating and analyzing EDI data stored in the data lake, organizations can gain insights into warehouse operations and inventory levels. This enables real-time monitoring of stock levels, facilitates inventory replenishment, and supports efficient warehouse management practices.

- **Predicting seasonal sales or purchasing behavior**: Analysis of historical EDI data within the data lake enables organizations to forecast seasonal sales patterns and purchasing behavior. By identifying trends and patterns in transactional data, businesses can optimize inventory stocking, plan marketing strategies, and anticipate fluctuations in customer demand more accurately.

- **Predicting shipment ETAs for managing risk and customer satisfaction**: By leveraging EDI data stored in the data lake, organizations can predict estimated times of arrival (ETAs) for shipments. By analyzing historical data alongside external factors such as weather conditions and transportation routes, businesses can mitigate risks associated with delays, proactively manage logistics, and enhance customer satisfaction by providing accurate delivery timelines.

Use Case: Integrating EDI Data into SAP Systems

Integrating EDI data directly into SAP ERP systems enables line-of-business users to drive efficiency. The following are a few examples:

- **Optimizing sourcing of raw materials from suppliers**: By seamlessly integrating EDI data into SAP ERP systems, business users can access real-time information on supplier transactions, inventory levels, and pricing. This enables informed decision-making regarding the sourcing of raw materials, ensuring timely procurement, cost optimization, and maintenance of supply chain resilience.

- **Exchanging purchase orders and invoices**: Users can utilize integrated EDI data within SAP systems to streamline the exchange of purchase orders and invoices with trading partners. This eliminates manual data entry and facilitates electronic communication, improving accuracy, reducing processing times, and enhancing collaboration with suppliers and customers.

- **Submitting bills of lading and customs declarations**: Integrating EDI data directly into SAP systems enables users to submit bills of lading and customs declarations seamlessly. This automated process ensures compliance with shipping regulations, expedites customs clearance procedures, and enhances visibility into the movement of goods across international borders.

Architecture

Figure 4-9 illustrates a sample architecture pattern that can be leveraged to implement both of the preceding use cases. A trading partner or automated pipeline loads the data into Amazon S3. AWS B2B Data Interchange automatically transforms the EDI data into XML or JSON, with optional mapping of the EDI data, and makes this transformed data available in a separate S3 bucket. AWS Glue Crawler classifies this data into AWS Glue Catalog, which makes the data ready to be queried using Amazon Athena or Amazon QuickSight. Alternatively, the transformed data can be fed into the SAP system using Amazon AppFlow for the second use case.

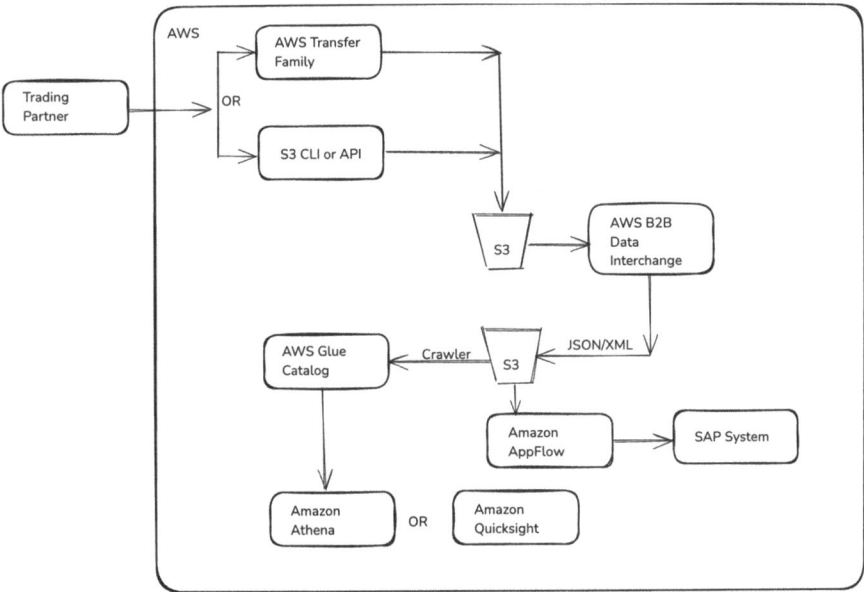

Figure 4-9. *Architecture pattern for creating a data lake with data from EDI transactions, integrating it with the SAP system, and gaining insights from it via analytic services*

Summary

In this chapter, we explored the transformative potential of SAP BTP services that are powered by AWS as well as the AWS native services and the AWS SDK for SAP ABAP. The journey through this chapter has illuminated the ways in which these services can modernize and streamline business processes, enabling innovation with minimal coding effort and fostering a more agile and responsive business environment.

We started by introducing an evaluation framework to identify which business processes are most suitable for modernization. Next, we explored how to decide between using SAP BTP services or AWS native services, depending on the specific use case. Then, we presented the AWS and SAP Joint Reference Architectures, demonstrating their application across various scenarios. You also learned about the AWS SDK for SAP ABAP and how, with the use of familiar ABAP syntax, developers can modernize their SAP business process. The AWS SDK for SAP ABAP is a tool that stands at the intersection of SAP's robust business process management and AWS's dynamic cloud services. We then discussed how using AWS B2B Data Interchange, customers can transform their EDI processes, which a majority of SAP customers use. We discussed how using AWS B2B Data Interchange, customers can effortlessly onboard and administer their trading partners while automating the conversion of EDI documents into standard data formats like JSON and XML. You learned how customers can reduce the time, complexity, and cost associated with preparing and integrating EDI data into their business applications and purpose-built data lakes, enabling them to focus on driving business insights using the AWS suite of analytics services.

Throughout the chapter, we also discussed real-life use cases that not only adapt existing processes but also create new opportunities for business process innovation.

You are encouraged to take the insights and methodologies detailed in this chapter and apply them to your own business scenarios. The integration of SAP with BTP services and AWS services through the AWS SDK for ABAP offers a pathway to significant benefits, including cost savings, increased efficiency, and the ability to quickly respond to changing market demands.

The potential for innovation is vast, and the tools are at your disposal. Whether you are looking to optimize existing processes or embark on entirely new projects, SAP BTP services, powered by AWS and AWS native services, provide a robust foundation for your endeavors. You now should be well equipped to embark on a strategic and informed innovation journey, making the right decisions, leveraging the right technologies and drawing inspiration from proven success stories. Embrace the opportunity to become a leader in business process innovation by leveraging the power of SAP and AWS together. Next, Chapter 5 explores how to optimize manufacturing processes by using these services.

CHAPTER 5

Smart Factory

Smart factory is your gateway to a data-driven, intelligent, and connected manufacturing ecosystem.

In recent times, there has been a growing acknowledgement across all industries of the imperative need for digital transformation within manufacturing operations to uphold competitiveness and resilience. The COVID-19 pandemic highlighted weaknesses in global supply chains and vulnerabilities within various sectors. The pandemic has emphasized the necessity for more adaptable, agile solutions that are fully integrated digitally. Additionally, evolving consumer expectations are driving the advancement of smart factory technologies and reshaping the concept of future manufacturing facilities. Commonly called the "Amazon effect," the demand for next-day delivery has been steadily increasing, significantly boosting the requirement for smart factory technology as outdated systems struggle to meet the logistical demands of this trend. Moreover, manufacturers are tired of facing increased risks and operational disruptions. Therefore, it is essential to implement digital factory technologies to improve efficiency and visibility, especially in today's dynamic business environment.

The following topics will be covered in this chapter:

- What is a smart factory?

- Digital Manufacturing and Its Key Tenets - which are Industrial analytics and building an industrial data lake, Predictive maintenance, Predictive quality, Digital twin and Voice technology

© Bidwan Baruah, Krishnakumar Ramadoss and Abarajith Vivekanandha 2024
B. Baruah et al., *Evolve from Infrastructure to Innovation with SAP on AWS*,
https://doi.org/10.1007/979-8-8688-0890-6_5

- AWS solutions for manufacturing and industry

- SAP solutions to enable digital manufacturing

- Secrets to a successful implementation

What Is a Smart Factory?

A *smart factory* is a manufacturing facility that uses technology to share information between machines, sensors, and people. The smart factory is becoming a reality thanks to the cloud and the Internet of Things (IoT).

We are currently in the midst of the fourth industrial revolution, popularly known as *Industry 4.0*, and the smart factory is the cornerstone of Industry 4.0. This revolution is characterized by the integration of advanced technologies such as artificial intelligence (AI), machine learning (ML), robotics, IoT, and big data analytics into various industries, particularly manufacturing. Chapter 2 discussed event-driven architectures. IoT is simply an event-driven architecture wherein we are creating, consuming, and reacting to events of interest.

Edge and Edge Devices

The following two terms frequently arise when discussing smart factories: *edge* and *edge device*.

- **Edge**: The location where the devices transmitting and/ or computing data are physically located. Sometimes, the edge is classified into near edge (plant floor, for example) and far edge (such as vehicles, etc.).

- **Edge devices**: Pieces of equipment that transmit data between the local network and the cloud. The role of an IoT edge device extends beyond data transmission to the cloud; it also facilitates data exchange with nearby edge devices, fostering localized and intelligent data management tailored to each device's needs. IoT edge devices serve as filters for data, enabling onsite storage and minimizing the expense associated with migrating data to cloud services.

Factories often house a cluster of servers dedicated to storing, analyzing, and managing locally generated data. Referred to as a *fog*, this operates independently within the facility. Data stored and processed in the fog, or a refined version thereof, may also be transmitted to the cloud for further analysis, storage, and distribution. The presence of a fog infrastructure enables uninterrupted factory operations, even in scenarios where connectivity to the cloud is disrupted.

Often, there is ambiguity on how edge and fog is perceived by different people. As shown in Figure 5-1, from the perspective of a networking professional, who focusses only on infrastructure, everything within a factory, even if it's thousands of miles away from its cloud counterpart, may be considered as part of the edge (a).

Looking deeper into the factory, we might encounter various edge devices such as sensors. These devices could be connected to a local or client server, which, in turn, might be linked to a group of servers forming a local fog. In this scenario, some individuals might view "the edge" as encompassing both the edge devices and the edge server (b). Alternatively, a distinction could be made between the edge devices and the edge server by categorizing them as *far edge* and *near edge*, allowing the edge server to be situated at the near edge while remaining separate from the edge devices (c).

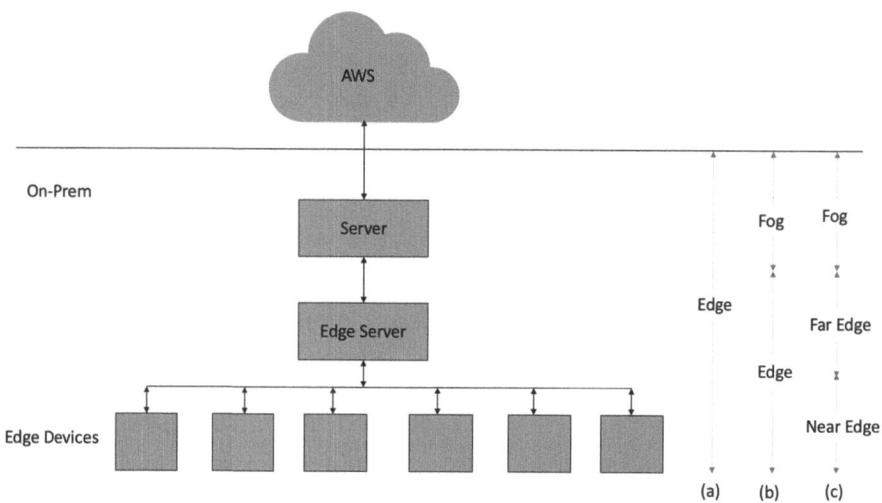

Figure 5-1. *Various definitions of edge*

Importance of Smart Factories

Companies are adapting smart factories for a variety of reasons, as outlined in this section.

Enable Digital Manufacturing

Digital manufacturing is about using digital technologies and data-driven processes to optimize and improve factory operations. The key tenets of digital manufacturing are covered in the upcoming section "Digital Manufacturing and Its Key Tenets."

Create Smart Products

Product performance and usage data improve product and process design. Smart products redefine customer experiences and lead to new business models that increase customer value. Data from smart products at customer premises facilitates the development of the next generation of the product. Ensuring that shipped products are "smarter and connected" enables a comprehensive understanding of their usage. This information is then accessed by the research and development (R&D) team to provide customers with new features via software updates and an improved overall user experience.

A compelling example of a company with successful product innovation is iRobot, famous for its Roomba vacuuming robot, which revolutionized the world of vacuuming. Its first challenge was to have its vacuums connect to a highly scalable, high-availability cloud application and an IoT back-end platform. To support the web applications connecting to Roomba vacuums, iRobot implemented a serverless architecture built on AWS IoT, AWS Lambda, and over 20 other AWS services. This strategic solution allowed iRobot to avoid the complexities associated with managing servers at scale. Remarkably, fewer than ten employees were needed to operate this solution, contributing to the realization of the next generation of smart homes.

Another example of a company achieving successful product innovation is Mercedes-Benz, which has implemented digital services via the foundation of "connected car." This involves the seamless transmission of data between vehicles and back-end applications, as well as third-party services. Through a mobile application, users can receive vehicle-specific alerts and updates, as well as execute remote commands. This feat is accomplished through an event-driven architecture leveraging SAP Advanced Event Mesh within SAP Business Technology Platform, described in Chapter 2.

Optimize Supply Chain Management

In a seamlessly integrated supply chain, tracking is enabled throughout the entire journey from the point of origin to the destination. This is achieved by capturing data points throughout the value chain using IoT, SAP, and other IT systems that are commonly used in planning and execution. This comprehensive approach establishes a sense-and-respond system, ultimately creating a close-knit connection among suppliers, manufacturing, distributors, customers, and their end consumers.

Attain Sustainability Goals

Organizations can achieve sustainability goals through the implementation of smart factories and IoT by

- Collecting energy usage data

- Understanding consumption patterns

- Implementing a control plan for optimization

- Monitoring production processes in real time to minimize waste generation

Now, let's dive deep into the key tenets of digital manufacturing.

Digital Manufacturing and Its Key Tenets

Digital manufacturing holds immense importance because it aligns not only with SAP's vision of an intelligent enterprise but also the fundamental principles of Industry 4.0, which center on utilizing advanced digital technologies to revolutionize conventional manufacturing methods. This is achieved through various technologies and strategies, including implementing real-time visibility and insights, operational excellence, predictive maintenance and asset optimization, predictive quality, digital twins, and the integration of voice technology.

The following are the key tenets of digital manufacturing:

- Industrial data lake and analytics

- Predictive maintenance

- Predictive quality

- Digital twin

- Voice technology

Industrial Data Lakes and Analytics

As discussed in Chapter 3, data is an enabler, and manufacturing companies have recognized that they have a wealth of data that can be leveraged to improve operations. With the emergence of AI/ML, companies are eager to identify their AI/ML use cases. For those who already have a mature data platform and are deriving business from it, incorporating AI/ML is a logical next step. However, for other companies, they need to start with the basics first—extracting and using the data to address various business problems. For these companies, creating an industrial data lake is the first logical step.

Many manufacturers rely on data from multiple systems to carry out their operations. Over the years, IT departments have been responsible for deploying and managing these systems. However, with the advent of cutting-edge technology, various departments now engage with SaaS applications, and there is an influx of data from sources such as IoT sensors, camera feeds, programmable logic controllers (PLCs), and more. This proliferation of systems poses a challenge in management and, most importantly, hinders the extraction of maximum value from the accumulated data.

To address this issue, *industrial data lakes* have emerged as a viable solution. These are highly scalable, available, secure, and flexible data stores capable of handling exceptionally large datasets. A data lake allows manufacturers to capture structured, semi-structured, and unstructured data, employing the open data format of their choice. The data can then be organized and tagged in a central, searchable catalog. Subsequently,

215

AI and ML technologies can be applied to analyze the data within this centralized data lake, enabling predictions of future business outcomes. Figure 5-2 provides a high-level overview of an industrial data lake.

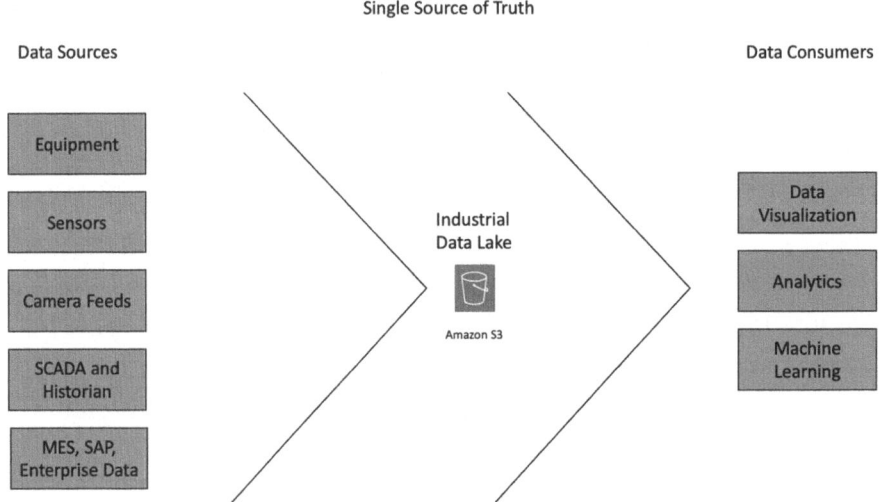

Figure 5-2. *Industrial data lake*

There are many prerequisites to build an industrial data lake— connectivity to the equipment, building digital model assets, ingesting the data, creating operational dashboards, and so forth. Instead of investing time on all this heavy lifting, customers can take advantage of AWS services to address these challenges and instead invest their time in figuring out which data and which key performance indicators (KPIs) make the most sense to gain insights, optimize processes, and make data-driven decisions.

The primary AWS services that can help in this heavy lifting are AWS IoT SiteWise and AWS IoT Core. As depicted in Figure 5-3, data from the industrial assets can be made available on AWS via AWS IoT SiteWise and AWS IoT Core. Once the data is on AWS, apart from monitoring the equipment, this data can be further used for other use cases such as predictive maintenance, asset condition monitoring, and predictive quality, all of which are discussed in detail later in this chapter.

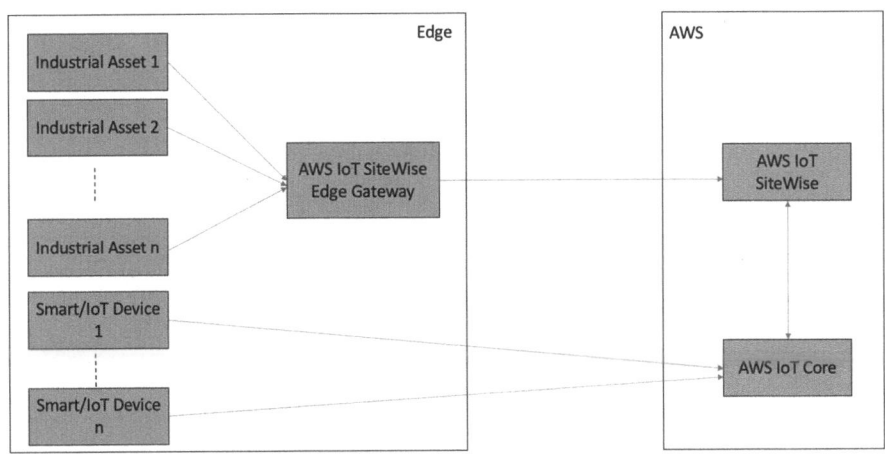

Figure 5-3. *High-level data ingestion patterns to AWS*

Let's take a look at both these services in more detail.

AWS IoT SiteWise

As introduced in Chapter 1, AWS IoT SiteWise is a managed service provided by AWS that helps in the collection, organization, and analysis of industrial data from various sources. It is designed to help industrial customers monitor and manage their equipment and processes more effectively by providing insights into asset performance, operational efficiency, and overall productivity. The capabilities of AWS IoT SiteWise can be summarized in three key areas:

- **Ingest the data**: AWS IoT SiteWise facilitates the ingestion and processing of data from various sources, allowing for the seamless collection of information. It processes and stores the data in an AWS IoT SiteWise–managed time series storage. We are going to dive deep into this process in a bit.

- **Create digital model assets**: AWS IoT SiteWise
  enables the creation of digital models for assets,
  establishing a mapping between the collected data
  and the corresponding real-world assets. This digital
  representation helps organize and contextualize the
  information. What are these models? Think about
  models as classes in computer programming. Just like
  you create classes and instances of classes, you create
  models and instances of models. Every asset in the
  "real life" industry will have a model in IoT SiteWise—a
  model for the plant site, a model for the production
  line, a model for the machine, and so on. Instances of
  the model Plant Site would be, for example, Plant Site
  Chicago, Plant Site Austin, and so on. Instances for
  Production Line would be, for example, Production
  Line 1, Production Line 2, and so on. Why do we need
  to create these models and instances of models? To link
  measurements to real-world assets, as illustrated in the
  similar example shown in Figure 5-4.

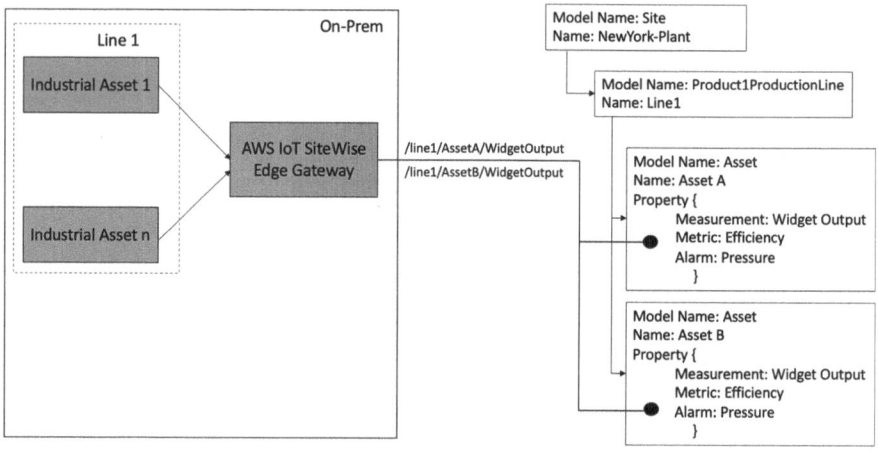

Figure 5-4. *Digital model assets*

- **Monitor the data**: AWS IoT SiteWise provides tools to monitor and analyze the ingested data, allowing for real-time insights and informed decision-making based on the performance of assets and other relevant metrics. AWS IoT SiteWise Monitor, an integral feature of AWS IoT SiteWise, enables the development of managed web applications and dashboards for effective monitoring of this data. Some customers prefer visualizing and monitoring their data through dashboards on the widely used data visualization platform, Grafana. To achieve this, they utilize the AWS IoT SiteWise plug-in for Grafana and configure AWS IoT SiteWise as the designated data source for seamless integration and display of data in Grafana dashboards.

One notable feature of AWS IoT SiteWise is IoT SiteWise Edge, which extends the preceding three key capabilities of AWS IoT SiteWise to the edge.

IoT SiteWise Edge and Its Capabilities

AWS IoT SiteWise Edge operates on premises, where it locally collects and monitors equipment data before transmitting it to AWS. AWS IoT SiteWise Edge can be installed on various local hardware, including industrial gateways, computers, AWS Outposts, and devices within the AWS Snow Family—basically, any device that supports AWS IoT Greengrass, which is an open source local software runtime and cloud service designed to assist in building, deploying, and managing software on these devices. The device on which AWS IoT SiteWise Edge is deployed is referred to as the AWS IoT SiteWise Edge Gateway. Due to its on-premises functionality, local applications utilizing data from AWS IoT SiteWise Edge can remain operational even in situations of intermittent cloud connectivity.

Let's look at the capabilities of AWS IoT SiteWise Edge in detail:

- **Collect and process industrial data**: AWS IoT SiteWise Edge capabilities are delivered via two *packs*. The *Data Collection pack* offers support for reading data through common industrial protocols like OPC UA, Modbus, and Ethernet /IP and pushes this data to AWS IoT SiteWise. The other pack is the Data Processing pack. Equipment data frequently exists in an unstructured format and lacks the necessary context for application and end-user comprehension. Processing this raw data is essential to derive metrics that can inform business decisions. For instance, numerous pieces of equipment convey their status through numerical values, necessitating conversion into meaningful states such as running, idle, or stopped. Additionally, there may be a need to down-sample high-frequency data, such as reducing data points from one per second to one per minute. All of these tasks fall under the umbrella of data processing.

- **Monitor the data locally**: There are two options to monitor the collected and processed data locally:

  - **AWS IoT SiteWise data plane APIs**: AWS IoT SiteWise Edge offers the same AWS IoT SiteWise query APIs both locally and in the AWS Cloud. You may develop local web applications or dashboards designed for the OT team, who would leverage these APIs to retrieve near-real-time data and metrics directly from your on-premises gateway.

- **Utilize AWS IoT SiteWise Edge dashboarding capability**: Develop monitoring dashboards effortlessly, without the need for coding or SQL queries, and access them locally.

- **Tools to manage gateways locally**: AWS offers tools for local monitoring of gateways, allowing you to assess the health of the gateway device and troubleshoot connectivity issues with associated data sources (edge devices) without relying on the cloud. A downloadable local application designed for Windows PCs enables users to monitor and debug gateway issues effectively. This application facilitates monitoring gateway health, including aspects such as CPU usage and available space, as well as evaluating connectivity with data sources and addressing potential challenges.

Now that you understand the capabilities of AWS IoT SiteWise and AWS IoT SiteWise Edge, let's look at the process of data ingestion. This is the most important as well as the most problematic process for many customers to get started with IoT because they have diverse heterogenous industrial data sources. They struggle with ingesting the data from these diverse sources, which require unique communication protocols or translation.

Data Ingestion Using AWS IoT SiteWise

AWS IoT SiteWise Edge can ingest equipment data using the following protocols:

- **OPC United Architecture (OPC UA) natively**: OPC UA is a specification for a popular protocol used in industrial automation systems. It facilitates communication between machines, sensors, PLCs, SCADA systems, and various applications, including cloud-based solutions.

221

- **Ethernet/IP and Modbus via built-in-connectors**: Ethernet/IP and Modbus are both communication protocols used in industrial automation and control systems.

- **MQTT via custom AWS IoT SiteWise MQTT connector**: MQ Telemetry Transport is a commonly used industrial communication protocol that follows the pub-sub pattern. Refer to this blog (`https://aws.amazon.com/blogs/iot/aws-iot-sitewise-adds-support-for-10-new-industrial-protocols-with-domatica-easyedge-integration/`) for more details.

- **Modbus (TCP & RTU), Ethernet/IP, Siemens S7, KNX, LoRaWAN, MQTT, Profinet, Profibus BACnet, and Rest interfaces via AWS Partner Domatica's EasyEdge software**: All the customer needs to do is connect the devices supporting the above protocols to EasyEdge and set up EasyEdge as a data source in the AWS IoT SiteWise Edge gateway. Refer to the AWS blog at `https://aws.amazon.com/blogs/iot/aws-iot-sitewise-adds-support-for-10-new-industrial-protocols-with-domatica-easyedge-integration/` for detailed step-by-step instructions.

Figure 5-5 shows all the options. Alternatively, using the AWS IoT SiteWise API, your applications at the edge or in the cloud can directly send data to AWS IoT SiteWise.

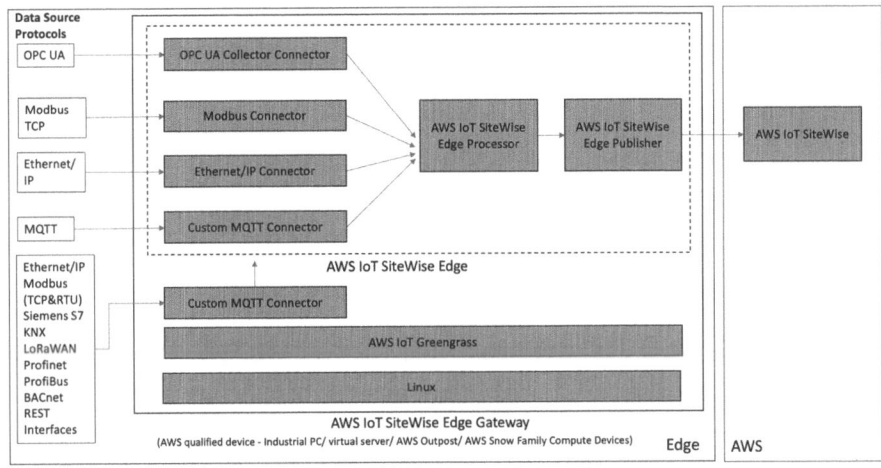

Figure 5-5. *Data ingestion patterns using AWS IoT SiteWise for data sources using different protocols*

AWS IoT Core and Data Ingestion

As explained in Chapter 1, AWS IoT Core supports connections from devices via MQTT/HTTP/LoRaWAN to the AWS Cloud. Through AWS IoT Core, data can be forwarded to other AWS services.

Let's dive deep into how AWS IoT Core processes and routes this data forward.

AWS IoT Core provides a comprehensive solution for managing IoT data, offering secure connectivity, flexible message routing, reliable message queuing, device shadowing, and robust security features. Primarily, AWS IoT Core acts as a message broker, receiving data from IoT devices and routing it to the appropriate destination based on predefined rules (called IoT rules) and policies. It also provides message queuing capabilities, allowing messages from IoT devices to be buffered and queued before being processed by downstream applications or services. This helps to decouple the ingestion and processing of data, ensuring reliable message delivery and scalability.

Figure 5-6 shows how MQTT messages received from IoT devices are processed by AWS IoT Core. As shown in the figure, the process consists of the below stages,

- MQTT messages are usually in JSON format. If not in JSON format, IoT rules also support binary payloads and there are options to decode base64 as well as AWS Lambda can be used for more complex transformations.

- IoT topics serve as communication channels for IoT devices to publish and subscribe to these messages. MQTT topics are hierarchical in nature and follow a topic-based publish-subscribe model.

- IoT rules define the logic for processing incoming messages from IoT devices based on predefined criteria. These rules use SQL-like syntax to filter, transform, and route messages based on message content, device attributes, or other metadata.

- IoT actions specify the actions to be taken when a rule is triggered by incoming messages. AWS IoT Core supports a variety of built-in actions, like storing messages in Amazon Simple Storage Service (Amazon S3), invoking AWS Lambda functions, sending notifications via Amazon SNS, and so forth.

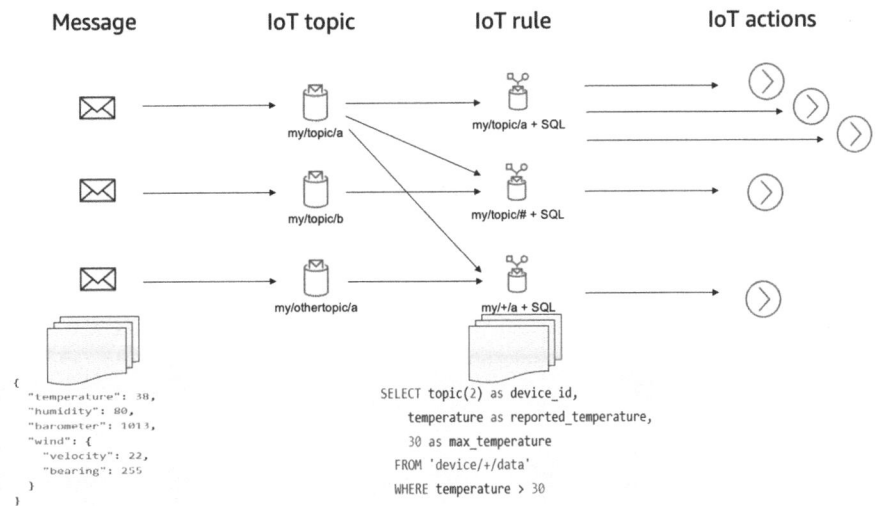

Figure 5-6. *AWS IoT core message processing*

Knowing how various protocols can be managed during the data ingestion process, you are prepared to look at the end-to-end process. In Figure 5-7, you can observe the complete journey of industrial data from the source to the data lake in Amazon S3. In the initial phase of this journey, Figure 5-7 illustrates three distinct patterns, which consider variations in setup and preferences that may differ from one customer to another.

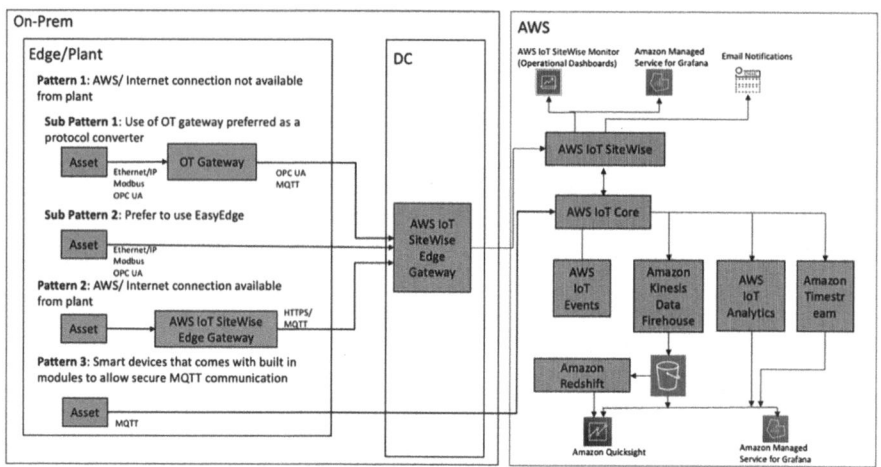

Figure 5-7. *End-to-end architecture patterns for AWS industrial data lake*

Regardless of which pattern you choose, your data may be transmitted to either AWS IoT SiteWise or IoT Core.

If your data resides in AWS IoT SiteWise, you have the option to utilize its native features for identifying production performance issues with your equipment or processes. Alternatively, you can make the data available in AWS IoT Core. Once your data is in AWS IoT Core, you have several options:

- Send the data to an Amazon S3 data lake using Amazon Kinesis.

- Make the data available in AWS IoT Analytics for data analysis through an IoT rule.

- Push the data to Amazon Timestream using a built-in IoT rule.

- Forward the data to an AWS Lambda function, which, using AWS AppSync GraphQL mutation and Amazon DynamoDB, can broadcast real-time streaming data to subscribers. You may find the detailed solution architecture here: https://aws.amazon.com/blogs/mobile/iot-with-aws-appsync/.

Irrespective of which of the preceding options you choose, you can achieve the final step of data visualization by using either Amazon QuickSight or Amazon Managed Grafana dashboards.

A great customer story of this architecture pattern is from Apollo Tyres, India's leading tire manufacturer, who has used AWS to digitally transform. By moving all of its IT infrastructure, including its SAP workloads, to AWS, Apollo Tyres uses AWS's broad portfolio of services to innovate its business processes. Leveraging AWS IoT SiteWise, Apollo Tyres engineered an IoT-in-a-box solution. This solution facilitated the swift connection of production machines on the factory floor to AWS in just five days. Once integrated, the solution gathers data from various machines, including mixers, tire building equipment, and curing presses, and streams it to the data lake. Apollo Tyres utilized Amazon Redshift, a cloud data warehouse, to construct a comprehensive dashboard for visualizing production insights sourced from the data lake. This dashboard offers business teams and plant managers real-time visibility into the manufacturing process, thereby enhancing production efficiency and productivity. For instance, it has led to a 50% reduction in the idle time of curing presses, which shape the tire in a mold. More details are available in this press release, https://press.aboutamazon.in/news-releases/news-release-details/apollo-tyres-goes-allaws-make-factories-smarter-iot-and-machine.

Overall Equipment Effectiveness Metric

One of the basic ways to enhance production efficiency is to use the data derived from equipment to calculate the very popular and widely adopted metric *overall equipment effectiveness (OEE)*. OEE plays a pivotal role in systematically enhancing manufacturing processes, providing valuable insights that enable organizations to identify and address inefficiencies, optimize equipment performance, and ultimately improve overall production efficiency. OEE serves as a measure of how efficiently a manufacturing operation utilizes its full potential during scheduled operational periods. OEE is calculated using the following formula:

$$OEE = Equipment\ Availability \times Performance \times Quality$$

AWS IoT SiteWise helps you get the values for equipment availability, performance, and quality.

To understand OEE, consider the following example scenario. A production line in a plant ran for 360 minutes today but was supposed to run for 480 minutes, and it produced 2,880 pieces of the product, out of which 2,736 pieces were actually good, useable products and the rest (144) had to be discarded. By looking at these numbers, you would know that the line is running inefficiently but you may not be able to quantify it accurately. This is where OEE comes in handy. So, OEE for this asset (production line in this example) can be calculated as shown in Figure 5-8.

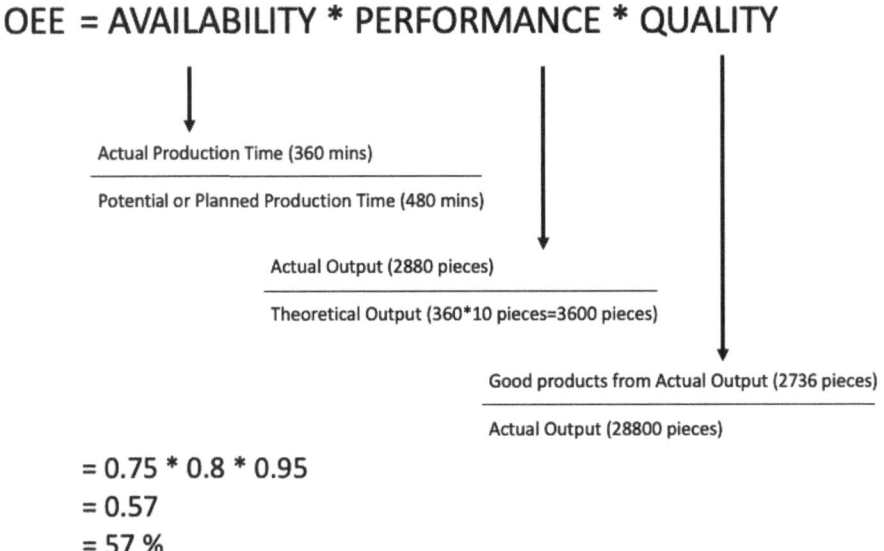

OEE = AVAILABILITY * PERFORMANCE * QUALITY

Actual Production Time (360 mins)
───────────────────────────────
Potential or Planned Production Time (480 mins)

Actual Output (2880 pieces)
───────────────────────────────
Theoretical Output (360*10 pieces=3600 pieces)

Good products from Actual Output (2736 pieces)
───────────────────────────────
Actual Output (28800 pieces)

= 0.75 * 0.8 * 0.95
= 0.57
= 57 %

Figure 5-8. *Example of calculation of OEE*

As you can see, the OEE for this asset is 57%. You might not have known that this line is running only a bit above 50% efficiency had you not used OEE. That is why OEE is the gold standard to measure manufacturing efficiency.

OEE results are usually analyzed by supervisors, such as plant managers, using reports via tools like Amazon QuickSight and AWS IoT SiteWise Monitor. Those reports can help expose areas to improve factory performance, increase utilization of equipment, increase confidence that equipment is always operating at its peak productivity, and reduce maintenance costs.

Once organizations understand OEE in real time, they can optimize production processes. For example, consider Georgia-Pacific (`https://aws.amazon.com/solutions/case-studies/georgia-pacific/`), one of the world's leading makers of tissue, pulp, packaging, and building products, which has optimized the speed of its paper production lines. Every day, Georgia-Pacific manufacturing facilities across North America produce hundreds of papers and tissue parent rolls. The paper production process is intricate and sensitive: tears or breaks may occur both during the manufacturing of parent rolls and during the conversion of these large rolls into consumer-ready bath or tissue products. Frequent tears or breaks can lead to downtime in paper machines and converting lines, incurring substantial costs for Georgia-Pacific—potentially millions of dollars per year per line, especially considering the company's extensive network of over 150 converting lines.

To extract valuable insights from manufacturing data collected at paper production plants, Georgia-Pacific faced the challenge of relying on disparate sources to analyze data related to material quality, moisture, temperature, and other critical features. To address this issue, the company adopted an AWS advanced solution using AWS IoT and analytics services. This solution facilitated the collection and analysis of data from equipment at manufacturing facilities throughout North America, resulting in:

- Anticipation of equipment failure 60–90 days in advance

- Improved ability to run more production lines predictably

- Assurance of delivering the highest quality products

Another SAP on AWS customer, Volkswagen Group, embarked on a transformative journey to enhance its automotive manufacturing and logistics processes. Central to this initiative was the development of the Volkswagen Industrial Cloud on Amazon Web Services (AWS), an integral component of the AWS-based Digital Production Platform (DPP). This platform seamlessly consolidates data from over 120 factory sites, creating a central industrial data lake on Amazon S3. DPP also helped Volkswagen establish a cloud-based architecture that facilitates the scaling of applications globally. Following an architecture pattern very similar to the one outlined above, in Figure 5-7, Volkswagen connects data originating from various machines, plants, and systems across multiple sites, thereby constructing a robust data lake on Amazon S3. AWS's IoT services play a pivotal role in delivering valuable insights into manufacturing operations across facilities.

This approach optimizes production processes and enhances overall operational efficiency. The establishment of a company-wide data lake on Amazon S3 enables Volkswagen to analyze data comprehensively. This analysis yields insights that pinpoint operational trends, improve forecasting accuracy, and streamline operations by identifying inefficiencies in production and minimizing waste. Furthermore, Volkswagen has seamlessly integrated its SAP S/4HANA system with the DPP and plans to improve manufacturing outcomes with DPP by 30% by 2025. To learn Volkswagen's entire story, refer to this YouTube video: `https://www.youtube.com/watch?v=t9ED1BseanA`.

Once you have a data lake on Amazon S3 with industrial data, you can enrich it with SAP data. This involves contextualizing the data with information such as production orders, bills of material, equipment details, inventory supplier data, and other relevant data stored in the SAP environment. If you're utilizing the SAP Plant Maintenance (PM) module, you can further enrich the data lake with maintenance records and asset settings. The same is true if you are using SAP Quality Maintenance (QM), which enables you to enrich the data lake with quality records like inspection results, work orders, and so on.

With all this diverse data consolidated in a single repository, you can introduce machine learning to derive insights for predictive maintenance and predictive quality. This can be achieved through tools like Amazon Lookout for Equipment and Amazon Lookout for Vision (see Figure 5-9). You may also develop custom predictive maintenance or quality models using Amazon SageMaker or Amazon Bedrock. The combination of real-time data contextualized with historical records enables the ML model to more effectively detect anomalies and improve its accuracy over time.

Moreover, integrating with SAP PM contributes to the model's maturity, enabling it not only to predict but also to prescribe actions. By examining maintenance records from previous occurrences of a specific failure or event, the system can prescribe actions based on historical resolutions, enhancing the overall effectiveness of maintenance strategies. This brings us to our next topic, predictive maintenance.

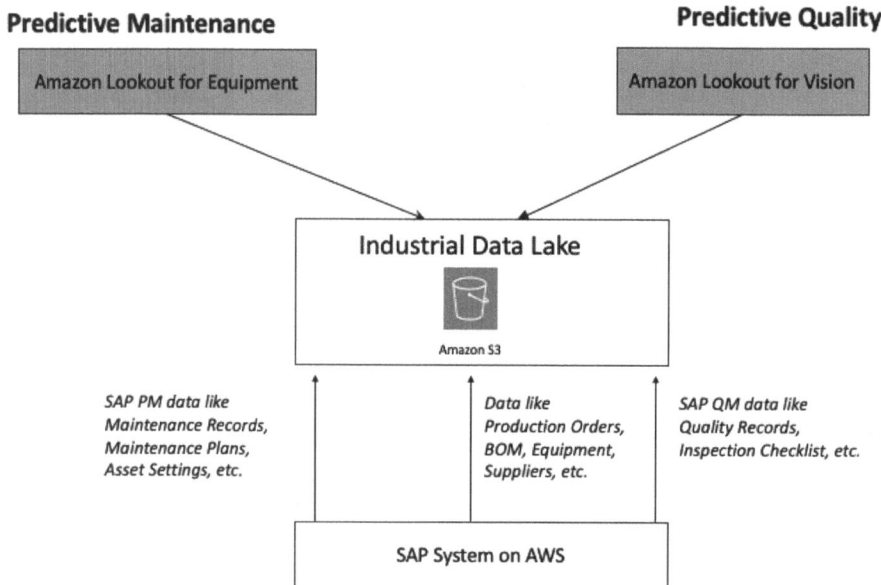

Figure 5-9. *High-level architecture of integration of Amazon Lookout for Equipment and Amazon Lookout for Vision with an SAP system on AWS*

Predictive Maintenance

Presently, there are two predominant approaches to machine maintenance, reactive and preventive. In the reactive model, maintenance occurs after a machine failure, leading to elevated maintenance costs and extended downtimes. Preventive maintenance relies on scheduling maintenance tasks periodically, based on either time or usage, which can lead to excessive maintenance if performed too frequently or potential machine breakdowns between maintenance cycles if done less frequently.

Predictive maintenance, where maintenance is executed precisely when needed based on machine data, is an emerging alternative. Predictive maintenance is the activity of monitoring and evaluating the condition of equipment, detecting developing faults, and planning specific corrective maintenance activities for a time when it is most cost effective. However, implementing predictive maintenance using machine learning is intricate and demands access to scarce resources. In an ideal scenario, your would have the knowledge of when industrial machinery requires maintenance, allowing for systematic removal of equipment from service to manage downtime and prevent critical failures. Reality, however, is more complex due to the diverse types of machines, each exhibiting distinct symptoms, signals, and failure modes. But, with continuous monitoring via AWS IoT services, you can quickly identify potential failure events before they happen. Using IoT sensors to monitor equipment performance, coupled with AI and ML, allows companies to predict equipment issues proactively.

Implementing predictive maintenance is not a one-click process. It depends on the complexity of the machinery and equipment involved, the availability and quality of data, the level of expertise within the organization, and so forth. In essence, predictive maintenance follows a maturity curve. Figure 5-10 illustrates the conventional maturity curve for maintenance capability. Typically, organizations commence with reactive maintenance, progress toward scheduled maintenance, and ultimately adopt condition-based maintenance, with preventive maintenance as the ultimate goal.

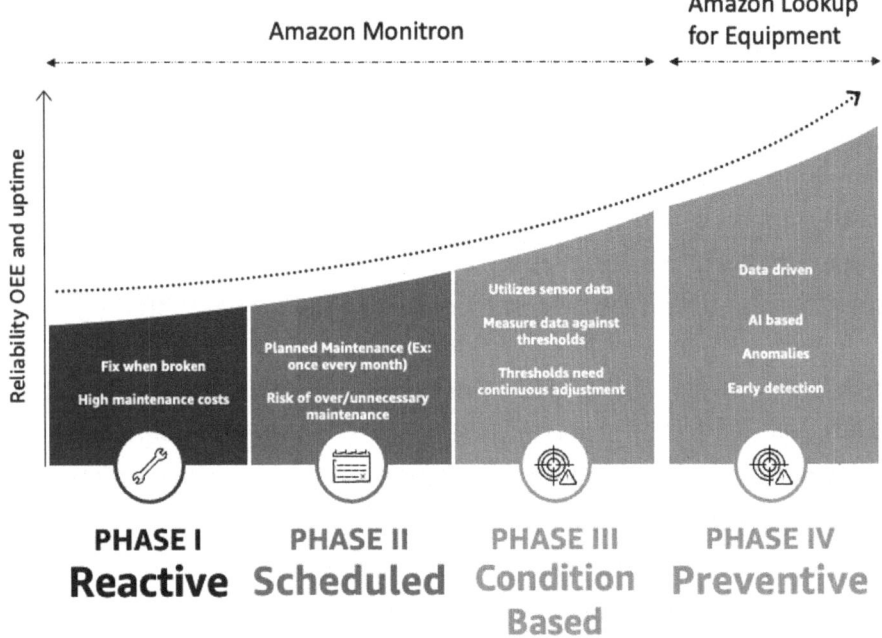

Figure 5-10. *Maintenance capability maturity model*

The following are a few common challenges for customers to get to phase II or phase III from phase I:

- Installing sensors is an expensive initiative.

- Analyzing sensor data to detect potential failures and setting up a correct mechanism to send alerts are not easy.

AWS addresses these challenges through AWS Monitron, briefly introduced in Chapter 1.

Amazon Monitron

In this solution, cost-effective Amazon Monitron sensors are affixed to the asset, capturing vibration and temperature data. This data is then transmitted to AWS via a Monitron gateway, which establishes a Bluetooth connection to the sensor. Once the data is in AWS, the service utilizes both ML algorithms and International Organization for Standardization (ISO) standards. This dual methodology provides the operations technology (OT) team with valuable insights into the health of the asset. Monitron also includes a user-friendly app through which notifications and alerts can be received. Importantly, users have the ability to provide feedback on the alerts, and this feedback is integrated back into the service for continuous improvement.

Let's look at the end-to-end architecture:

1. The equipment's sensor data is captured and transmitted to AWS, where the Amazon Monitron service conducts an analysis using vibration ISO standards and ML techniques.

2. The measurement data and ML inference outputs from Amazon Monitron are exported to an Amazon Kinesis data stream and, through Amazon Data Firehose (formerly Amazon Kinesis Data Firehose), delivered to Amazon S3.

3. Upon the availability of a new set of sensor data and inference results in the Amazon S3 bucket, SAP customers experience a seamless process whereby either a maintenance document is generated for changes in temperature or vibration, incorporating the new values for these parameters, or, in the case of anomaly detection, a maintenance notification is created.

235

Every time a new set of sensor data and inference results is available on the Amazon S3 bucket, you can use AWS Lambda to orchestrate the triggering of API call via SAP Business Accelerator Hub (as shown in the architecture diagram at `https://aws.amazon.com/solutions/guidance/predictive-maintenance-with-amazon-monitron-and-sap/`) to create the maintenance document or maintenance notifications in the SAP ERP Central Component (ECC) or SAP S/4HANA system. This notification prompts maintenance planners or maintenance technicians to investigate and address the identified anomalies.

A notable success story involving Amazon Monitron comes from Koch Ag & Energy Solutions (KAES), a subsidiary of Koch Industries. KAES implemented Amazon Monitron on crucial assets, enabling predictive maintenance activities that prevent equipment failures, thereby eliminating unplanned outages.

The KAES approach can be summarized as "Think Big, Start Small, Scale Fast." Here's a brief timeline of KAES's implementation of Amazon Monitron:

- 2020: Initially deployed five sensors.

- 2021–2022: Expanded to 122 sensors across five sites.

- 2023: Deployment increased to over 500 sensors, detecting hundreds of anomalies that prompted process adjustments or equipment repair work orders.

As shown in Figure 5-11, the monitoring data is utilized through the Amazon Monitron monitoring app, empowering operators to take necessary actions. Additionally, KAES leverages a Kinesis stream to feed this data to a data historian, a system that collects and stores time-series data from industrial equipment and processes it for real-time monitoring and historical analysis, providing further context. This data feeds into AI/ML models, offering analysts deeper insights into ongoing operations.

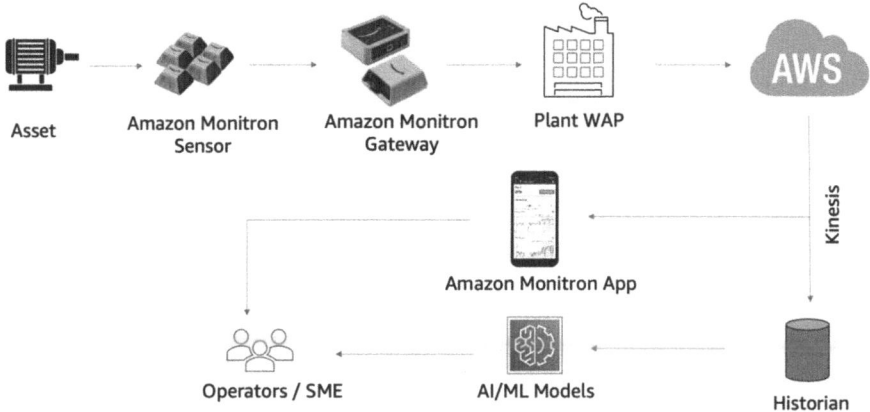

Figure 5-11. *Architecture for predictive maintenance use case at KAES*

Refer to the YouTube video "Predictive Maintenance at Scale: KAES's Journey with Amazon Monitron" (`https://www.youtube.com/watch?v=FYfXj_rcVzs`) to learn more about KAES's implementation journey along with specific instances when Amazon Monitron came to its rescue.

Once organizations are in condition-based maintenance (phase III in Figure 5-10), the transition to preventive or predictive maintenance (phase IV) is a big one because predictive maintenance is based on ML, and building ML pipelines to provide inference output is difficult, costly, and a time-consuming process. It involves the identification of critical assets and the location of appropriate sensor data before handing it off to a data scientist. The data cleansing and preparation phase, typically managed by a data scientist, also requires a significant amount of time. Subsequently, selecting the proper algorithm, tuning hyperparameters, and refining the model through multiple iterations add further complexity. Once deployed in production, there may be a need to retrain models for improved accuracy and to address issues like data drift. To mature to phase IV, AWS offers Amazon Lookout for Equipment.

Amazon Lookout for Equipment

Amazon Lookout for Equipment adopts a different approach by automating the generation, analysis, and deployment of models for individual assets without the need for extensive data preprocessing. This substantially reduces the time required to build ML models from months to hours. The differences in the workflows between traditional AI/ML models and Amazon Lookout for Equipment are illustrated in Figures 5-12 and 5-13, while the detailed approach adopted by Amazon Lookout for Equipment is depicted in Figure 5-14.

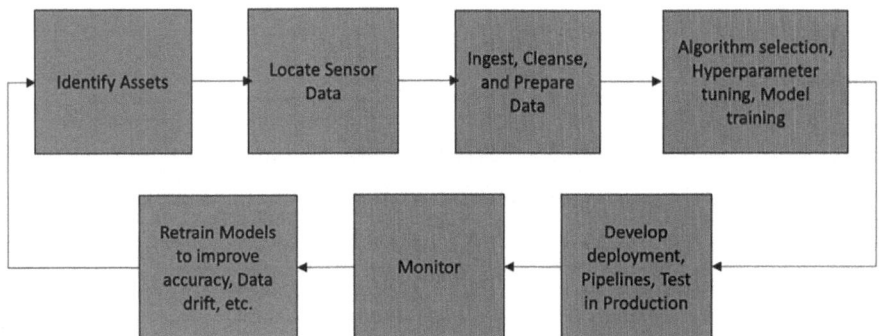

Figure 5-12. *Traditional AI/ML workflow*

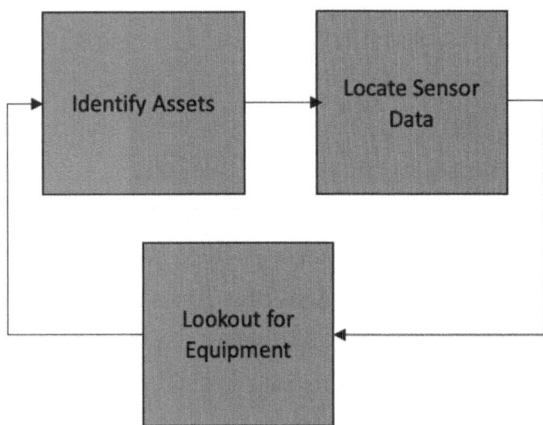

Figure 5-13. *Amazon Lookout for Equipment workflow*

238

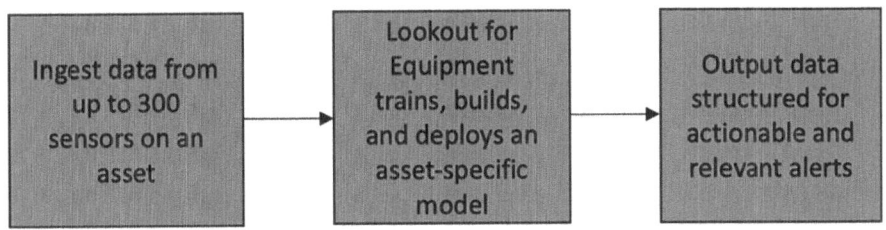

Figure 5-14. How Amazon Lookout for Equipment works

Amazon Lookout for Equipment leverages historical equipment data obtained from existing sensors, coupled with historical maintenance events, usually available in the IoT/industrial data lake (as discussed earlier) to construct a tailored ML model. This model learns the typical behavior patterns of the machine. When operational data deviates from the established normal behavior, Amazon Lookout for Equipment identifies and flags the deviation. Users can receive actionable and relevant alerts concerning impending failures and abnormal ML behavior, enabling accurate detection of equipment abnormalities, swift diagnosis of issues, and the ability to take proactive measures to reduce downtime, prevent critical failures, and optimize overall machine performance.

Collaborating with AWS enabled Toyota Motors to create a predictive maintenance system, shifting from unexpected machine breakdowns to a proactive maintenance schedule even during weekends and holidays. Toyota's predictive maintenance system utilizes AWS IoT SiteWise and Amazon Lookout for Equipment for machine learning. Both services facilitated the development of dashboards for Toyota's team members to visualize equipment anomalies and anticipate breakdowns before they occur. Toyota found the system to be user-friendly, requiring no data science background for implementation in its factory. Within the initial months of adopting AWS, Toyota experienced over 16 instances where early anomaly detection resulted in savings exceeding $80,000 in equipment downtime and spare parts costs. This approach has empowered Toyota's team to address issues during planned downtime efficiently. Watch the customer testimonial at https://www.youtube.com/watch?v=ymvz7UT5fhg.

Figure 5-15 is a sample end-to-end architecture pattern. So far, we have covered how to include the asset data in the data lake on Amazon S3. This data lake is enriched with maintenance history data from the SAP system on AWS. The Amazon Lookout for Equipment anomaly model is trained from this data. Then, a Lambda function gets triggered based on the inference results of the model. If an anomaly is detected, similar to the Amazon Monitron use case, you can call an SAP BTP API endpoint to create a maintenance notification so that maintenance planners or maintenance technicians can act on it. This is also the basis of the architecture pattern suggested in this prescriptive guidance at `https://aws.amazon.com/solutions/guidance/amazon-lookout-for-equipment-integration-with-sap-plant-maintenance/`.

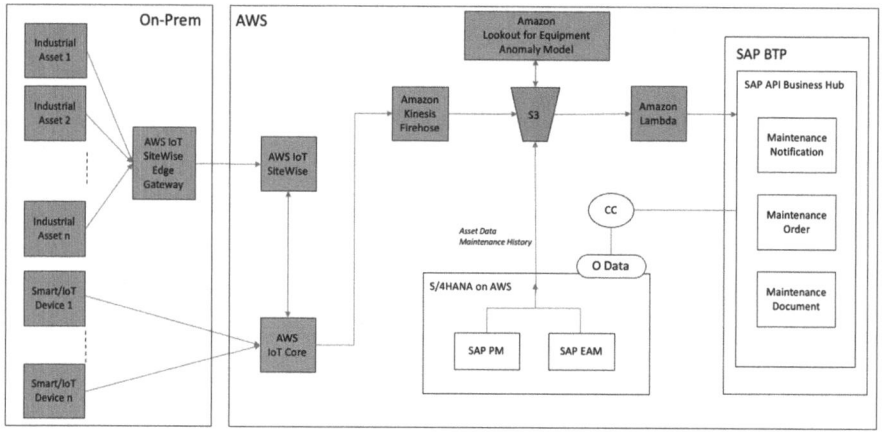

Figure 5-15. *Amazon Lookout for Equipment architecture pattern*

Predictive Quality

Quality control primarily relies on sampling methods within manufacturing organizations. However, by incorporating cutting-edge technology like computer vision, specifically through advanced tools like Amazon Lookout for Vision, manufacturing entities can revolutionize their quality assurance processes.

The capabilities of Amazon Lookout for Vision extend beyond traditional quality inspection methods, offering real-time analysis of large datasets with enhanced accuracy. This not only ensures that each product is thoroughly examined but also helps in identifying deviations or defects that might be missed in conventional sampling approaches. It enables manufacturing organizations to establish a more comprehensive and efficient quality control system. This, in turn, leads to a substantial reduction in the likelihood of defective products reaching consumers, thereby bolstering customer satisfaction and minimizing returns.

The continuous monitoring and analysis capabilities of Amazon Lookout for Vision contribute to a proactive approach, allowing manufacturers to address potential issues before they escalate, fostering a culture of continuous improvement in product quality. Overall, embracing computer vision in quality control processes can significantly elevate the standards and effectiveness of manufacturing operations.

There are several compelling reasons to use Amazon Lookout for Vision:

- **Enhanced efficiency**: Amazon Lookout for Vision significantly improves the speed, consistency, and accuracy of inspections when compared to manual inspection processes. By seamlessly integrating this with SAP Quality Management, manufacturers can reduce reliance on manual inspections, significantly improve accuracy, and automate creation of notifications, defects, measurements, and orders in the SAP system.

- **Near 100% sampling**: Quality inspection is usually done in samples on batches. By using Amazon Lookout for Vision, you can make sure all products pass through the automated gate check.

- **Rapid inference results**: The model can swiftly analyze an image, deliver an inference result, and present the outcome in less than 30 seconds, contributing to rapid decision-making.

- **Continuous model improvement**: Operators, quality managers, and process engineers have access to feedback tools, enabling them to enhance models based on practical insights and domain knowledge, ensuring ongoing improvement.

- **Cost-effective quality control**: Amazon Lookout for Vision offers a cost-effective solution for quality control inspections compared to traditional machine vision systems, providing organizations with an economical approach to maintaining high quality standards.

- **User-friendly and accessible**: Amazon Lookout for Vision is designed to be user-friendly, allowing users to get started without requiring any prior ML experience. This accessibility ensures a smooth onboarding process for users of varying technical backgrounds.

Quality management encompasses two primary aspects: quality planning and quality execution (which is basically inspection). Material master attributes for quality, attributes for quality itself in the form of characteristics, classes, and codes are defined. Inspection plans specify the materials, attributes, and sample sizes for inspection. With Amazon Lookout for Vision, the constraint of human resources is eliminated, allowing for nearly 100% sampling, which ensures a higher first pass yield (FPY) and rolled throughput yield (RTY) because every product and batch undergoes a quality phase gate.

Measurement documents evaluate quality in terms of characteristics and codes, and quality information records set thresholds for supplier quality. Quality inspection begins with an inspection lot, serving as the starting document for inspection transactions. Results are recorded, and defects may lead to notifications, quality orders, or corrective action plans. Similar to Amazon Lookout for Equipment, explained in the previous section, Amazon Lookout for Vision employs ML algorithms to detect anomalies in objects. The detection process involves training on a predefined set of "good" images of those objects, enabling the system to recognize and flag any deviations or defects during operational processes.

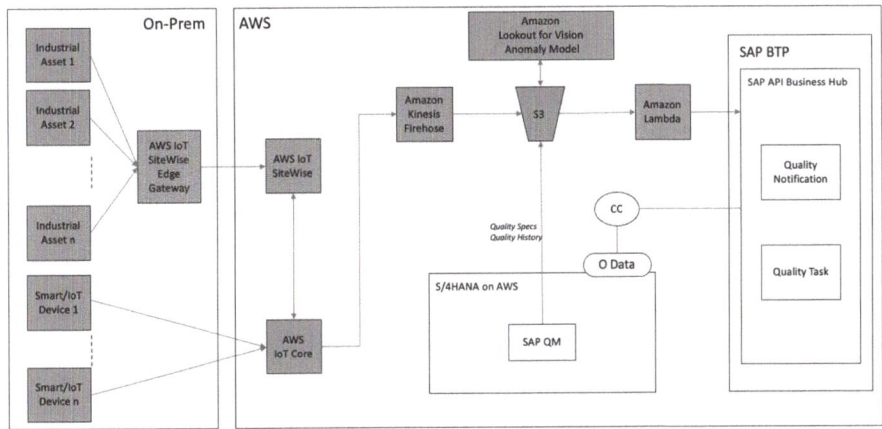

Figure 5-16. *Amazon Lookout for Vision architecture pattern*

Once images from a manufacturing facility camera are captured, they are transferred to Amazon S3. Once a new image is uploaded in the S3 bucket, an Amazon Lambda function is triggered, which calls the DetectAnomalies API. If an anomaly is detected, an SAP QM Notification or a Defect or a Measurement or a QM Order can be created in the SAP ECC or SAP S/4HANA system by using the appropriate APIs via SAP Business Accelerator Hub, just like you can create a maintenance notification when using Amazon Lookout for Equipment. This process is depicted in Figure 5-16.

By leveraging Amazon Lookout for Vision, Invista, a subsidiary of Koch Industries and a global manufacturer of chemical intermediates, polymers, and fibers, successfully automated visual inspections across its production lines, resulting in faster response time to issues and enabling proactive interventions that significantly enhanced production efficiency. This also eliminated the need for certain reactive responses, consequently reducing operational complexity and costs.

The deployment of Amazon Lookout for Vision proved instrumental in improving the reliability of detecting process anomalies and unusual conditions within Invista's production processes. This not only contributed to a reduction in waste but also empowered technicians to proactively identify potential process issues. With the ability to take earlier corrective actions, Invista experienced a notable improvement in overall operational efficiency. The complete case study can be found at `https://aws.amazon.com/solutions/case-studies/invista-case-study/`.

Digital Twin

A *digital twin* is a living digital replica of a physical system that is driven and updated by the data from the physical machines continuously to mimic the true structure, state, and behavior of the physical system to understand and predict its behavior in order to drive business outcomes. The simplest example is Google Maps or Apple Maps, each of which is a digital twin of the earth's surface.

Consider an industrial scenario where you have numerous industrial motors on your manufacturing floor, each vital for maintaining smooth operations. To gain insights into their operational aspects, such as speed and efficiency, creating a digital twin becomes crucial. This involves constructing a 3D model, serving as a visualization object, allowing for the simulation of the physical motor. Through the analysis of data collected from sensors, using AI and ML models, you can drive business outcomes like streamlined operation, optimized resources, and anticipated issues. Leveraging a digital twin can also play a crucial role in advancing sustainability goals for the organization.

By utilizing operational data within the context of a digital twin, organizations can focus on enhancing efficiency and reducing the consumption of resources like power, water, or gas in their operational processes. This targeted approach toward resource optimization contributes directly to achieving sustainability objectives, promoting eco-friendly practices, and implementing responsible resource management.

As shown in Figure 5-17, one of the prerequisites to creating a digital twin is that data from all the industrial equipment must be available/ stored in AWS. You have already learned about different ways to create an industrial data lake on Amazon S3, so assuming this prerequisite is taken care of, there are still a few challenges:

- Bringing together data from various sources (like IoT data from sensors, video feeds from cameras, etc.)

- Asset modeling

- Adding insights such as simulations using modeled data

- Creating 3D visualizations

- Publishing web applications to be used by operators

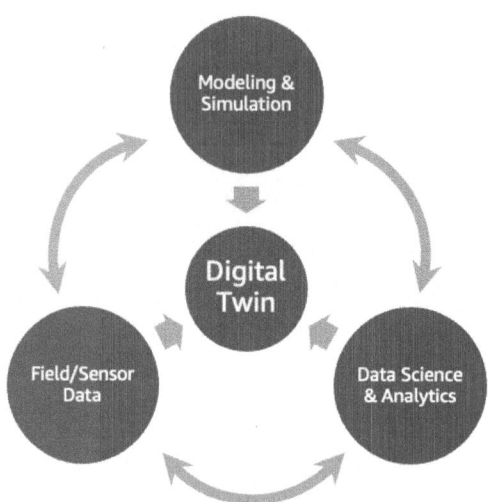

Figure 5-17. *How a digital twin works*

AWS IoT TwinMaker helps you with these challenges. The entire end-to-end process involves the following five stages, as depicted in Figure 5-18:

1. Create a workspace to serve as a container for all artifacts necessary for digital twin creation, such as asset models, data source connectors, files, documents, and so on. Data ingestion is handled by built-in connectors. If your data is already stored in AWS natively, AWS IoT TwinMaker can effortlessly access it using its built-in connectors. For data in third-party databases on AWS, you need to create custom connectors.

2. Conduct asset modeling, a process very similar to the process of creating digital model assets described earlier in the section "AWS IoT SiteWise."

3. Add streaming insights. A mechanism for self-learning within the system is crucial, and this is where ML simulations play a vital role. Building an ML model involves incorporating these learnings.

4. Build scenes, which are essentially 3D models serving as visual representations of your physical systems.

5. Create a web application to visualize these scenes. AWS IoT TwinMaker offers a low-code experience for constructing the required web application, enabling plant operators and maintenance engineers to access and interact with the digital twin. AWS IoT TwinMaker includes a plug-in for Amazon Managed Grafana featuring a 3D scene viewer, allowing the creation of 3D-enabled application dashboards.

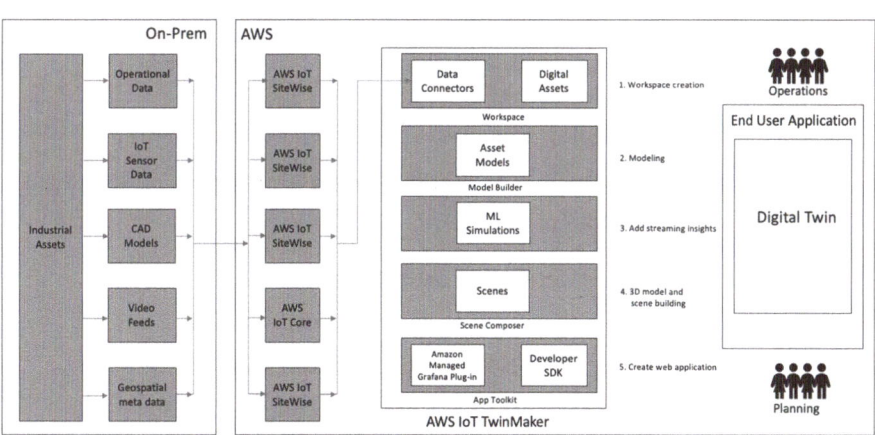

Figure 5-18. *Implementing a digital twin using AWS IoT TwinMaker*

Voice Technology

Voice technology is in place already in multiple warehouses for picking, replenishment, put-away, packing, material movement, and so forth. Any business process wherein users have to do some physical activity, take notes, and enter them in an application could use voice to improve productivity and accuracy. We are increasingly seeing customers wanting to do more with conversational apps using Amazon's cloud-based voice service, Alexa, in the workplace. Let's discuss the picking process as an example.

A distribution center employee, upon signing in, receives instructions on the specific location within the warehouse to commence picking. In the course of their activities, pickers validate each instruction, complete the pick simultaneously, and update SAP in real time as each pick is completed. This is way more efficient than a manual process. Benefits of voice-based picking include

- Significantly improved operational efficiency by eliminating the need for manual scans, task verification, and data entry, leading to streamlined workflows.

- Faster processing times compared to traditional mobile device usage, surpassing expectations set by paper-based operations.

- A significant reduction in error rates. By enabling pickers to focus solely on task execution without the need for interruptions caused by cross-checking content on a screen device, the likelihood of errors is greatly minimized.

- Substantial savings in end-user training costs and time. The intuitive nature of voice commands simplifies the learning curve, allowing users to quickly acclimate to the new system with minimal training requirements.

Figure 5-19 depicts the high-level end-to-end architecture pattern for integrating Alexa with SAP.

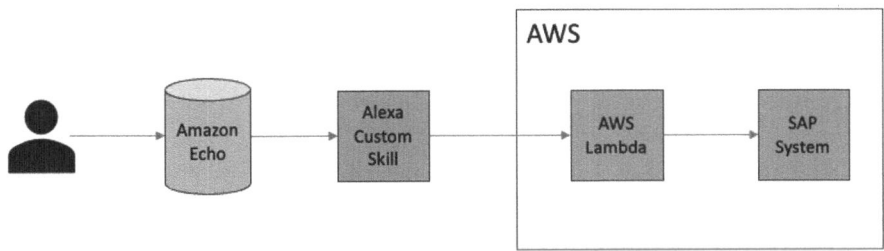

Figure 5-19. *Architecture pattern for Alexa integration with SAP*

Amazon Alexa allows intuitive interaction with everyday technology. To enhance the user experience, Amazon provides the Alexa Skills Kit, a set of tools and APIs for creating voice-based experiences on Alexa. There are multiple hosting options for Alexa skills; the option shown in Figure 5-19 is through an Amazon Lambda function triggered by voice commands via an Amazon Echo device.

Alexa skills consist of voice commands enabling operations. Let's say we are building Alexa skills for Smart Factory 1 to get the inventory of Product 1 in Warehouse 1. Every Alexa skill needs the following resources:

- An *invocation*, which signals Alexa to start the interaction. For example, in this example, invocation can be "Smart Factory 1."

- *Intents* represent the tasks to Alexa. Each intent denotes an action that fulfills a user's spoken request. These intents are defined within a JSON structure referred to as the *intent schema*. In our example, "Get Inventory" might be an intent.

- *Slots*, which narrow the tasks requested. Think about slots as arguments or parameters of intents or as the where clause in a SQL statement. In this example, "Warehouse 1" might be a slot.

249

Suppose the user issues the following voice command: "Alexa, ask Smart Factory 1 to get inventory of Product 1 in Warehouse 1." The resources would be

- Invocation name: Smart Factory 1

- Intent: GetInventory

- Slot: Warehouse1

This is the Alexa custom skill that is represented in Figure 5-19.

After you have built your Alexa skill, you must link it with an Amazon Lambda function in order to pass the intent schema to be processed. You would need to link the ARN of the Lambda function in the Endpoint tab of the Alexa Developer Console for that particular skill. You need to save the skill, make a note of the skill ID, and link this skill ID in the "Add trigger" configuration of the Lambda function.

As soon as the user makes the request via voice, the skill understands the intent and triggers the specific Amazon Lambda action. The Lambda function built with Python can connect to the SAP ECC or SAP S/4HANA system to make the appropriate SAP BAPI/function module call using PyRFC. SAP returns the result of the execution with error codes and additional data. If you prefer to make OData calls to the SAP system using the SAP Gateway service, instead of calling BAPIs or function modules, you may have the Amazon Lambda function call the SAP OData endpoint present in the SAP system. Another option is to expose the OData service in the SAP system as an API in SAP BTP through the SAP Integration Suite and have the Lambda function call it. Authentication is not factored in this architecture pattern; you may use an Amazon Cognito user pool for authentication or the SAP Authorization and Trust Management (https:// api.sap.com/package/authtrustmgmnt/overview) service from SAP BTP. The Amazon Lambda function may access AWS Secrets Manager for credential information.

Having covered the fundamental principles of digital manufacturing that have paved the way for the introduction of numerous AWS services in this domain, let's now explore thoughtfully curated solutions leveraging these AWS services, designed to address typical use cases within the industrial sector.

AWS Solutions for Manufacturing and Industry

According to *Avoiding Pilot Purgatory: How to Choose the Right Use Cases to Accelerate Industrial Transformation*, a 2020 e-book published by LNS Research,

- 70% of manufacturers have started their smart manufacturing journey.

- <30% see the business value expected.

- >40% identify data integration issues as the primary driver.

To address the challenges identified by customers, AWS offers purpose-built solutions to solve industry-specific business challenges and assist these customers in achieving tangible results. These solutions utilize AWS services, AWS solutions, and AWS Partner Solutions, along with guidance such as reference architectures. For more information, refer to the Solutions for Manufacturing & Industrial page of the AWS Solutions Library (`https://aws.amazon.com/solutions/industrial/`), which groups the solutions under different categories. The solutions listed in the "Smart Manufacturing / Production & Asset Optimization" category are most relevant for this chapter. The following sections describe each of those solutions in turn.

Asset Maintenance & Reliability

These solution packages help to integrate disparate data sources including operational systems, sensors and others, leverages this data to remotely oversee the health of factory assets, forecast potential failures, implement corrective measures, prevent unexpected downtime, and consolidate data for enhanced reliability within a unified platform.

Asset Performance Management

Asset Performance Management (APM) helps in attaining operational excellence by facilitating data-driven decision-making regarding asset performance and the risk of failure. By leveraging insights derived from data analysis, APM assists in mitigating operational risks and reducing costs while optimizing the balance between efficiency and capital investment. APM involves proactively monitoring data silos across various applications to offer intelligent business analytics, enabling customers to visualize and integrate operations seamlessly, thereby fostering operational excellence.

Automation Software Management

Automation Software Management solutions streamline the organization of PLC code versions, device configurations, and more, empowering manufacturers to automate their software management procedures. These solutions aid in reducing unexpected downtime by preventing error-prone PLC code updates and promoting systematic version management of automation code. They support comprehensive implementation of DevOps methodology, covering development, testing, and other facets, while also offering features such as automated tracking of version changes and notification of modifications to users. Through the utilization of these solutions, manufacturers can securely execute over-the-air updates of firmware, patches, and PLC code, resulting in notable time and cost efficiencies. Additionally, these solutions enhance productivity by enabling remote monitoring of hundreds of PLCs, eliminating the need for manual visits to each one individually.

Cloud Manufacturing Execution System (MES)

Cloud MES on AWS empowers manufacturing clients to swiftly digitize their operations on a large scale. This solution integrates cutting-edge features from chosen AWS Partner Solutions with the scalability, accessibility, security, and reliability of the AWS Cloud, facilitating essential MES functions including order dispatching and execution, process enforcement, and comprehensive management and visibility of data, events, and alerts. Moreover, Cloud MES extends its functionality to encompass scheduling and sequencing, maintenance, and energy management, providing a holistic solution for manufacturing optimization.

Connected Worker

The Connected Worker solutions boost workforce productivity by equipping employees with tools to comprehend vital KPIs, pinpoint root causes of incidents, and foster improved collaboration across teams. They expedite training and onboarding for standard operating procedures (SOPs) via virtual and simulated scenarios, enhancing employee skills and diminishing errors and inefficiencies. Additionally, they reinforce adherence to safety protocols and promptly alerts workers to real-time hazards, ensuring a safer work environment.

Composable Operations Applications

What we often observe among our customers is that as they grow, their operations tend to become more intricate. This complexity typically leads to the formation of organizational silos, whether it's in terms of an increased number of departments/functions or the associated data and applications. Ultimately, this impedes their ability to deliver value and adapt to the rapidly changing business landscape, driven by both internal and external factors. Consequently, stakeholders face significant pressure not only to continually benchmark themselves against competitors and industry standards but also to modernize their data and application infrastructure to foster innovation and leverage emerging disruptive technologies. In this context, we think the vision that ultimately all of our customers are thinking about is how can they become more of what we call a composable enterprise.

A *composable enterprise* is one that structures itself around modular business capabilities, enabling quick and seamless interchangeability of these building blocks based on evolving needs and circumstances. Composable Operations Applications can be used to empower both the business and the stakeholders so that they are able to effectively not only collaborate but also rapidly start composing both the applications and the experiences across the engineering technology, information technology, and operations technology landscapes. Composable Operations Applications are built, partly or wholly, as cloud-native and vendor-agnostic, interchangeable building blocks of business capabilities. These modules can be leveraged in conjunction with existing core systems like Product Lifecycle Management (PLM), Manufacturing Execution Systems (MES), Enterprise Asset Management (EAM), SAP, and so on to close gaps, extend processes, or enable manufacturing enterprises to rearrange and reorient as needed depending on external or internal factors. The foundation and the pillars of a composable enterprise are shown in Figure 5-20.

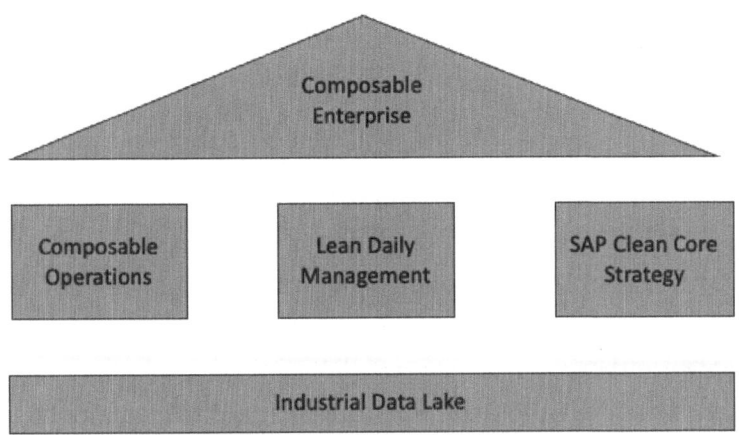

Figure 5-20. *Foundation and pillars of a composable enterprise*

With composability, organizations can achieve digital acceleration, greater resiliency, and the ability to innovate through disruption. AWS has partnered with market-leading, low-code, rapid application development platform Mendix to offer cloud-native, vendor-agnostic applications to compose your manufacturing operations with modular, interchangeable blocks addressing desired business capabilities.

Industrial Data Fabric

Industrial Data Fabric (IDF) solutions on AWS is designed to empower customers in achieving smart manufacturing goals, whether they are starting from scratch or have encountered challenges with existing projects. IDF offers buyer and builder solutions that make it easier to ingest, store, contextualize, and act on manufacturing data across the value chain. Very often industrial customers start with a single use case having two or three data sources where context is created, thereby making data useful for consumption for only that use case. Scaling the use case for production and incorporating additional data sources to the same use case or creating completely new formats (context) from existing and new data sources is a challenge that customers face today in managing industrial data on the cloud.

The difficulty comes down to creating and maintaining different formats from different industrial data sources to be consumed by multiple use cases in a scalable manner coupled with robust data governance. Solutions from AWS partners having native integration on AWS allows information technology (IT) and operations technology (OT) users to contextualize and normalize data into rich information for analytics and other business systems. The solutions are designed to be maintained and scaled across the enterprise as the number of use cases that rely on industrial data grow exponentially.

Lean Daily Management

The Lean Daily Management solution aims to assist personas like plant managers in gaining comprehensive insights into their lean manufacturing processes and supporting them in their daily management routines. They can know on a daily basis whether they are on track or off track in meeting their goals via the visualization of updated statuses, alerts, and actions across transactional and operational data from their operations, thereby enabling the identification and addressing of potential issues or red flags. By automating the aggregation and interpretation of metrics from various data sources, Lean Daily Management solution reduces manual efforts significantly. Moreover, it decreases reliance on IT teams for delivering and scaling multiple views for different users and managing evolving requirements. Additionally, it enhances productivity by fostering cross-functional collaboration and minimizes the time required for new hires to grasp key operational metrics.

Computer Vision for Quality Insights

We have discussed this topic in-depth earlier, but if you are interested in exploring AWS Partner Solutions in this area, Computer Vision for Quality Insights solutions offer an excellent starting point to assess your options. These solutions leverage data and images from vendor-agnostic vision sources to streamline process orchestration, derive valuable insights, and empower customers to achieve zero defects on a large scale. These insights enable manufacturers to establish closed-loop integration with root-cause analysis solutions and deploy countermeasures effectively to minimize the cost of quality. Benefits include providing a ready-to-deploy solution in days versus months, automating the visual inspection process, providing insights to reduce rework and scrap, eliminating human error, and reducing inspection time for increased throughput and productivity.

Operational Technology (OT) Cybersecurity

OT Cybersecurity solutions can play a critical role in safeguarding industrial processes and systems, ensuring operational continuity, and defending against threats that could potentially cause production disruptions, safety hazards, or data breaches. In today's interconnected and digital landscape, these solutions are indispensable for maintaining the integrity and security of industrial operations.

Predictive Quality

We have also discussed the topic of predictive quality in detail. AWS Predictive Quality solutions provide manufacturers leading KPIs to identify quality before defects occur. They integrate manufacturing process data using AWS IoT services, combine it with in-process and end-of-line quality data, and use AI/ML computations to predict quality defects before they occur.

In addition to these solutions, SAP offers its own packaged solutions that can run on AWS. Let's proceed to discuss those next.

SAP Solutions to Enable Digital Manufacturing

SAP provides its own products and solutions to enable digital manufacturing in conjunction with the AWS native services discussed thus far. SAP Manufacturing Integration and Intelligence (SAP MII) has been serving as SAP's flagship product for integrating shop-floor operations with business process in SAP systems like SAP ERP (SAP ECC) for many years. It offers a versatile solution for integrating various shop-floor applications, equipment, and sensors with SAP systems. SAP MII is depicted in Figure 5-21.

Beyond integration capabilities, SAP MII offers services for organizing, distributing, and contextualizing manufacturing data through dashboarding and analytics features. Dashboarding and analytics functionalities are achieved by developing reports with data sourced from systems like SCADA, Historian, and MES, enabling manufacturing personnel to visualize and leverage accurate and timely data for critical decision-making.

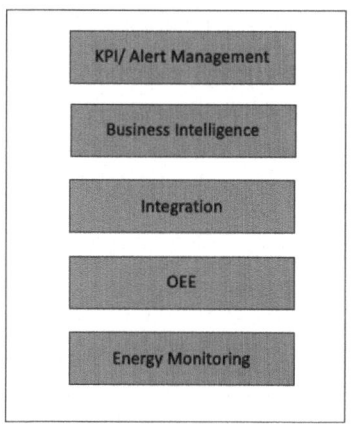

SAP MII

Figure 5-21. *SAP MII and its components*

So, how does SAP MII communicate with these data sources that use different protocols? This challenge was discussed in detail earlier in the context of using AWS IoT SiteWise. This communication is taken care by a software component called SAP Plant Connectivity (PCo). SAP PCo facilitates data exchange between various industry-specific standard data sources from different manufacturers, such as process control systems and Historian, using protocols like OPC UA, DA, Modbus, MQTT, and others, with SAP systems like SAP MII, SAP Manufacturing Execution (SAP ME), SAP ERP, SAP Extended Warehouse Management (SAP EWM) or databases like SAP HANA and Microsoft SQL Server. The significant advantage of SAP MII lies in its ability to provide both integrations, encompassing data connectivity and transformation, and intelligence, including user interface and manufacturing analytics, within a real-time environment.

While SAP MII serves as a framework for delivering light MES functions, it does not offer a full-fledged MES application. For this purpose, another product called SAP Manufacturing Execution (SAP ME) is available, which is a prebuilt application containing a wide range of solutions for managing orders, work instructions, routings, rework, documentation, and quality control.

SAP MII typically requires very low latency with on-premises equipment, devices, and IT systems, so it is usually hosted on prem; however, latency has not been a problem for several SAP on AWS customers who are hosting SAP MII on AWS with SAP PCo and the integrated devices being on prem. SAP PCo can directly interface with SAP ECC or S/4HANA, but incorporating SAP MII in between, as shown in Figure 5-22, simplifies the process of calling RFCs/BAPIs without needing to expose them as web services and also serves as a buffer during maintenance periods when SAP ECC or S/4HANA is unavailable.

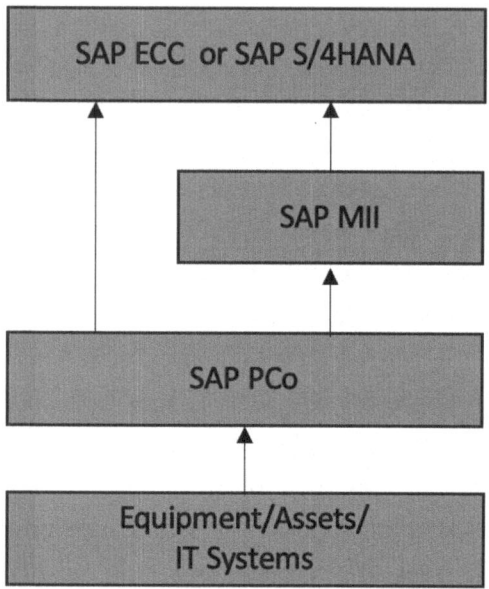

Figure 5-22. *Integration of SAP MII with SAP ECC or SAP S/4HANA and industrial assets*

SAP's next-generation products for managing and maintaining your physical assets throughout their life cycle is delivered under the umbrella of Enterprise Asset Management (EAM), which uses IoT, ML, mobility, GIS, GPS, and other applications. Apart from those, which are a part of the SAP S/4HANA system, EAM contains the following cloud solutions (previous versions of which were earlier offered as SAP on-prem or AWS native deployment), which can be easily integrated with the SAP S/4HANA system using the guided procedures via Cloud Integration Automation Service (CIAS).

- **SAP Asset Performance Management**: Based on Industry 4.0 technologies such as the SAP AI Core infrastructure, SAP Asset Performance Management powers dynamic, intelligent maintenance strategies. SAP Asset Performance Management helps asset owners, managers, plant supervisors, and reliability engineers in measuring and enhancing asset performance, as well as optimizing maintenance strategies. SAP Asset Performance Management is the next-generation offering that refines the capabilities from SAP Asset Strategy and Performance Management/SAP Predictive Asset Insights toward a better alignment with SAP S/4HANA. It offers greater control, simplifies planning processes, and enhances accuracy, thereby enabling more effective asset management and maintenance planning.

- **SAP Business Network Asset Collaboration**: The SAP Business Network Asset Collaboration (formerly SAP Asset Intelligence Network) solution establishes a unified platform for efficiently exchanging asset data among stakeholders and coordinating work orders between maintenance and service partners. This solution enables operators, manufacturers, and service providers to collaborate seamlessly on a single digital network, thereby enhancing resilience, transparency, and operational reliability throughout the asset life cycle. Moreover, it seamlessly integrates with SAP S/4HANA to facilitate business-to-business integration for asset data and maintenance processes.

- **SAP Service and Asset Manager**: The SAP Service and Asset Manager (formerly SAP Asset Manager) mobile app empowers maintenance and field service technicians to work safely, productively, and autonomously. It offers both online and offline access to context-rich visualizations and actionable insights, facilitating end-to-end enterprise asset and service management processes. This enables technicians to efficiently address issues and make informed decisions while on the go, ultimately enhancing overall operational effectiveness. SAP Service and Asset Manager uses a three-tier architecture:

- SAP Service and Asset Manager mobile client

- SAP Business Technology Platform Mobile Services as Middleware

- Mobile Add-On for SAP S/4HANA or SAP Mobile Add-On as primary business back-end system with integration services using Mobile Application Integration Framework (MAIF)

- **SAP Field Service Management**: The SAP Field Service Management solution enables the optimization of field service operations, empowering managers, technicians, and dispatchers to deliver exceptional customer experiences while improving overall service quality.

Figure 5-23 shows how all the preceding solutions fit together in the bigger scheme of things.

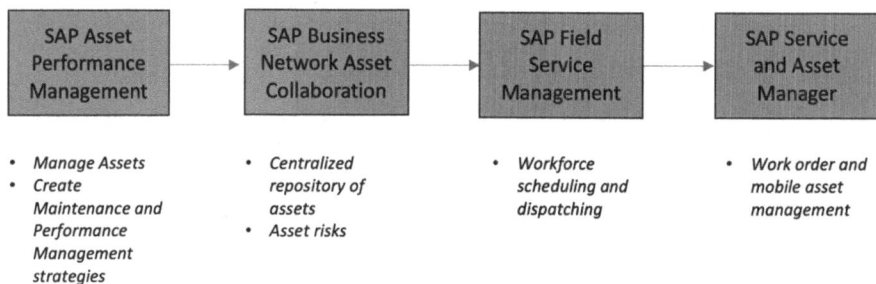

Figure 5-23. *How all the different solutions fit together*

All of these solutions integrate seamlessly with SAP
S/4HANA. Figure 5-24 shows a sample architecture pattern of integration
of one of the solutions with SAP S/4HANA.

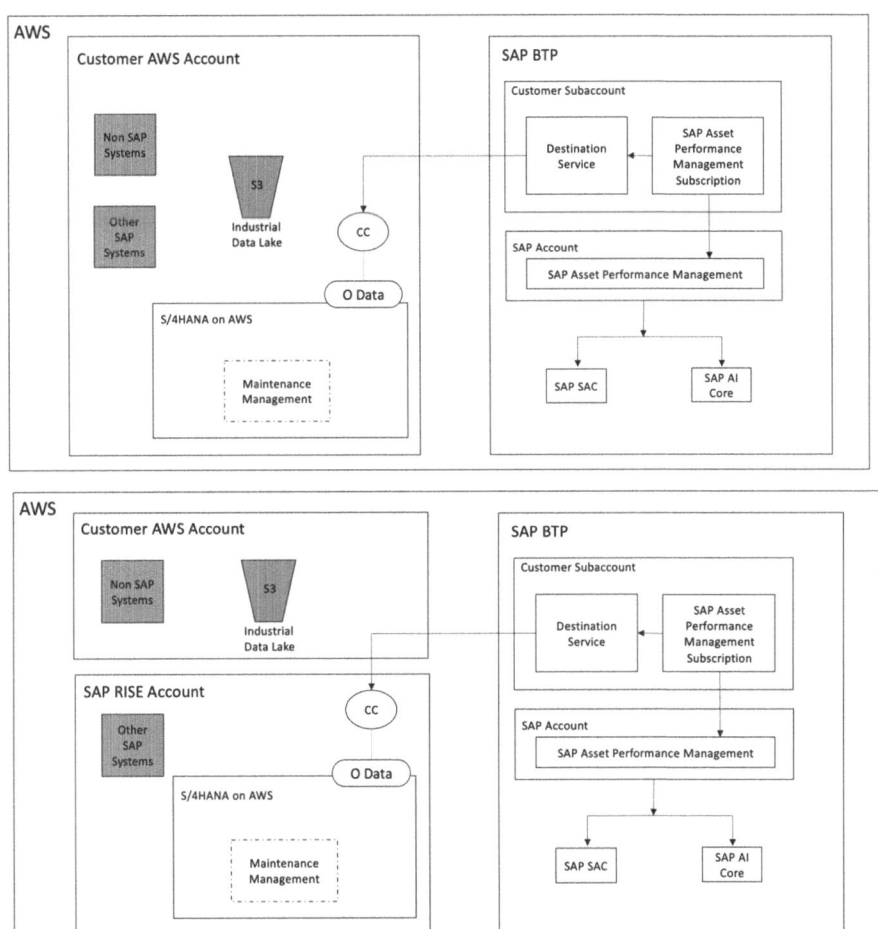

Figure 5-24. *Example architecture pattern of integration of one of
the solutions with SAP S/4HANA (both AWS native and SAP RISE
scenarios)*

SAP now provides capabilities similar to those of SAP ME and SAP MII but with more extensive analysis and reporting capabilities via SAP Digital Manufacturing Cloud (SAP DMC). These capabilities include a broad range of tools and services, allowing for scalability and integration with other cloud services via SAP Integration Suite. Additionally, SAP DMC incorporates advanced technologies such as AI and ML. SAP DMC has two primary components, SAP Digital Manufacturing Cloud for Execution (SAP DMCe) and SAP Digital Manufacturing Insights (SAP DMCi).

SAP DMCe offers a solution for managing manufacturing control, monitoring production processes, and managing resources and jobs. It also facilitates technical integration of machinery and equipment. SAP DMCi serves as a one-stop shop for performance management tailored for manufacturing, independent of various sites a customer might have. It ensures seamless data integration across SAP and non-SAP environments, spanning various production systems. This integration enables real-time collection and visualization of KPIs, which provides a global platform to take decisions that can lead to enhancing production performance. The below Figure 5-25 shows how SAP DMC fits in a sample modern industrial SAP architecture pattern.

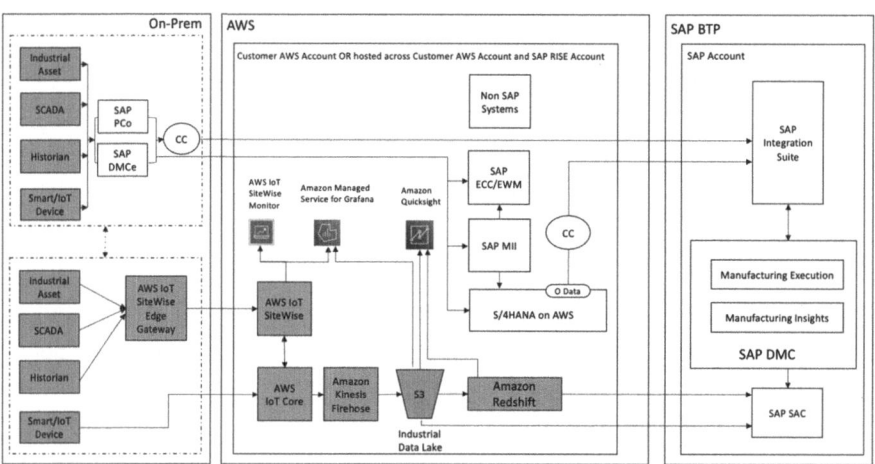

Figure 5-25. *End-to-end integration of SAP DMC with SAP S/4HANA and industrial assets*

SAP Solutions That Complement AWS Offerings

SAP solutions such as SAP Digital Manufacturing Cloud (SAP DMC), SAP Manufacturing Execution (SAP ME), and SAP Manufacturing Integration and Intelligence (SAP MII) can effectively complement AWS offerings, including the creation of an industrial data lake on Amazon S3 and other capabilities of AWS IoT and analytics services. These integrations enable organizations to capitalize on the strengths of both platforms, combining SAP's extensive manufacturing expertise with AWS's scalable cloud infrastructure and advanced data management capabilities.

AWS IoT SiteWise, particularly, AWS IoT SiteWise Monitor, is ideal to gain simple, scalable, real-time monitoring and visualization into equipment performance. However, you would use SAP DMC to get a broader view of manufacturing-specific functionalities, including SAP MES capabilities, which are essential for managing the full life cycle of production processes within a manufacturing plant. Also, you would use Amazon S3 to store data from industrial assets for long-term storage, advanced analytics, and large-scale processing. AWS IoT SiteWise's scalability complements SAP DMC's operational insights, ensuring that shop-floor data is seamlessly transferred to the cloud, where it can be aggregated, analyzed, and visualized.

Furthermore, data from SAP ME, SAP MII, and SAP DMC can be enhanced using AWS's machine learning services, such as Amazon SageMaker and Amazon Bedrock. For instance, predictive maintenance models can be developed using Amazon SageMaker, or Amazon Bedrock can be directly used, with data stored in Amazon S3 and then integrated back into SAP DMC for real-time operational decision-making. Similarly, data from AWS IoT SiteWise, stored in Amazon S3 or Amazon Redshift, can be fed into SAP Analytics Cloud (SAP SAC) for further analysis, enabling more informed business decisions.

The SAP solutions discussed in this chapter are specifically designed to manage, monitor, and optimize manufacturing processes. In contrast, AWS IoT is a versatile, general-purpose platform designed to connect IoT devices, manage data, and enable the development of IoT applications. AWS IoT offers services for device connectivity, data processing, analytics, and integration with other AWS cloud services. While SAP solutions are tailored for manufacturing, AWS IoT is applicable across various industries, including automotive, healthcare, and agriculture, making it a more flexible IoT platform.

Secrets to a Successful Implementation

Implementing an IoT project successfully requires careful planning, strategic decision-making, and effective execution. Here is a five-stage process that might help you for a successful implementation:

1. **Focus on business outcomes**: Identify the use cases that align with your business goals and address real challenges or opportunities. Scope your project based on the answers to these questions:

   - What are the desired business outcomes?

   - What defines the success criteria in terms of business outcomes?

2. **Focus on a proof of concept (POC)**: Select one of your plants as your pilot site. Create a comprehensive end-to-end integration for production, ensuring high quality, by gathering data from equipment and presenting it on cloud dashboards for a production line. This will be the Minimum Viable Product (MVP). This solution would intentionally focus on specific aspects and

won't cover all customer use cases or connect all assets. Its limited scope will aim to showcase tangible business value as a proof of value (POV).

3. **Visualize operational metrics in near real time and historical views**: Ingest, process, and store data from different types of equipment/PLCs and Historians for visualizing the actionable operational KPIs on dashboards in both near real time and historical views. Based on the desired business outcome, the primary goal should be to optimize operations for the production line, enhance productivity, and improve overall availability and see what this improvement percentage looks like.

4. **After POC is successful, prepare for rapid deployment at scale**: Based on the success of the POC, create the foundation of an industrial data lake ready for self-serve and advanced analytics. Create a production-grade technical deployment template designed to automate the deployment of the solution across similar production lines spanning various plants, divisions, and locations. This template should aim to ensure consistency and efficiency in implementing the solution on a broader scale.

5. **Evaluate third-party and/or generative AI solutions**: Most of the concepts and architecture patterns described in this chapter are simplified by third-party solutions. This again goes back to the conversation of build vs. buy. Generative AI is revolutionizing all aspects of technology, including IoT. These

advancements are reshaping the way developers construct applications and how users interact with them. Although IoT-specific generative AI is in its early stages, two overarching categories of use cases have emerged. Firstly, it can help IoT solution developers in building more efficient solutions swiftly and with enhanced quality. Secondly, it can felicitate how end-users' interact with IoT devices or solutions to get both insights from the data and recommendations. We are going to visit this as well as all other aspects of gen AI in Chapter 6.

Summary

A smart factory represents the next evolution in manufacturing, where advanced technologies like IoT, AI, and ML converge to create a highly automated, flexible, and data-driven production environment. At the core of a smart factory is industrial analytics, which involves building an industrial data lake to aggregate, store, and analyze vast amounts of data from machines, sensors, and other sources. This data powers predictive maintenance, allowing manufacturers to anticipate equipment failures before they occur, and predictive quality, which helps in identifying potential defects in the production process. Digital twin technology further enhances these capabilities by creating virtual models of physical assets, enabling real-time monitoring and optimization.

AWS offers a range of solutions for manufacturing and industrial operations, such as AWS IoT and AWS ML services, which help in building and managing these advanced capabilities. SAP complements these offerings with its digital manufacturing solutions, which provide the tools needed to manage and optimize the end-to-end manufacturing business processes. Voice technology, another emerging tool, can be leveraged to streamline operations and improve worker efficiency. Successful implementation of these technologies requires careful planning, a successful POC, and a strong focus on change management to ensure that the digital transformation delivers tangible business value.

CHAPTER 6

Business Transformation with AI/ML

Leveraging the power of AI/ML is no longer optional—it has become essential for optimizing inefficient business processes.

Several of the previous chapters have already discussed artificial intelligence (AI) and machine learning (ML) and AI/ML services from AWS. Chapter 1 introduced the basics of AI and ML and how they are related to each other. It discussed the capabilities of two AWS AI/ML services, Amazon SageMaker and Amazon Bedrock, and how they are architected. In Chapter 4 discussed SAP-related use cases using AWS AI services like Amazon Textract. Chapter 5 explored how predictive maintenance and predictive quality can be implemented using AI services like Amazon Monitron, Amazon Lookout for Equipment, and Amazon Lookout for Vision. It also discussed how voice technology can be implemented in the shop floor by using Amazon Alexa. In this chapter, we will explore how all the AWS AI and ML services fit together holistically and we will dive deep into the features of generative AI. We will start with the basics of generative AI and then discuss use cases with example architecture patterns.

© Bidwan Baruah, Krishnakumar Ramadoss and Abarajith Vivekanandha 2024
B. Baruah et al., *Evolve from Infrastructure to Innovation with SAP on AWS*,
https://doi.org/10.1007/979-8-8688-0890-6_6

The following topics will be covered in this chapter:

- Why AI and ML matter now

- AI and ML offerings from AWS and use cases

- Generative AI

- Foundation models

- How large language models work and how to make them smarter

- Amazon Q

- Patterns to use generative AI capabilities

- Generative AI use cases for SAP

Why AI and ML Matter Now

The potential of AI and ML to drive digital transformation in the enterprise has gained a lot of traction in the past year. It's interesting to note that machine learning has been around for more than 50 years. Most of the common ML techniques that are used today essentially were invented decades ago. What has changed recently is that with cloud computing, AI and ML have become accessible to all businesses—not limited to just the major tech giants and hardcore academic researchers. Cloud computing has removed so many of the barriers to experimenting and innovating with AI that even risk-adverse businesses are making it part of their strategies. Adding to the mix the massive mature sets of data to train the models on and the graphics processing units (GPUs) that can significantly accelerate training of the ML models brings us to a point where all the hype is transitioning to real impact on businesses. For example:

- Customer experience is being transformed via capabilities such as conversational interfaces, smart biometric authentication, and personalization and recommendations.

- In retail, sophisticated demand planning and forecasting models are dramatically improving accuracy.

- Automation is making supply chain management more efficient.

- Backoffice manual processes like invoicing and check processing are now automated, resulting in faster processing times, better accuracy, greater visibility and control.

Let's explore all the major services and use cases.

AI and ML Offerings from AWS and Use Cases

Successful business outcomes from the application of AI/ML technology are strongly associated to the careful selection of its use case. This section discusses common and impactful AI/ML use cases and identifies which specific AWS service to leverage according to the scenarios. Table 6-1 lists the AWS AI services mapped to the uses cases that are most commonly used in conjunction with SAP business processes.

Table 6-1. *Use Cases Mapped to AWS Services*

Use Case	AWS Service to Explore
Extract information from documents, such as e-mails, forms, invoices, etc.	**Amazon Textract**: ML service that automatically extracts text, handwriting, and data from scanned documents.
Translate text	**Amazon Translate**: Neural machine translation service that delivers fast, high-quality, affordable, and customizable language translation.
Computer vision	**Amazon Rekognition**: Provides pretrained and customizable computer vision (CV) capabilities to extract information and insights from your images and videos to solve your most pressing computer vision needs with no ML skills required and at a lower cost. Works well for large assets.
Detect visual defects	**Amazon Lookout for Vision**: Spots product defects using computer vision to automate quality inspection. Works well for small assets.
Predictive maintenance	**Amazon Lookout for Equipment:** ML industrial equipment monitoring service that detects abnormal equipment behavior so you can act and avoid unplanned downtime.
Predictive maintenance	**Amazon Monitron:** End-to-end system that monitors equipment by capturing vibration and temperature data from equipment through wireless sensors and sends alerts when it detects anomalies or potential failures.

(*continued*)

Table 6-1. (*continued*)

Use Case	AWS Service to Explore
Search information contained in documents, audio recordings, or videos easily	**Amazon Kendra**: Intelligent enterprise search service that helps you search across different content repositories with built-in connectors.
Convert audio files into text	**Amazon Transcribe**: Automatic speech recognition service that uses machine learning models to convert audio to text.
Build a chat application (with execution of scripted workflows based on identified user intent)	**Amazon Lex**: Fully managed AI service with advanced natural language models to design, build, test, and deploy conversational interfaces in applications.
Forecasting	**Amazon Forecast**: Fully managed service that uses machine learning to deliver highly accurate forecasts
Voice technology	**Amazon Alexa**: Voice service available on hundreds of millions of devices from Amazon and third-party device manufacturers that offer an intuitive way to interact with the technology.

Chapter 1 covered in detail the capabilities of the AWS ML service Amazon SageMaker; let's look at a couple of case studies.

SAP on AWS customer TC Energy (formerly TransCanada) is a North American energy company headquartered in Calgary. TC Energy faced challenges with manually identifying anomalies in its pipelines due to the sheer volume of data and 88,000 points that required real-time monitoring and assessment. Its pipeline capacity process needed to be optimized. To address this need, TC Energy developed and introduced ORBIT, an operational business intelligence application that uses machine learning to aid customers in optimizing gas throughput. Leveraging Amazon

SageMaker for training and deploying ML models, ORBIT consolidates various data sources, including historical pipeline data and operator insights, into a single platform. This integration optimizes data to achieve maximum operational efficiency on a daily basis. With ORBIT, TC Energy customers can streamline the integration of diverse pipeline data and enhance it for optimal operational performance each day. Additionally, operational planners can easily share this data with teams such as gas control and field operations, facilitating the optimization of maintenance and asset utilization. More details are available here: `https://aws.amazon.com/solutions/case-studies/tc-energy/`.

Another SAP on AWS customer, Apollo Tyres (introduced in Chapter 5), utilizes an automated tire inspection program that examines tire defects through photos captured as tires move along the production line. Powered by Amazon Rekognition, the program automates image and video analysis to enable factory supervisors to promptly address manufacturing irregularities, ensuring the delivery of high-quality tires that adhere to stringent safety regulations. The complete case study is available at `https://press.aboutamazon.in/news-releases/news-release-details/apollo-tyres-goes-all-aws-make-factories-smarter-iot-and-machine`.

Generative AI

Transformative and innovative technologies are enabling organizations to solve complex problems and, more importantly, helping them reimagine the way they usually solve any problem. One such technology is generative AI, or gen AI as it is commonly known. But why gen AI and why now? Gen AI is a technology that can produce content, that are very close to those generated by humans, which makes a huge impact in terms of simplifying a lot of tasks that we do in our daily life. The true power of gen AI goes beyond the search engine and Chatbot—it can change the way organizations run. Adoption of gen AI has picked up speed recently because of the convergence of three technological advancements: widespread availability of vast amounts of data and information, advancements in ML, and the accessibility of highly scalable computing power required to train the models.

Advancements in ML technology have led to *foundation models (FMs)*. FMs have evolved through a series of innovations and breakthroughs in ML research. The start of foundation models can be linked back to early studies in how humans understand and use language. These early models, like simple neural networks for language, paved the way for smarter ways to understand language. Then came word embeddings, like Word2Vec, which were a big step forward in understanding natural language. Word embeddings work by figuring how words relate to each other in sentences by giving them special codes, making language models better at understanding context. The *transformer*, a new type of system, changed how we look at language sequences. It did away with old methods and used a new approach called *self-attention*, which helped models understand sentences better by looking at all the words at once. This made a big impact on how we build models that understand language.

Large language models (LLMs) are a type of foundation model that are pretrained on language. They are designed to understand, generate, and manipulate human language. Hence, LLMs are used for gen AI use cases because of their ability to understand and generate coherent, contextually relevant human-like text. LLMs are ML models trained on massive datasets (which is what "large" in LLM refers to), enabling them to understand and respond to numerous natural language queries. They are pretrained on colossal text-based datasets, often spanning petabytes, resulting in models with tens to hundreds of billions of parameters. Typically, LLMs undergo initial pretraining on vast text corpora, followed by fine-tuning for specific tasks.

Foundation Models

There are three types of foundation models, as depicted in Figure 6-1 and described in the following list:

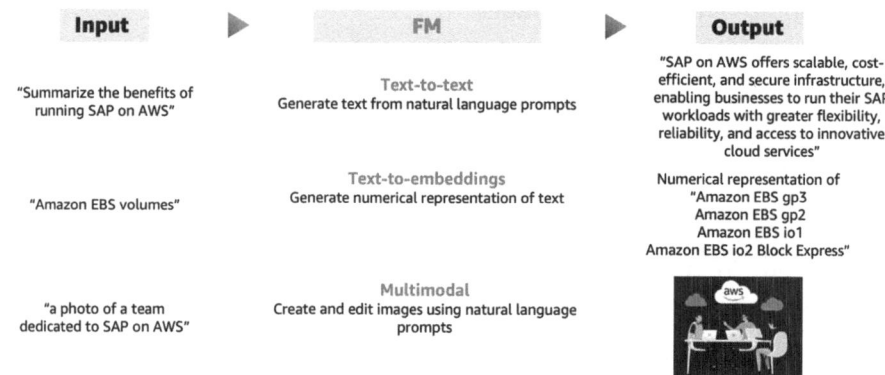

Figure 6-1. *Types of foundation models*

- **Text-to-text**: This type of FM accepts text input and produces text output. It is commonly employed in generative AI applications, aiding tasks like summarizing lengthy text documents or providing answers to questions based on document content.

- **Text-to-embedding**: In this type of FM, text input is converted into an embedding, which is simply an array of numbers (a vector), as the output. Embeddings are numerical representations of words, sentences, images, and audio and video data that reside in a multidimensional vector space. Machines understand only numerical data, hence we need to translate the text into this format. By assigning a numerical representation to a different feature of each word, we can find the distance between them. For example:

Word: Dog: Vector 1 = [89.7, 202, 23.7]

Word: Puppies: Vector 2 = [99.7, 302, 13.7]

These are two vectors, each with three embeddings. Again, a vector is an array of numbers. Embeddings are numbers. As you might expect, dogs and puppies have vectors closer to them, which will help machines understand them better and output a better response. This technique is achieved through the use of ML algorithms that enable the understanding of the meaning and context of data (semantic relationships) and learning of complex relationships and patterns within the data (syntactic relationships). You can use the resulting vector representations for a wide range of applications, such as information retrieval, image classification, natural language processing, and many others.

- **Multimodal:** This type of FM accepts text as input and generates outputs in alternative modalities, such as images.

The Need for Amazon Bedrock

Organizations can leverage foundational models to construct their applications, but they typically encounter several challenges. Firstly, there isn't a single optimized foundation model that is suitable for every task. As foundation models evolve over time, it often necessitates the use of multiple models. Moreover, FMs have their own limitations, as they are unable to tackle complex tasks requiring interaction with external systems.

FMs are very powerful but cannot execute the tasks by themselves. To make these kinds of capabilities possible, developers need to provide definitions and instructions to FMs, configure the FMs to access company data sources, write custom code to execute tasks via API calls, and set up cloud policies for data security. Also, organizations need to make sure application integration is seamless without the need to maintain infrastructure.

To address these challenges, AWS announced the general availability of Amazon Bedrock in September 2023. Amazon Bedrock is a fully managed service with which enterprises can experiment with a wide variety of top foundation models from leading AI companies, customize them privately with their own data using techniques like fine-tuning and retrieval-augmented generation (RAG), and interact with the FMs using APIs. Customers do not need to control the model deployment or scalability; they just choose the model and interact with it through the API. This makes Amazon Bedrock appealing. Amazon Bedrock uses Agents that invoke APIs dynamically to execute tasks. Let's explore what Amazon Bedrock Agents are.

Amazon Bedrock Agents

Gen AI applications often require the capability to execute multistep tasks across various data sources. Amazon Bedrock Agents enable you to automatically orchestrate and analyze a request and break it down into the correct logical sequence of tasks using the FM's reasoning capabilities. All that's required is setting up an Agent with access to the organization's enterprise systems, processes, and knowledge bases. The Agent then determines the logic, identifies which APIs to call and when to call them, and executes the tasks in the correct sequence.

If the Agent requires additional information from the user, it automatically asks the user for those details using natural language. One of the key advantages of Amazon Bedrock Agents is that they leverage the most up-to-date information available to provide relevant and answers securely.

Figure 6-2 shows how Amazon Bedrock Agents work.

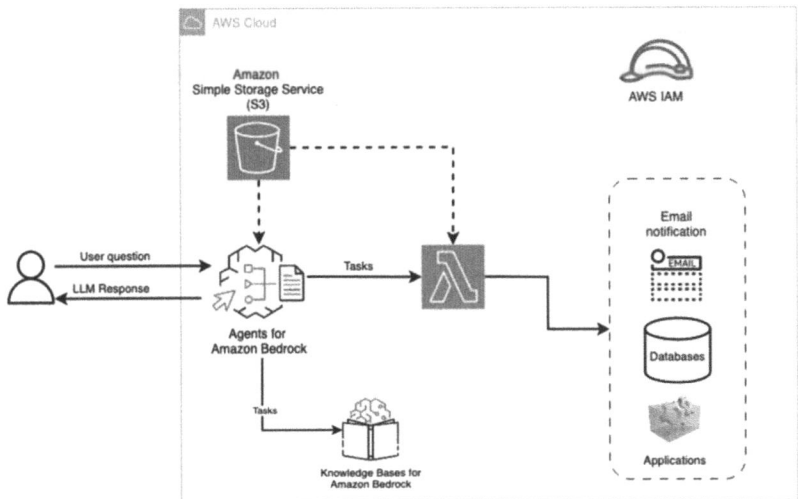

Figure 6-2. *Amazon Bedrock Agents*

Now, let's explore how LLMs work and how we can make them smarter by using prompt engineering, fine-tuning, and RAG.

How LLMs Work and How to Make Them Smarter

From the capability perspective, what do LLMs really enable? Previously, it was always difficult to programmatically understand natural languages, and turn it into structured data. Although it was doable with traditional task-specific natural language processing (NLP) models, there were limitations. While it was also possible to turn structured data back into text using a simple template method, LLMs make it possible to generate dynamic text in different styles in natural languages that seems to understand user intention or intent from natural language, and communicate to users like a real human.

Another important advantage of LLMs compared to traditional NLP models is that LLMs are generic. This allows users to use one model (or a limited number of models) to do many tasks, as opposed to training and operating multiple models, which is usually costly and requires specialized knowledge. LLMs enable user experiences that were not possible or practical before.

However, LLMs have a couple of drawbacks. An LLM can answer incorrectly, also called *hallucination*, if it is used as is. To avoid that, you need to provide proper context so that the LLM answers correctly. Also, an LLM might not be able to answer because it has no customer data. In that case, you need to provide a Knowledge Base, an Amazon Bedrock capability, which enables the LLM to answer. For example, if you ask an LLM how much you spent on food in your last hotel stay, it won't be able to answer; but if you provide your hotel invoice as a data source to a Knowledge Base, it would answer correctly.

Vector Embeddings

To understand LLMs, you first need to understand the transformer architecture, which is the fundamental building block of LLMs. It has led to the development of LLMs such as generative pretrained transformers (GPTs) and bidirectional encoder representations from transformers (BERT). But to understand transformers, you first need to understand vector embeddings.

Figure 6-3 illustrates a simple example at a very high level to help you understand how LLMs work without having to know technical terms like transformers and vector embeddings. Suppose that you ask an LLM to predict the next word in the following sentence:

"The creamer is bad, the coffee tastes <blank>"

The preceding sentence is a prompt in generative AI terminology. A *prompt* is simply a sentence, question, or any text provided to the model, for which a response or a prediction task is expected. You pass this prompt to the tokenizer, which translates words into specific numerical values based on a predefined dictionary. As introduced in 'Text to Embeddings' type of FM in the section "Foundation Models," LLMs primarily operate with numbers rather than words. Hence this step. The tokenizer places the numerical values in the context.

In the diagram shown in Figure 6-3, although natural language text is depicted as the content in the context, tokens actually are used. Natural language text is shown instead of tokens for your easy understanding. This context is passed to the LLM. Once the LLM receives the tokenized words, it performs the necessary statistical computations to forecast the subsequent token. This predicted token is integrated back into the context. Subsequently, this updated context is relayed back to the LLM, prompting it to generate the next token. This iterative process continues until a termination condition is met, such as reaching the maximum token setting within the LLM or the generation of an end-of-sequence token, signifying completion. At this point, the context is passed to the de-tokenizer, which reverses the tokenization process by converting the numerical tokens back into words. The resulting output is then provided to the user.

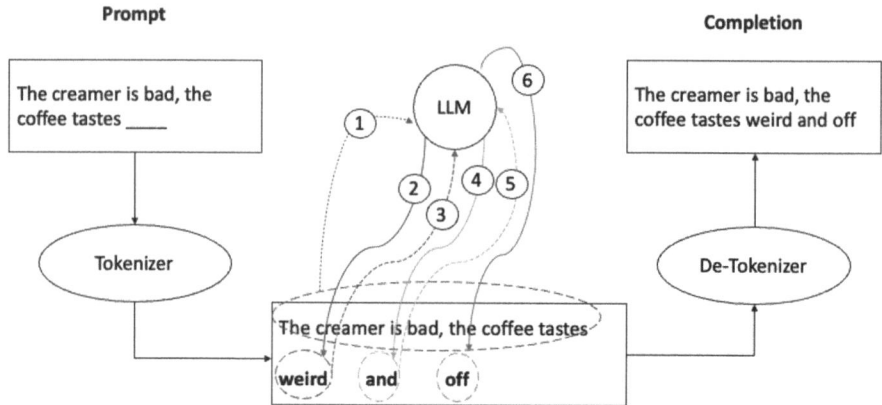

Figure 6-3. *High-level architecture of how LLMs work*

Now that you have some understanding of how LLMs work, let's go a level deeper and see how a piece of text is converted into a vector and what these "statistical computations" are.

There are multiple techniques to convert a sentence into a vector. The following are two popular methods:

- Use word embedding algorithms, such as Word2Vec, and then aggregate the word embeddings to form a sentence-level vector representation.

- Use pretrained language models, like BERT or GPT, which can provide contextualized embeddings for entire sentences. These models are based on deep learning architectures such as transformers, which can capture the contextual information and relationships between words in a sentence more effectively.

What Is a Transformer?

To start with, assume that a *transformer* is a black box that converts an input into an output magically, as depicted in Figure 6-4.

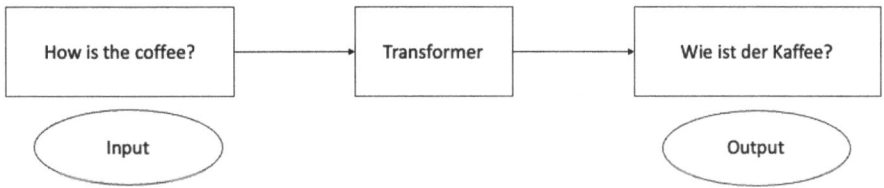

Figure 6-4. *What a transformer does*

Drilling down a level, this black box is composed of two main parts, an encoder and a decoder, as shown in Figure 6-5.

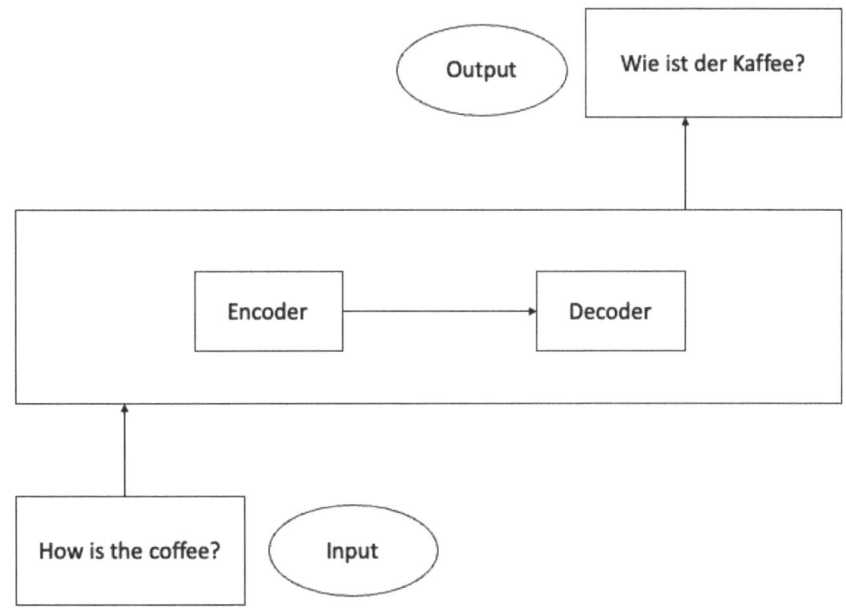

Figure 6-5. *Components of a transformer*

However, both the encoder and the decoder are actually a stack with multiple layers (same number for each), as shown in Figure 6-6. All encoders present the same structure, and the input gets into each of them and is passed to the next one. All decoders present the same structure as well and get the input from the last encoder and the previous decoder. The journey of data through the transformer model culminates in its passage through a final layer, which functions as a classifier.

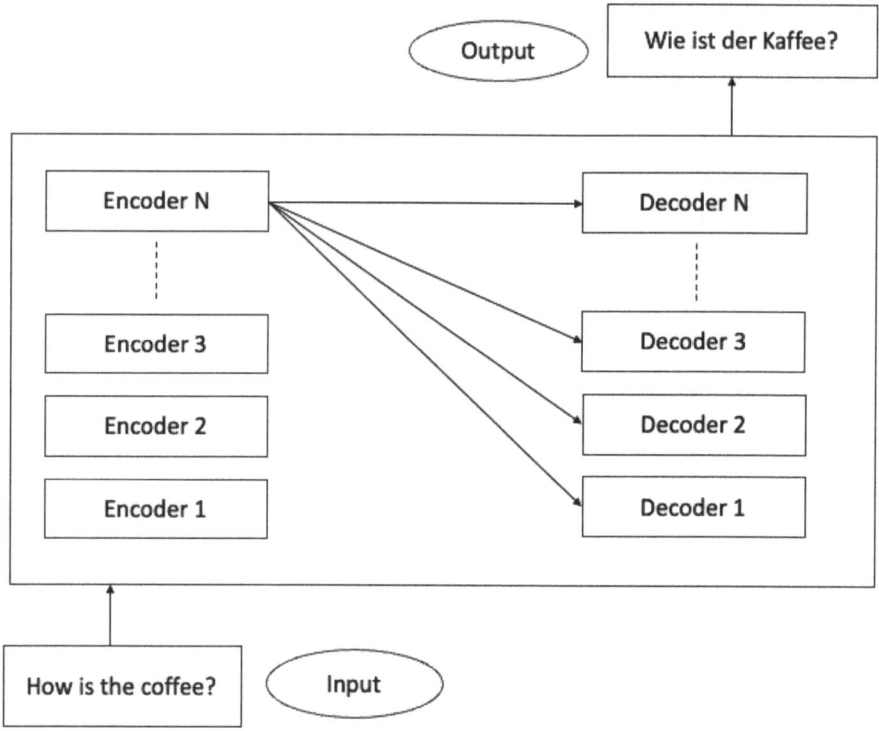

Figure 6-6. *Architecture of transformer*

Back to our example, after the tokenizer converts each word of the sentence into tokens, one of the transformer layers converts tokens and positions the tokens into vectors. All these vectors are stored in a databased called a *vector database.*

When these vectors are plotted in space, synonyms tend to cluster together. This clustering aids machine learning models in identifying similarities among items that are sparsely distributed. For the sake of simplicity, instead of a multidimensional space, consider a three-dimensional space, as shown in Figure 6-7. In this context, when the LLM processes input data, it learns to associate words based on their proximity in this three-dimensional space. For example, it might learn that "coffee" is closer to "creamer" than "taste" is. Similarly, "good" and "weird" might be closer to each other than "coffee" is to either of them.

Essentially, training an ML model involves establishing associations between words to determine their relative proximity or similarity in the dimensional space. During the model training process, efforts are directed toward constructing these word associations to indicate which word is closer to another. While this explanation simplifies the complexities inherent in training an LLM, it offers a foundational understanding of how the model learns and represents word associations.

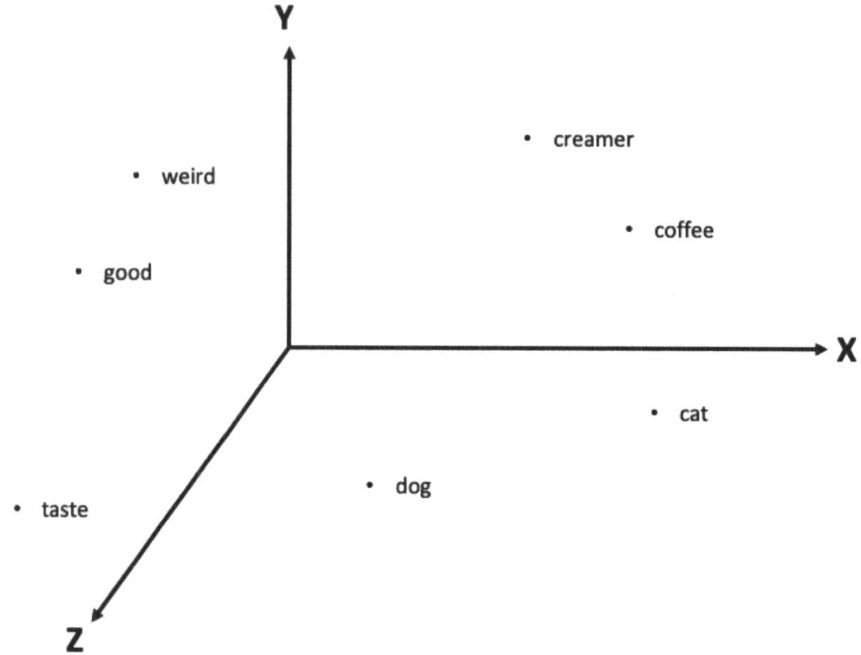

Figure 6-7. *Words plotted in three-dimensional space*

Embeddings can have hundreds or thousands of dimensions—too many for humans to visualize. Now imagine that each vector has hundreds of embeddings, and there are hundreds of millions of vectors.

Customization of LLMs

A standard LLM may not suffice for organizational needs because it lacks specific knowledge about a particular domain and/or an organization. Hence, customization of LLMs becomes necessary. There are four primary techniques for customization:

- Prompt engineering (with context)

- Retrieval-augmented generation (RAG)

- Fine-tuning an LLM

- Training your own LLM

These techniques vary in complexity, required skill level, and cost, as depicted in Figure 6-8, Figure 6-9, and Figure 6-10, respectively.

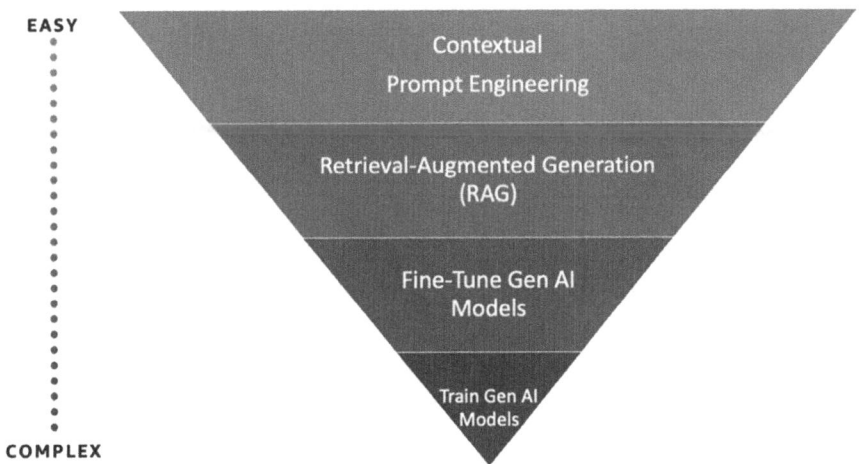

Figure 6-8. *LLM customization patterns and required skills*

	Pattern	Skills required	Organization development maturity
EASY	Prompt Engineering	API Integration	Basic development capability
	RAG	Data Engineering on documents Embedding Vector DB tuning	Strong development capability Mature engineering practices
	Fine-tune	Experience training, tuning, and hosting ML Models ML Ops	ML (LLM), data, and development capability Mature engineering practices
COMPLEX	Build own model from scratch	LLM Model training and operation	Strong LLM and data capability Access to large amount of training data Mature engineering practices

Figure 6-9. *Required skills and organizational development maturity*

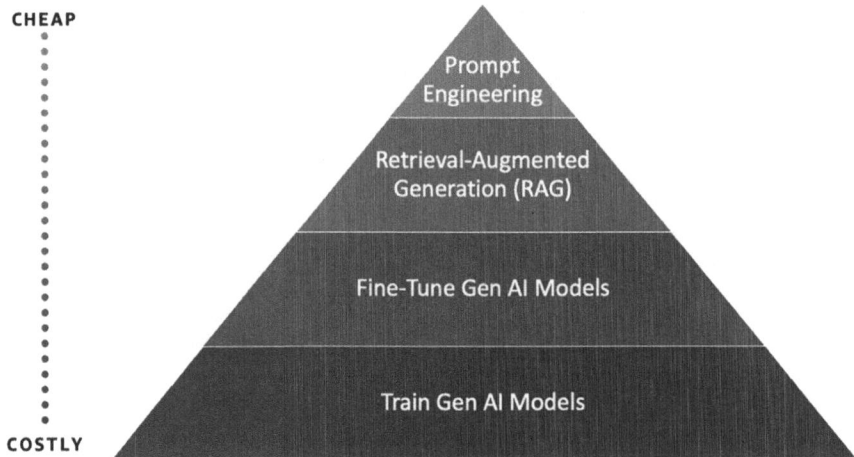

Figure 6-10. *LLM customization patterns and associated costs*

Prompt Engineering

In the world of generative AI (gen AI) applications, the typical workflow involves providing a prompt to a model and receiving a completion in return. However, enhancing the performance of LLMs without resorting to additional data or retraining poses a challenge. To address this, researchers have developed a technique known as *prompt engineering*, which is the process of designing and optimizing input prompts to guide an LLM's behavior and improve its performance on specific tasks.

A type of prompt engineering called *in-context learning (ICL)* involves manipulating the content within the context window, which essentially is the input provided to the LLM in the chat window. By refining how prompts are formulated or questions are posed, users can guide the LLM to generate better responses without the need for costly retraining. Retraining LLMs entails significant compute cost and time investment, so organizations prefer to not do that. As the name *foundational* models implies, you don't want to tweak these models too much.

ICL can be classified into zero-shot, one-shot, and few-shot inference techniques. To differentiate these techniques, let's go back to our coffee and creamer example and have the LLM model classify the sentiment as positive, negative, or neutral. In this example, the result of which is shown in Figure 6-11, the word "okay" is used intentionally to make determining the sentiment difficult for the LLM. Since no examples have been provided (*zero-shot inference*), the LLM may provide a correct sentiment analysis or may get confused. The model is expected to make predictions based solely on its understanding of the task and the context provided.

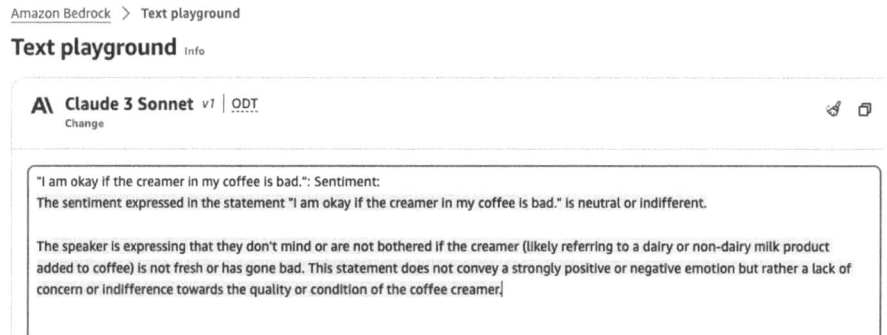

Figure 6-11. *Zero-shot inference technique*

You also can train an LLM by providing one or more examples and guiding it to understand how sentiment should be classified. In a *one-shot inference* scenario, you give the LLM a single example to learn from, such as labeling a statement with positive sentiment, and guide the LLM on how to approach these kinds of tasks. For a *few-shot inference* scenario, you give a couple of examples, such as labeling a statement with positive sentiment as "positive" and labeling another statement with negative sentiment as "negative." Then, you prompt the LLM to classify a new review based on these examples.

In the one-shot example shown in Figure 6-12, the LLM has been instructed that an "okay" experience is in fact a negative sentiment. So, now the LLM knows what "okay" means and correctly predicts the sentiment.

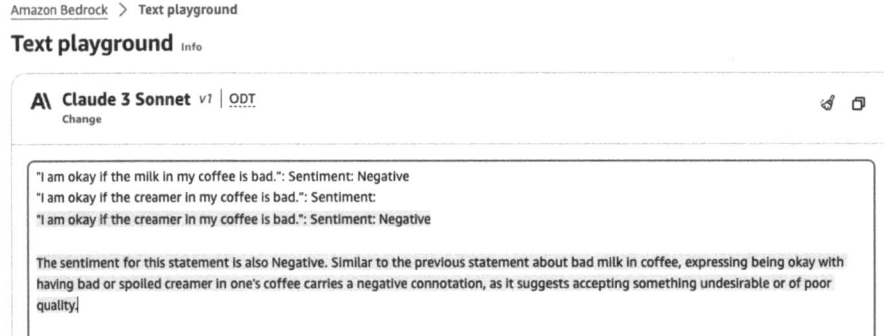

Figure 6-12. *One-shot inference technique*

By employing these techniques, users can effectively guide the LLM to understand and perform specific tasks without the need for extensive retraining, making prompt engineering a valuable strategy for leveraging existing models.

There is another technique called *chain-of-thought (CoT) prompting*, which is utilized to break down complex reasoning tasks into intermediary steps. CoT prompts are typically tailored to specific problem types. Users can initiate CoT reasoning by using the trigger phrase "(Think Step-by-Step)." Figure 6-13 shows two screenshots that provide a great example of a few-shot CoT prompting. It breaks down the task into logical steps to answer the question.

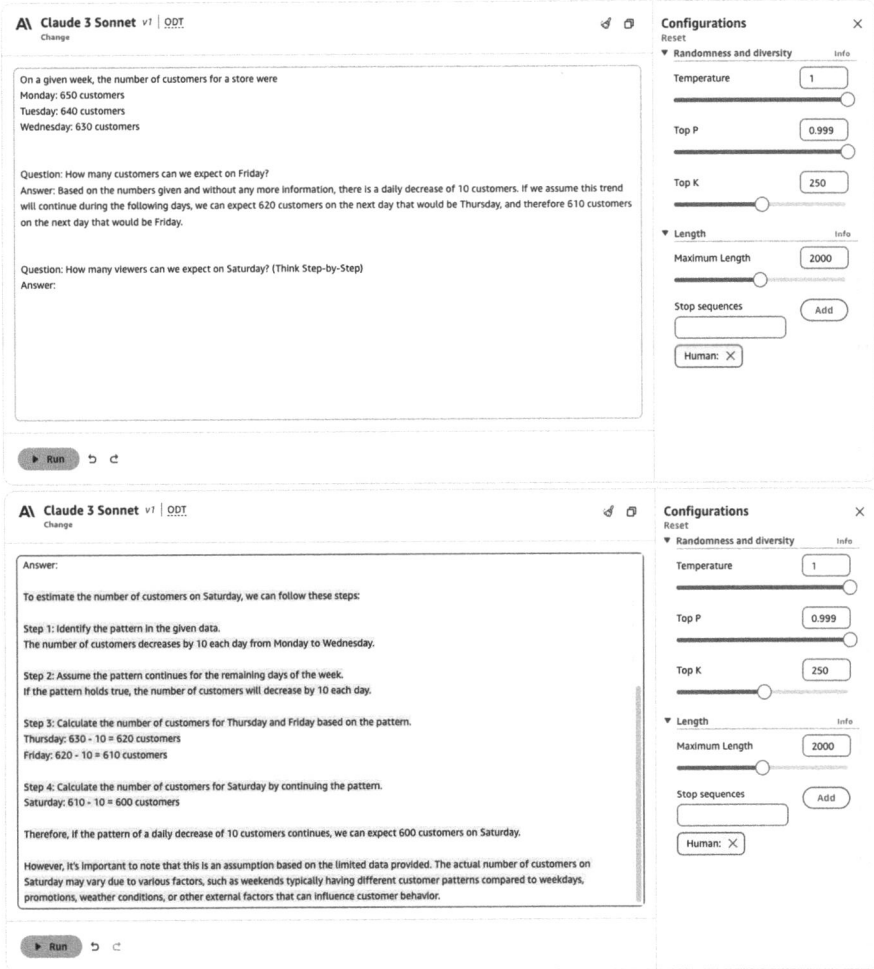

Figure 6-13. *Chain-of-thought prompting*

In-context learning offers several advantages:

- It enables immediate adjustments to specific prompts.

- It allows flexible adaptation without needing to retrain the entire model.

- It is computationally efficient, requiring fewer resources.

However, there are situations where in-context learning may not be suitable:

- The company context cannot be adequately expressed through prompts.

- Inference decisions rely on vast amounts of data, which may not align with an in-context approach.

- Adaptation is limited to the provided context, potentially constrained by the model's context size.

- Responses require a company-specific context.

- There are numerous use cases that may not be addressed with in-context learning, and the organization possesses the resources and expertise to fine-tune the large language model.

In such cases, fine-tuning becomes a valuable alternative.

Fine-Tuning

Fine-tuning is the process of adapting the pretrained LLM to a specific task or domain using a smaller or more relevant dataset. This can improve the performance or accuracy of the LLM for the target task. To understand fine-tuning, you need to understand pretraining.

Continuing the previous discussion of transformers, Figure 6-14 illustrates a transformer model. It depicts tokens representing input sentences. For each token, embeddings are looked up, and the transformer blocks are employed to enhance and route information upward. Eventually, a summary is generated at the top layer, which is then utilized to feed into the classifier.

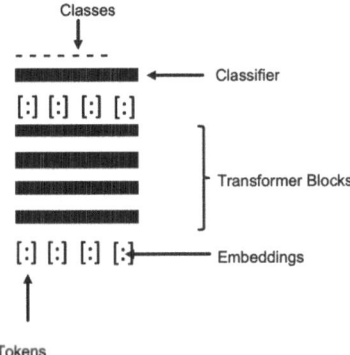

Figure 6-14. *Detailed architecture of a transformer*

When discussing pretraining and fine-tuning, they are essentially parallel concepts, not directly linked. As depicted in Figure 6-15, on one side, there's the task of next-token prediction. Conversely, on the other side is the task of sentiment prediction, where reviews are categorized as positive, neutral, or negative.

This divergence extends to the data utilized as well. For instance, on the sentiment analysis side, there might be only a limited dataset comprising perhaps 5,000 reviews or e-mails, manually labeled for sentiment. However, in self-supervised learning, a large corpus, potentially spanning the entire Internet, is employed, enabling the creation of labels. This means learning can occur from vast amounts of data, possibly petabytes in size, for next-token prediction.

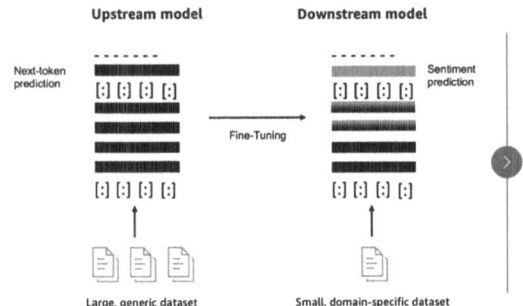

- Transfer learning of domain specific knowledge into a Foundation Model at reasonable cost
- Update of weights in the network, while architecture is kept
- Fine-tuning is task specific (e.g., MLM, CLM, PLM, translation, classification, …)

Figure 6-15. *Pretraining vs. fine-tuning*

The remarkable aspect here is that, on the left side, we may have gained insights into the intricacies of language features. Conversely, on the right side, we're limited to just 5,000 examples. When considering sentiment analysis, it's crucial to recognize that sentiment encompasses various nuances, such as understanding sarcasm, jokes, irony, or idioms. However, learning these nuances from just 5,000 samples is challenging. On the other hand, through pretraining, we potentially have access to tens of thousands or even millions of samples of irony, jokes, and sarcasm. Leveraging this extensive dataset allows us to capture a broader understanding of language nuances, even if they might seem irrelevant to our specific task. That is the importance of pretraining.

Sentiment analysis makes sense so far. There's a problem though. We have an upstream model where we learn about the world and then transfer this knowledge to start training our downstream model. Whenever we update the downstream model, there's a risk of inadvertently altering something crucial that we learned from the upstream model. This could happen without our awareness, and since there's no feedback mechanism to rectify it, we might lose valuable information. Thus, achieving a balance between updating the model and preserving task-specific knowledge becomes essential. So, fine-tuning is a balancing act and needs careful consideration. We would be very aggressive to make sure the top layers like Classifier have a higher learning rate. But each of the other layers below them could have smaller learning rates and the bottom layer, we can just totally freeze it. Then, we'll try training for fewer rounds (called *epochs*) and shorter sessions to avoid overfitting (when a model learns to perform really well on the data it was trained on but doesn't perform as well on other datasets) and potentially overwrite the transferred information. This is what fine-tuning is all about.

A common method for gauging the magnitude of large language models is by assessing their parameter count. You may be acquainted with descriptors such as a 1-billion, 10-billion, or even 100- or 200-billion parameter model. As these models increase in size, their abilities also expand. This growth is attributed to the parameters housed within, which encapsulate the essence of language comprehension. Throughout the pretraining stage, developers input extensive language data into the model, exposing it to numerous rounds of training. This iterative procedure fine-tunes and recalibrates the parameters until the model achieves the proficiency level that is good enough. However, when employing these foundational models in specific applications, challenges may arise. Despite employing various prompt engineering techniques and context learning strategies, if the model still fails to meet expectations, it could be due to the specialized nature of the data used in the application. This data might be pertinent to the application but might not have been part of the training data for the foundational model. To address this, fine-tuning the large language model with domain specific data as shown as Figure 6-16 becomes necessary.

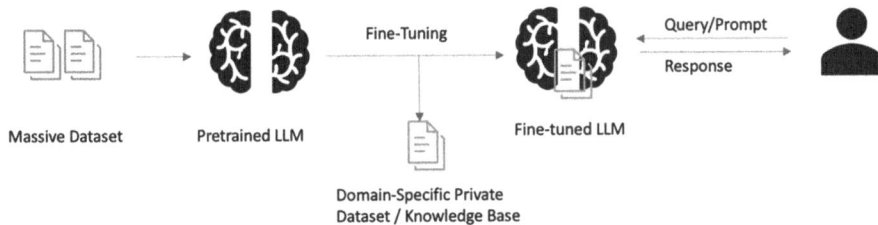

Figure 6-16. *High-level overview of fine-tuning*

Broadly, we can classify fine-tuning in two categories: supervised fine-tuning and reinforcement learning from human feedback.

Supervised Fine-Tuning

Supervised fine-tuning (SFT) involves refining a pretrained language model by utilizing labeled data to perform a specific task. The utilized data has been previously verified. This differs from unsupervised approaches, where data validation is not conducted. Typically, the initial training of the language model is unsupervised, while fine-tuning operates under a supervised framework.

There are few types of supervised fine-tuning, categorized by the extent to which parameters are adjusted during the learning phase: full fine-tuning, parameter-efficient fine-tuning (PEFT), and instruction fine-tuning, and reinforcement learning from human feedback (RLHF).

Full Fine-Tuning

With *full fine-tuning*, every parameter of the model is updated using labeled data, similar to pretraining but with a smaller dataset. Given that all model weights are susceptible to change, and considering there could be billions of them, substantial compute resources are necessary for this approach.

Through full fine-tuning, the model can grasp features and representations throughout all layers of its architecture. This results in maximal adaptability of large language models to specific tasks. Full fine-tuning frequently delivers notable performance enhancements compared to employing a pretrained model with more limited fine-tuning methods.

Parameter-Efficient Fine-Tuning

Full fine-training is computationally intensive and expensive, as it requires recalculating all parameters, which demands significant computing resources and time. Another challenge in fine-tuning LLMs for specific tasks is the potential issue of "catastrophic forgetting." This phenomenon occurs when the fine-tuning process modifies the weights of the original LLM, resulting in improved performance on the targeted task but a degradation in performance on other tasks.

To mitigate these challenges, researchers have developed *parameter-efficient fine-tuning (PEFT)* techniques. One such method is *low-rank adaptation (LoRA)*, which freezes the model's parameters and introduces additional trainable parameters specific to the dataset, as shown in Figure 6-17. This approach significantly reduces the computational overhead, as only a fraction of the model's parameters are updated. PEFT allows you to match the performance of full fine-tuning with only about 1% of the model's parameters.

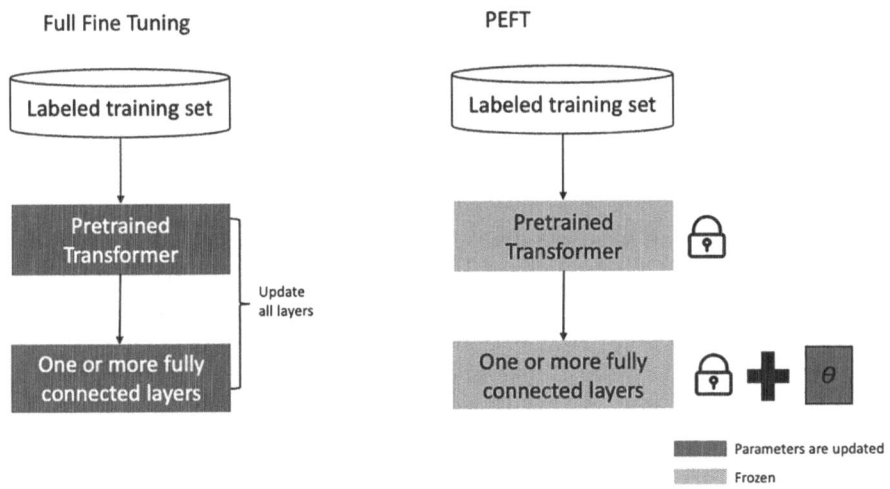

Figure 6-17. *Full fine-tuning vs. PEFT*

The following are the key advantages of PEFT, which can also be visualized via Figure 6-18:

- Low computational and storage costs

- Faster training time

- Small dataset (fewer examples)

- Suitable for multitask training

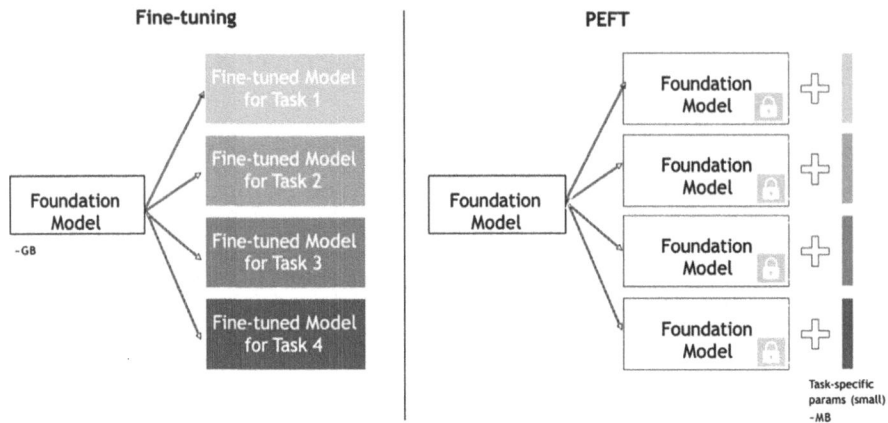

Figure 6-18. *Why parameter-efficient fine-tuning (PEFT) is useful*

Instruction Fine-Tuning

The core idea behind *instruction fine-tuning* is to provide the model with labeled examples illustrating the desired behavior or response. These labeled examples serve as directives for the model, guiding it to produce appropriate outputs in response to given input data or queries.

During the fine-tuning process, the model is trained using this instructional dataset, as shown in Figure 6-19, allowing it to grasp and generate responses aligned with the provided instructions. Through exposure to a diverse range of instruction-response pairs and adjustments to its parameters, the model improves its ability to understand and comply with specific instructions. The model can also be trained on multiple tasks simultaneously, improving its ability to generalize across various types of tasks, as shown in Figure 6-20. This is called .multitask instruction fine-tuning.

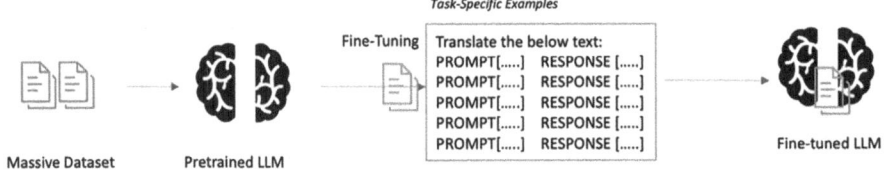

Figure 6-19. *Instruction fine-tuning*

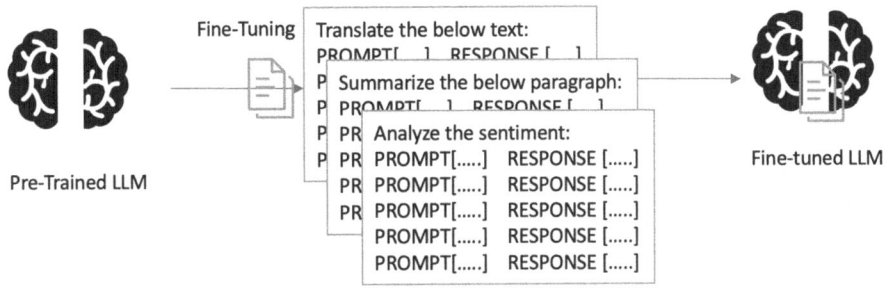

Figure 6-20. *Multitask instruction fine-tuning*

Reinforcement Learning from Human Feedback

Reinforcement learning from human feedback (RLHF) has been used to produce the best large language models we have today (e.g., GPT-4, Claude, and Llama 2). This technique holds great promise for developing AI systems that can learn complex behaviors while remaining aligned with human values. As shown in Figure 6-21, the process can be divided into multiple stages: language modeling for the base policy, modeling human preferences, and optimization via reinforcement learning.

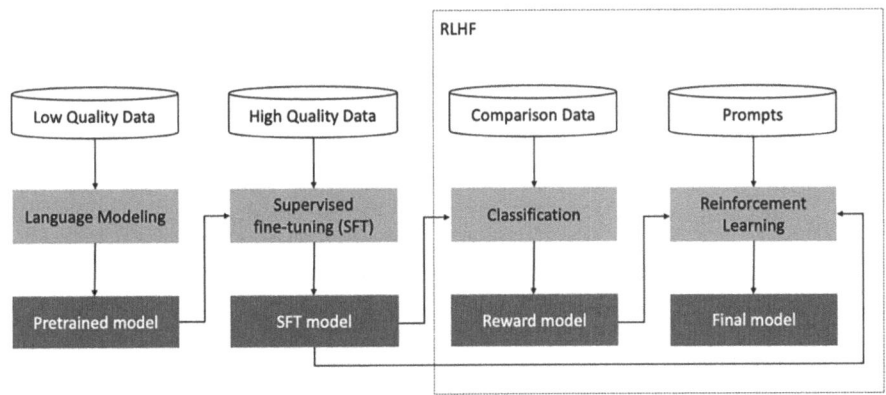

Figure 6-21. *Stages of RLHF*

You begin with a well-pretrained supervised model capable of performing the desired task. Subsequently, you gather outputs from the model for a specific prompt and ask humans to identify the superior response among them. This process results in the creation of a dataset comprising human preferences, which is then utilized to train the reward model. The trained reward model provides a score indicating the quality of any given response, facilitating standard reinforcement learning procedures.

Key considerations when deciding to use RLHF include the following:

- Use RLHF to better align LLM outputs with human preferences.

- The reward model is the most important part.

- Use a larger LLM to collect human preferences.

Key advantages of supervised fine-tuning (SFT) include

- SFT is effective for tasks where labeled data is available and where direct control over model behavior is desired.

- SFT allows for precise customization of the model's parameters to the target task, resulting in fine-grained adjustments tailored to specific task requirements.

- SFT allows the same pretrained model to be fine-tuned for different tasks.

Fine-tuning might not be advisable in the following scenarios:

- When a pretrained LLM already demonstrates high performance within your business context.

- When there are no or not enough labeled data that is specific to your task or domain, coupled with insufficient quality and reliability.

- When cost is a significant consideration, and the project's scale does not warrant the investment required for fine-tuning.

Fine-tuning is good, but it can be memory and/or compute intensive because

- **Entire model replicas need to be stored for each task**: If you have 100 different tasks and you need to fine-tune the model for each one of them, you need to maintain 100 copies of these fine-tuned weights. Even though the fine-tuning process itself is relatively lightweight compared to pretraining, it can still be memory intensive because you still need to train a large model, and just to be able to accommodate it, you might need many computational nodes.

- **A lack of batched inference opportunities**: In the case of inference, you're losing batching opportunities because each task has its own model replica. Consequently, executing requests on each replica necessitates traversing through separate model replicas. In contrast, batching these requests together could potentially save computational resources. Therefore, parameter-efficient fine-tuning aims to rectify these deficiencies.

Fine-tuning also can be costly. Fine-tuning a large language model leads to higher costs. The organization needs to have specialized AI/ML skills so that they can fine-tune the model. Also, it is impossible to keep up with the daily changing data because as the data changes, they will have to fine-tune the latest language model every day, which results in more cost. And then, as the latest language models come up with newer versions, it would result in additional cost.

303

These are the challenges with fine-tuning. All these problems are solved by RAG, and that's why we expect many customers to primarily use RAG or a combination of fine-tuning and RAG.

Retrieval-Augmented Generation

Retrieval-augmented generation (RAG) is an AI technique that allows a foundation model to incorporate knowledge from a repository to generate more accurate responses, without the overhead that is required by fine-tuning, although the level of customization available with fine-tuning may not be possible.

The best way to understand retrieval-augmented generation at a high level is to convert the three words in its name to their corresponding verb forms: retrieve, augment, and generate. Thus, the approach consists of three steps:

1. Retrieve data from an external source.

2. Augment the context of the prompt with the retrieved data.

3. Generate a response using the LLM, based on the prompt and retrieved data that is passed into the context.

As shown in Figure 6-22, the process begins with the user input text being converted into vector embeddings, as previously described. These embeddings are then used to query the vector database in order to identify all content that could potentially fulfill the task. This is achieved by requesting, for example, the five, seven, or ten nearest neighbors of the vector embedding. The vector database, which is already enriched by various data sources, then provides the relevant content in response. Subsequently, this content is passed as context, along with the instructions for completing the task, to an LLM. The LLM then generates a response based on the provided context and instructions.

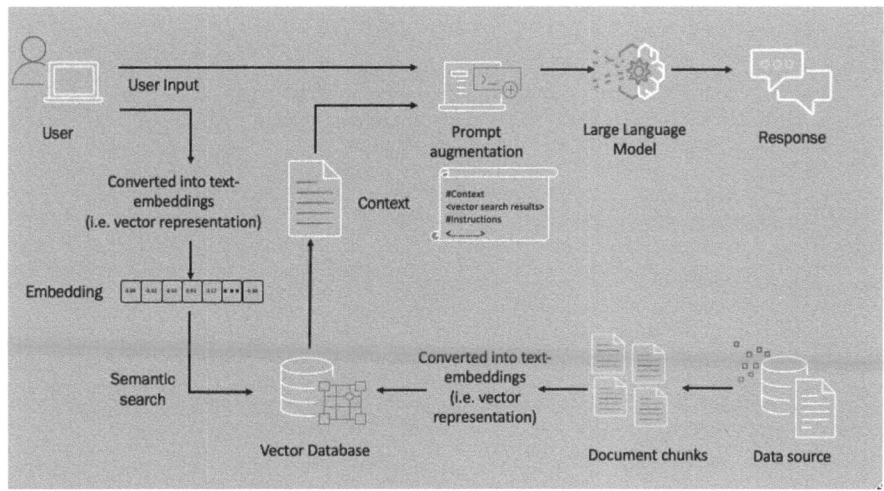

Figure 6-22. *How RAG works*

Now, in the world of Amazon, the conversion of user input into vector representations can be done by Embeddings LLMs like Titan Embeddings LLM. And the calls like the ones that request to generate vectors or return of the vectors, or the requests with the context and instruction that is sent to the LLM, can all be done using Amazon Bedrock APIs. Also, as shown in Figure 6-23, instead of the vector database, you may use Knowledge Bases for Amazon Bedrock, which gives FMs and agents contextual information from your own private data sources to deliver more relevant, accurate, and customized responses.

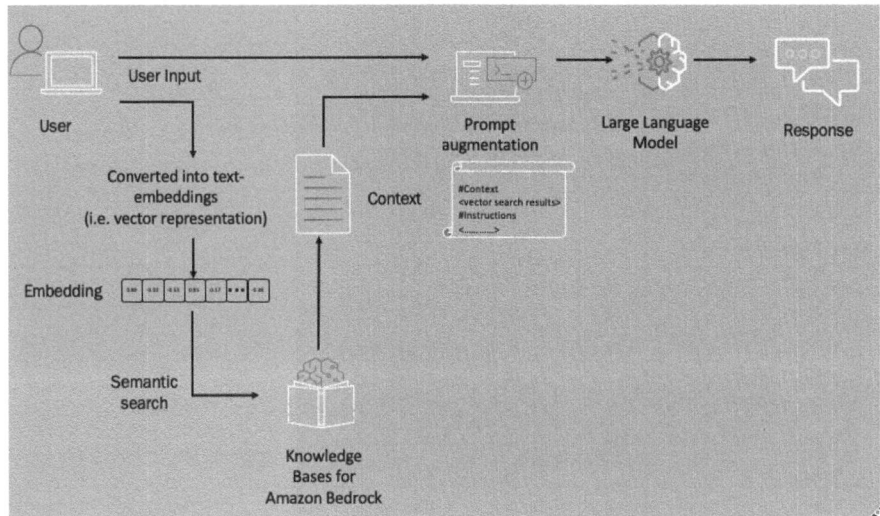

Figure 6-23. *How RAG works with Knowledge Bases for Amazon Bedrock*

The following are the key advantages of using RAG:

- RAG does not require labeled data, so it's more appropriate for cases where labeled data is insufficient or unavailable.

- RAG is beneficial for tasks that require incorporating external knowledge or context.

- RAG is particularly useful for cases where data is updated very frequently.

- Every developer who knows Python or basic coding skills can actually invoke the Amazon Bedrock APIs and build their gen AI application easily.

- As either data is changing in the databases or new documentation files are being created, RAG can adapt very easily to the changing data. And then of course,

- As new LLM versions or new LLMs are released, switching LLMs in a RAG application is very easy.

So, we expect most of customers to use RAG approach. And of course, they could use a combination of fine-tuning and RAG, as per their need. Building on the principles of RAG for enhancing information retrieval and knowledge generation, let's now explore how Amazon Q takes a different approach by leveraging its own advanced AI models to drive insights and decision-making in business processes.

Amazon Q

Amazon Q is a generative AI–powered assistant that is designed specifically for work and can be tailored to your business to have conversations, solve problems, generate content, and take actions using the data and expertise found in your company's information repositories, code bases, and enterprise systems. Figure 6-24 shows a screenshot of Amazon Q.

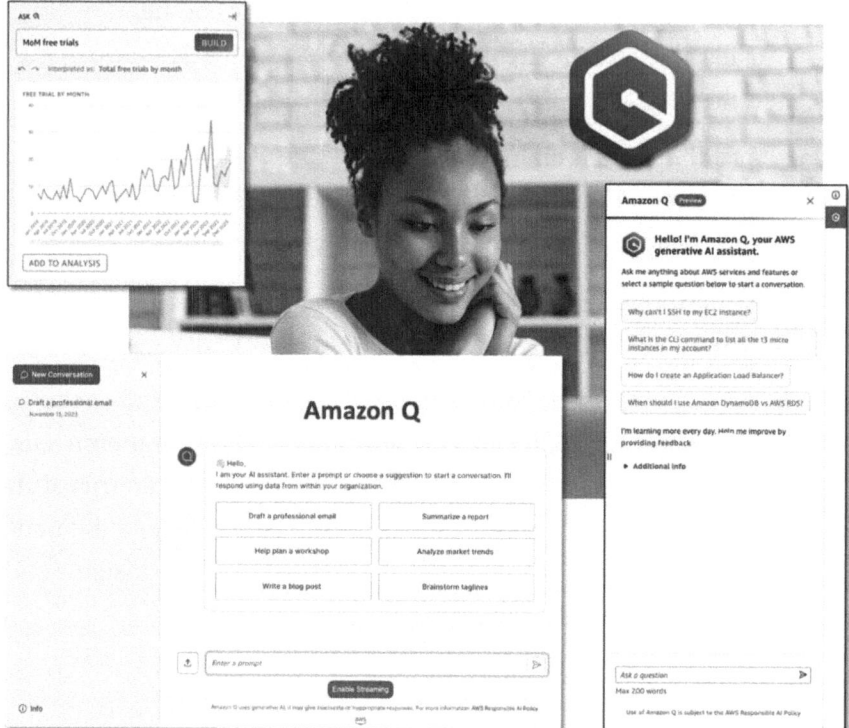

Figure 6-24. *Amazon Q*

Amazon Q is designed to provide accurate answers, solve problems, generate new content, and take actions using the data and expertise available in your company's information repositories, including your code and enterprise systems. With Amazon Q, you can streamline tasks, expedite decision-making, and foster creativity and innovation.

- **Engages in conversations to solve problems and generate content**: Amazon Q acts as the business expert, simplifying common tasks such as summarizing lengthy documents, drafting e-mails or articles, conducting research, or performing comparative analysis. It can also assist in completing tasks, reducing the time employees spend on repetitive work like filing tickets or creating new cases. Additionally, Amazon Q serves as the AWS expert. It is available in the AWS console to provide expert recommendations and step-by-step instructions when working with AWS services.

- **Understands your company information, code, and systems**: By connecting Amazon Q to your enterprise's systems, it can engage in tailored conversations, solve problems, generate content, and take actions that are pertinent to your business.

- **Personalizes interactions based on your role and permissions**: Amazon Q is able to identify users' roles and permissions, personalizing its interactions accordingly. Users without permission to access certain data cannot access it through Amazon Q either.

- **Built to be secure and private**: Amazon Q prioritizes security and privacy to meet the stringent needs of enterprises, ensuring that data is protected and confidential.

Patterns to Use Generative AI

As explained in Chapter 1, SAP is approaching the AI revolution by keeping in mind its core strength—business knowledge via business processes. So, SAP is calling its AI platform SAP Business AI. Aligned with its overall strategy, SAP is building its AI-related technical elements on the Business Technology Platform (BTP), naming it the AI Foundation. The AI Foundation serves as a one-stop shop for developers, allowing them to create AI- and generative AI–powered extensions and applications on SAP BTP. Additionally, SAP is developing ready-to-use services known as

SAP AI Services or SAP Business AI Services. These services are built upon the SAP AI Core, the fundamental AI runtime that facilitates the training, deployment, and monitoring of ML models.

SAP is advancing its gen AI capabilities through three initiatives:

- **SAP Joule**: SAP Joule is an integrated AI assistant, embedded across SAP's cloud enterprise portfolio. It serves as a co-pilot for SAP applications that helps users deliver work faster, with smarter insights and better outcomes through natural language prompts. It has three key advantages:

  - Faster access to transactions

  - Ability to ask for data insights quickly without running ad hoc queries

  - Personalized user experience by providing contextual recommendations and insights based on individual user behavior and historical data

- **SAP Business AI**: AI is infused in SAP applications and business processes with out-of-the-box use cases. A classic example is that now, using generative AI, you can create job postings in Success Factor or create an SOW in SAP Fieldglass. Key advantages include the following:

  - Reduction in manual processes and data errors

  - Improved decision-making with advanced analytics

  - Predictive insights

- **Enterprise Transformation**: Over the last few years, SAP has been working on new tools and capabilities to leverage gen AI to deliver SAP in a faster and more consistent way. These tools include the following features:

  - Faster application development via code generation

  - Business process automation

  - Advanced AI-driven analytics

There are three primary patterns to consume gen AI capabilities:

- **Pattern 1**: **Embedded**: AI capabilities are seamlessly integrated into SAP applications. This is in line with SAP Joule and SAP Business AI described in the previous list.

- **Pattern 2**: **Bolt on**: AI capabilities are added to existing processes as modular components via SAP BTP.

- **Pattern 3**: **Build it**: Build your own AI + gen AI solution on platforms like AWS.

  - **Subpattern 3.1**: Use a combination of SAP BTP and AWS services.

  - **Subpattern 3.2**: Use only AWS services.

Patterns 2 and 3 can be best described using the diagram shown in Figure 6-25.

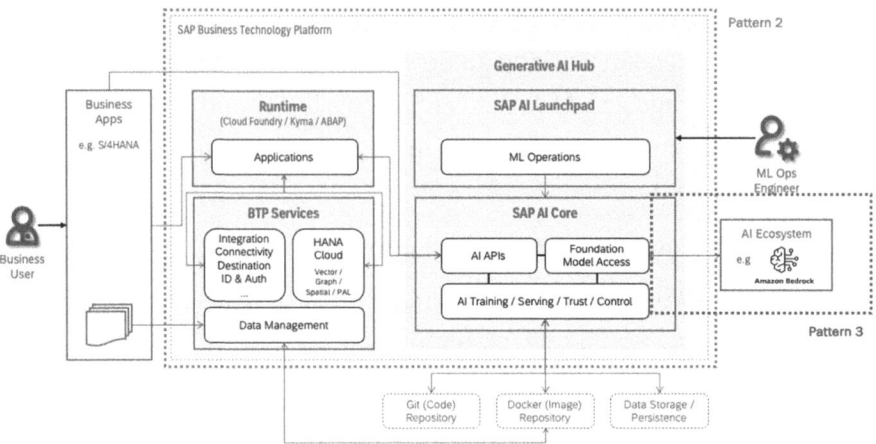

Figure 6-25. *Patterns 2 and 3 to consume gen AI capabilities*

As indicated in the previous list, with pattern 3, you have the option to either use a combination of SAP BTP services and AWS services (subpattern 1) or use AWS services only (subpattern 2). For an example of subpattern 1, you could build the gen AI application using SAP Cloud Application Programming Model (CAP) on SAP BTP itself (side-by-side extension) but utilize Amazon Bedrock to make calls to the LLMs. Alternatively, using subpattern 2, you may decide to develop a small custom ABAP application on the SAP S/4HANA system itself and call Amazon Bedrock via the AWS SDK for SAP ABAP, as depicted in Figure 6-26.

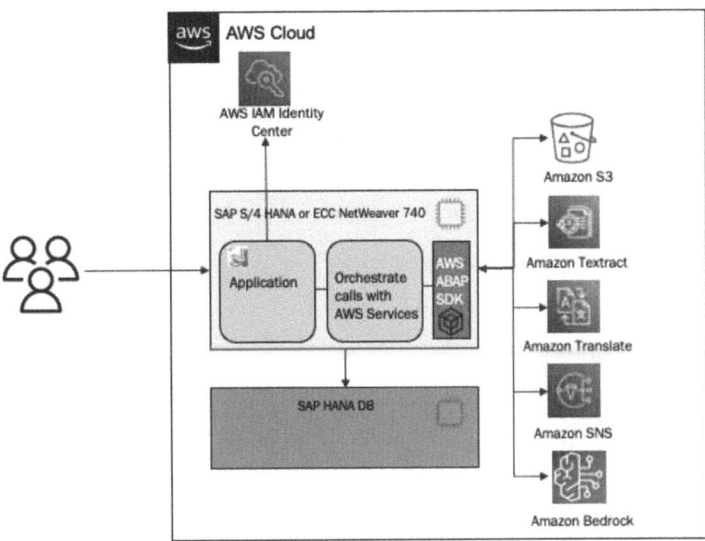

Figure 6-26. *Subpattern 3.2 using only AWS services to consume gen AI capabilities*

The diagram shown in Figure 6-25 also serves as an excellent transition to the diagram shown in Figure 6-27, illustrating the development of generation AI applications with RAG, a concept discussed in the previous section. Figure 6-27 highlights the utilization of SAP BTP and AWS services. This also exemplifies a classic instance of side-by-side extension. As depicted in the diagram, you can develop and deploy gen AI applications utilizing the SAP CAP framework, selecting either SAP BTP's Cloud Foundry, Kyma, or ABAP environment as the runtime. SAP HANA Cloud, powered by AWS Graviton, acts as the vector database. This architectural pattern demonstrates subpattern 3.1 on how to leverage Amazon Bedrock services from SAP BTP using SAP AI Core.

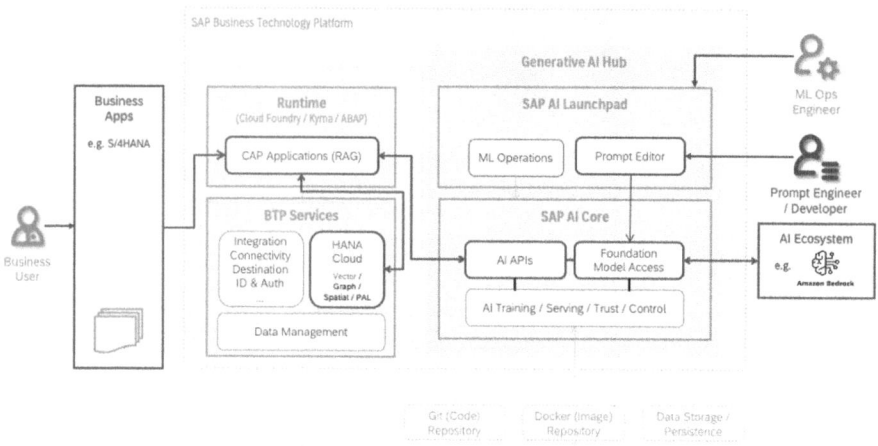

Figure 6-27. *Subpattern 3.1 using a combination of SAP BTP and AWS Services to consume Gen AI capabilities*

Now that we have discussed all the basics and patterns of implementing gen AI with SAP, let's look at some of the use cases.

Gen AI Use Cases for SAP

The use cases in this section are organized into two categories, technical and business, each of which relies on the following capabilities of generative AI:

- Code generation

- Writing assistance

- Question answering via information retrieval

- Invoking APIs to complete tasks

AWS's generative AI capabilities are now available to SAP on AWS customers, facilitating faster automation and innovation to unlock business value. The use cases leveraging AWS generative AI capabilities for SAP customers can be categorized as follows:

- **Technical use cases**: These include expediting development activities, particularly through the code generation capabilities of gen AI. Tools such as Amazon Q Developer and APIs for Amazon Bedrock facilitate this process. Accelerating code generation leads to swifter task and project completions.

- **Business use cases**: These encompass the development of applications and solutions aimed at enhancing productivity, gathering real-time insights, and improving decision-making processes. Specific examples include

  - **Enhancing productivity**: AWS generative AI capabilities can enhance productivity by summarizing and analyzing complex documents, translating documents, and generating e-mails using Amazon Bedrock together with the AWS SDK for ABAP.

  - **Gathering real-time insights**: AWS generative AI capabilities can gather real-time insights through natural language prompts using Amazon Bedrock, SAP BTP, and data warehousing solutions like SAP Datasphere.

  - **Improving decision-making:** AWS generative AI capabilities can improve decision-making by analyzing SAP data using generative BI capabilities with Amazon QuickSight.

Let's first look into the technical use cases.

Technical Use Cases

Code generation serves as the primary technical use case. From an SAP perspective, the following types of code generation use cases are possible:

- Accelerate your automation of SAP on AWS operational activities by creating shell scripts using Amazon Q Developer

- Also using Amazon Q Developer, generate code in Java or Python for solutions built in the Cloud Foundry environment of SAP BTP

- Build an Apache Spark ETL (extract, transform, and load) job using Amazon Q Developer for an SAP Analytics use case

- Generate ABAP code and code documentation by calling Amazon Bedrock via SE38 or Eclipse

The following sections dives deep into the first two use cases.

Amazon Q Developer to automate SAP Operational Activities

Let's look at how Amazon Q Developer works for the use case of accelerating your automation for SAP on AWS operational activities like SAP stop/start, kernel patching, and so forth.

One of the benefits of moving to AWS is increased staff productivity through purpose-built services by AWS. Various organizations have transitioned away from manual day-to-day operational tasks associated with the management of most of their systems and applications. Nonetheless, many organizations end up treating SAP as an exception when it comes to automation because SAP does not fit with the automation strategies commonly used for other applications, such as implementation of immutable infrastructure involving processes like immutable build, delete, and rebuild.

Often, the number of SAP administration or operational tasks exceeds the available team members' ability to handle them. As a result, critical strategic initiatives often remain unaddressed due to lack of time and resources. Automation solutions catering to SAP Basis functions have been accessible for quite some time, either through SAP itself or via third-party vendors; however, many organizations opt not to adopt them, often due to concerns such as those related to cost, alignment with their organization's DevOps automation strategy, complexities associated with managing multiple vendors, tools, technologies, and integration patterns, and so on. While some SAP Basis teams have successfully automated specific repetitive tasks and activities through the development of custom scripts, many have not been able to do so, primarily due to a prevailing skills gap in scripting and limited coding experience.

New technologies, such as generative AI coding assistants like Amazon Q Developer, are helping to bridge the skills gap. Amazon Q Developer acts as a bridge between logical comprehension and practical coding implementation. It enables SAP Basis administrators to translate their logical understanding into code by interpreting their logic, articulated in natural English. This approach significantly expedites the development process of SAP automation, making it more accessible and feasible for a wide range of organizations. Amazon Q Developer provides coding suggestions by converting simple natural English explanations of logic into operational code.

An SAP administrator is able to build a solution by developing an automation script for the operational activity -Start and Stop of SAP, using Amazon Q Developer.

The solution is orchestrated in two stages:

- Administrators, utilizing Amazon Q Developer, convey their logic in natural English to formulate a shell script capable of performing start and stop commands on a single Amazon EC2 instance.

- The embedded logic includes the verification of inputs, assessment of system installation, and execution of start/stop commands.

For the purposes of this section, we are using VSCodium (`https://vscodium.com/`) as the integrated development environment (IDE) with AWS Toolkit installed and Amazon Q Developer Auto-Suggestions turned on.

Prior to commencing the scripting process, it is imperative to construct a comprehensive model outlining the essential steps required to start and stop the SAP system successfully. When scripting for SAP start and stop procedures, the scope may encompass a number of conditions, checks, and logical considerations. Once you have established the structure of how you are going to call the script and what arguments you are going to pass, you can write in natural English what the shell script should do to start and stop the SAP system.

In an IDE such as VSCodium, create a new file in which to store the contents of the script you are going to write. Give the file a name and make sure it ends with `.sh` (e.g., `lab1.sh`). Figure 6-28 shows an example of the logic of the script in natural English that you may use. Copy and paste this to VSCodium or an IDE of your choice. Alternatively, you may write the logic you would like the script to follow in a similar fashion.

Once you are done, press Enter. The next step is to have Amazon Q Developer make suggestions about the code.

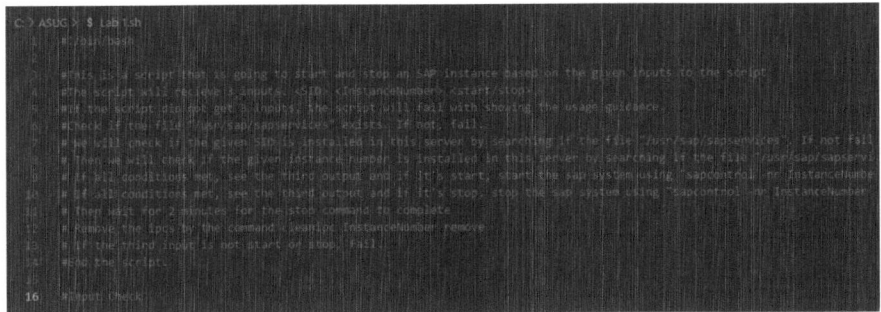

Figure 6-28. *Instructions in natural language for Amazon Q Developer*

The code suggestion prompt will look something like the screenshot shown in Figure 6-29. The indicator < 1/2 > means that there are two suggestions, and you may accept the one that makes the most sense to you. You can see the other suggestion by pressing the right-arrow key (>) on your keyboard. Toggle between the suggestions by the using right- and left-arrow keys.

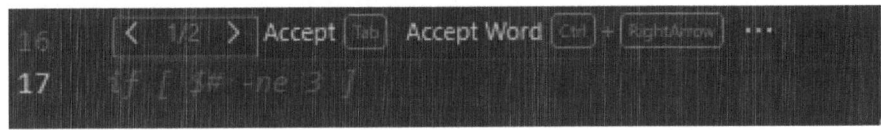

Figure 6-29. *Code suggestions from Amazon Q Developer*

To accept the proposed line of code, press the Tab key, as indicated in Figure 6-30 for line 18; else, write your own code. Once you accept the code, press Enter to move to the next line.

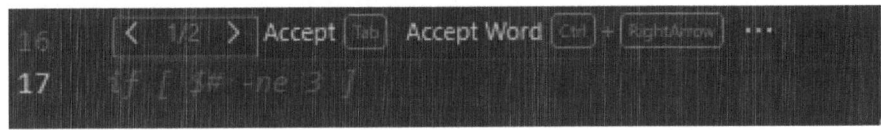

Figure 6-30. *Accepting the suggestion from Amazon Q Developer*

Based on your initial comment on what you would like to do, Amazon Q Developer should know what needs to be done next. Accept the lines of code, and once all sections of the code are completed, the script will be ready. It should look like the screenshot shown in Figure 6-31.

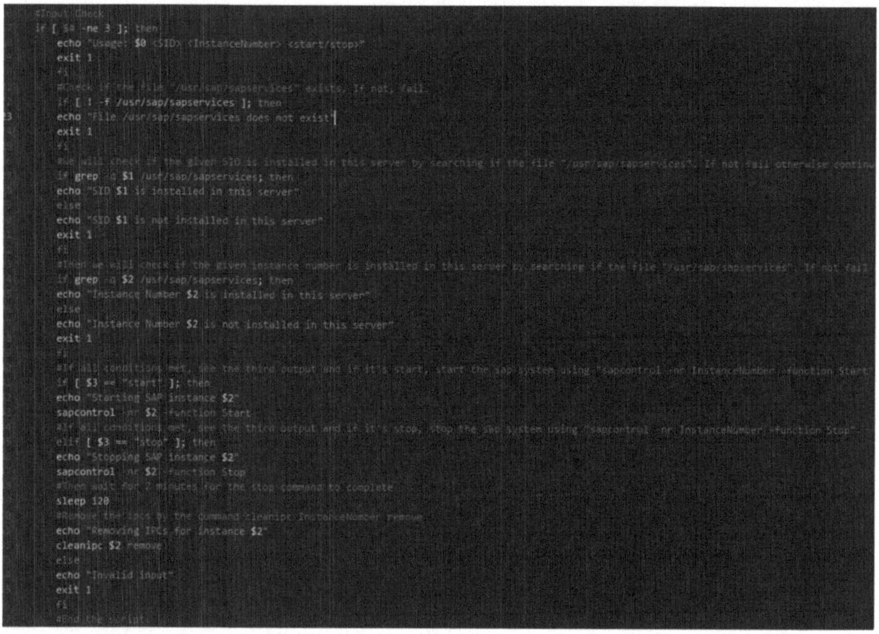

Figure 6-31. *Final output after accepting all suggestions from Amazon Q Developer*

It is important to recognize that the application of Amazon Q Developer is not limited solely to the start and stop of SAP systems but can be extended to encompass a broader spectrum of SAP operational scenarios. This use case is an illustrative example of the versatility of Amazon Q Developer. Its capabilities can be harnessed in diverse SAP-related use cases, such as kernel patching, database patching, and beyond. The flexibility of Amazon Q Developer opens up opportunities for enhancing efficiency and productivity across various aspects of SAP system management. It offers real-time code suggestions, enhancing developer workflow and efficiency. Additionally, it's adept at scanning for elusive vulnerabilities, fortifying application security.

Using Amazon Bedrock for Code Generation

First of all, as shown in Figure 6-32 and Figure 6-33, Amazon Bedrock can generate ABAP code directly in the AWS console.

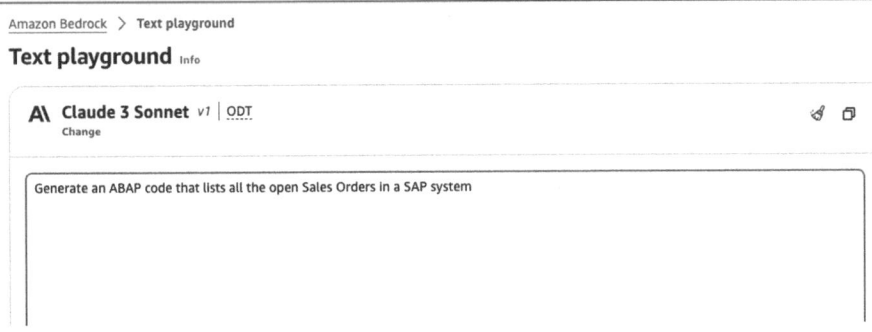

Figure 6-32. *Requesting Amazon Bedrock to generate ABAP code using the model Claude 3 Sonnet from Anthropic*

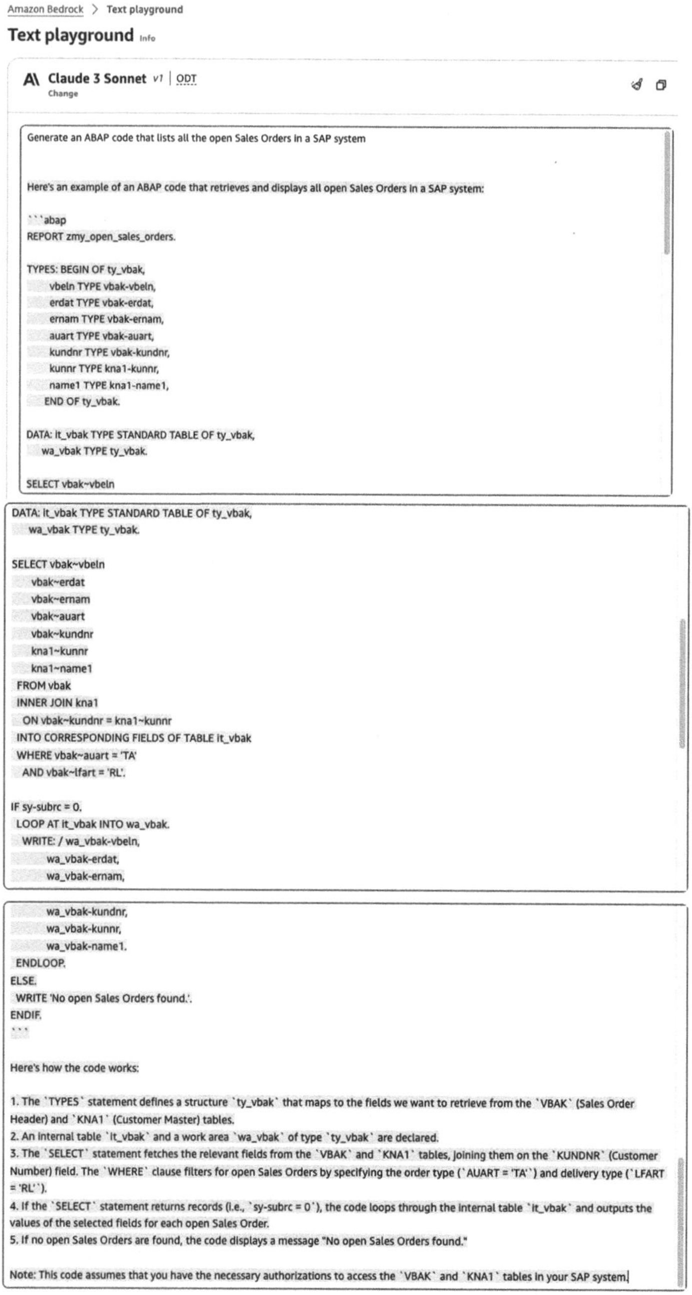

Amazon Bedrock > Text playground

Text playground Info

A\ **Claude 3 Sonnet** v1 | ODT
   Change

```abap
Generate an ABAP code that lists all the open Sales Orders in a SAP system

Here's an example of an ABAP code that retrieves and displays all open Sales Orders in a SAP system:

```abap
REPORT zmy_open_sales_orders.

TYPES: BEGIN OF ty_vbak,
       vbeln TYPE vbak-vbeln,
       erdat TYPE vbak-erdat,
       ernam TYPE vbak-ernam,
       auart TYPE vbak-auart,
       kundnr TYPE vbak-kundnr,
       kunnr TYPE kna1-kunnr,
       name1 TYPE kna1-name1,
       END OF ty_vbak.

DATA: it_vbak TYPE STANDARD TABLE OF ty_vbak,
      wa_vbak TYPE ty_vbak.

SELECT vbak~vbeln
```

```
DATA: it_vbak TYPE STANDARD TABLE OF ty_vbak,
      wa_vbak TYPE ty_vbak.

SELECT vbak~vbeln
       vbak~erdat
       vbak~ernam
       vbak~auart
       vbak~kundnr
       kna1~kunnr
       kna1~name1
  FROM vbak
  INNER JOIN kna1
    ON vbak~kundnr = kna1~kunnr
  INTO CORRESPONDING FIELDS OF TABLE it_vbak
  WHERE vbak~auart = 'TA'
    AND vbak~lfart = 'RL'.

IF sy-subrc = 0.
  LOOP AT it_vbak INTO wa_vbak.
    WRITE: / wa_vbak-vbeln,
           wa_vbak-erdat,
           wa_vbak-ernam,
```

```
           wa_vbak-kundnr,
           wa_vbak-kunnr,
           wa_vbak-name1.
  ENDLOOP.
ELSE.
  WRITE 'No open Sales Orders found.'.
ENDIF.
```
```

Here's how the code works:

1. The `TYPES` statement defines a structure `ty_vbak` that maps to the fields we want to retrieve from the `VBAK` (Sales Order Header) and `KNA1` (Customer Master) tables.
2. An internal table `it_vbak` and a work area `wa_vbak` of type `ty_vbak` are declared.
3. The `SELECT` statement fetches the relevant fields from the `VBAK` and `KNA1` tables, joining them on the `KUNDNR` (Customer Number) field. The `WHERE` clause filters for open Sales Orders by specifying the order type (`AUART = 'TA'`) and delivery type (`LFART = 'RL'`).
4. If the `SELECT` statement returns records (i.e., `sy-subrc = 0`), the code loops through the internal table `it_vbak` and outputs the values of the selected fields for each open Sales Order.
5. If no open Sales Orders are found, the code displays a message "No open Sales Orders found."

Note: This code assumes that you have the necessary authorizations to access the `VBAK` and `KNA1` tables in your SAP system.

Figure 6-33. *ABAP code generated by Amazon Bedrock using model Claude 3 Sonnet from Anthropic*

Now, it's just a matter of using APIs from the editor of your choice.

If you are using the ABAP Editor (Transaction: SE38) from your SAP system, follow these steps:

1. Install the AWS SDK for SAP ABAP.

2. Implement a menu exit for SE38 called 'Ask Bedrock,' as shown in Figure 6-34.

3. Create a custom code that uses the classes provided in the AWS SDK for SAP ABAP to interact with Amazon Bedrock. Define a structure to hold request parameters for the Bedrock service. Next, populate the request parameters, including the prompt, which is written in SE38. The structure can then be serialized into JSON format. Next, the code may invoke the LLM using an ABAP class provided by the AWS SDK for SAP ABAP for Amazon Bedrock. It would send the JSON request body and receive the response back from Amazon Bedrock. The response is deserialized into another structure and presented back to the SE38 screen.

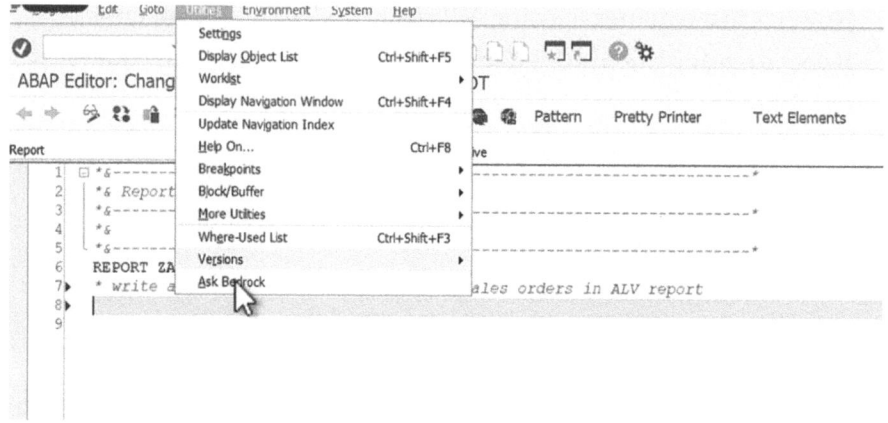

Figure 6-34. *ABAP Code Assistant in ABAP Editor*

If you are utilizing the Eclipse IDE, you have the option to install a plug-in called "SAP ABAP Assistant", as shown in Figure 6-35, with configuration as depicted in Figure 6-36. This plug-in helps integrate the Eclipse IDE with Amazon Bedrock. Next, the developer must authenticate and establish connections between the Eclipse IDE and one or more SAP systems. Similar to the SE38 use case, to initiate the code generation process, the developer writes a prompt in simple English within the ABAP program and then invokes the SAP ABAP Assistant plugin using the "Ask Bedrock" menu option available in the Eclipse IDE as shown in Figure 6-37. The SAP ABAP Assistant plugin sends a request with the selected prompts to Amazon Bedrock, which uses large language models (LLMs), such as the Anthropic Claude model, to generate the corresponding ABAP code as shown in Figure 6-38.

This plugin can also be used to generate documentation for existing SAP ABAP code. For more details, please refer to the Solutions Guidance, Guidance for Improving Application Development Productivity with the SAP ABAP Assistant on AWS (https://aws.amazon.com/solutions/guidance/improving-application-development-productivity-with-the-sap-abap-assistant-on-aws/).

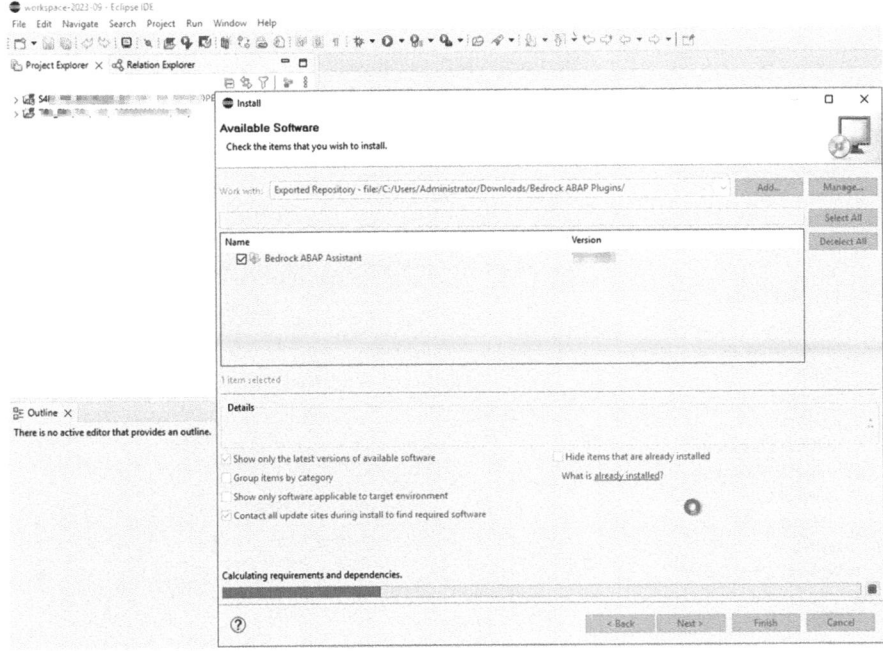

Figure 6-35. *ABAP Code Assistant in Eclipse*

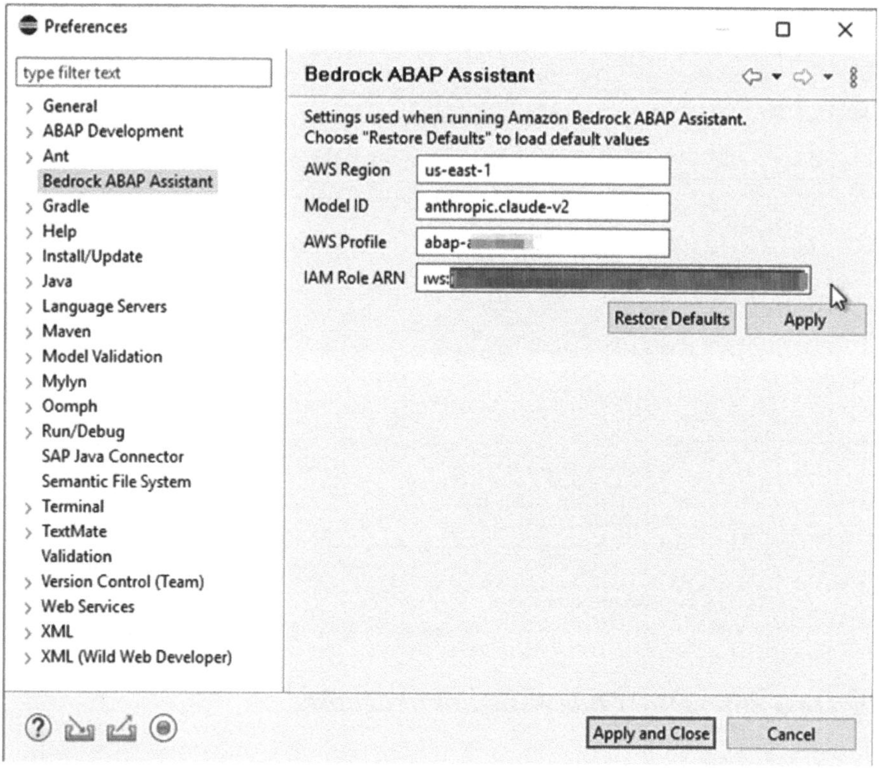

Figure 6-36. *Setting the preferences in ABAP Code Assistant in Eclipse*

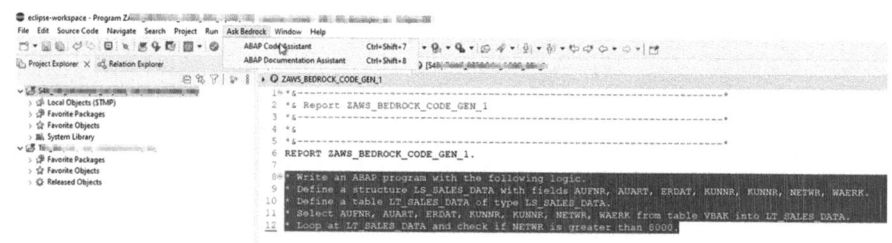

Figure 6-37. *Prompt for the LLM*

```
 1 * *&---------------------------------------------------------------------*
 2 *& Report ZAWS_BEDROCK_CODE_GEN_1
 3 *&---------------------------------------------------------------------*
 4 *&
 5 *&---------------------------------------------------------------------*
 6 REPORT ZAWS_BEDROCK_CODE_GEN_1.
 7
 8 * Write an ABAP program with the following logic.
 9 * Define a structure LS_SALES_DATA with fields AUFNR, AUART, ERDAT, KUNNR, KUNNR, NETWR, WAERK.
10 * Define a table LT_SALES_DATA of type LS_SALES_DATA.
11 * Select AUFNR, AUART, ERDAT, KUNNR, KUNNR, NETWR, WAERK from table VBAK into LT_SALES_DATA.
12 * Loop at LT_SALES_DATA and check if NETWR is greater than 8000.
13
14 TYPES: BEGIN OF ls_sales_data,
15          aufnr TYPE aufnr,
16          auart TYPE auart,
17          erdat TYPE erdat,
18          kunnr TYPE kunnr,
19          netwr TYPE netwr,
20          waerk TYPE waerk,
21        END OF ls_sales_data.
22
23 DATA: lt_sales_data TYPE TABLE OF ls_sales_data.
24
25 SELECT aufnr auart erdat kunnr netwr waerk
26   FROM vbak
27   INTO TABLE lt_sales_data.
28
```

Figure 6-38. *Response from the LLM*

Business Use Cases

SAP systems serve as the backbone for customers' business applications and processes, including those that set them apart from competitors. Generative AI and AWS services offer the potential to enhance and revolutionize these processes in ways previously unattainable.

The introduction to this "Gen AI Use Cases for SAP" section discussed some uses cases related to enhanced productivity, real-time insights, and improved decision-making. These use cases can be applied to various scenarios:

- A sales inventory analyst, procurement manager, or finance audit manager can now access real-time insights with natural language prompts from data stored in Amazon S3 or data warehouses like SAP Datasphere. Gen AI interprets the natural language and retrieves relevant data, eliminating the need for new ABAP or BW reports that could take days or weeks to develop and test. A great

327

example of this is building generative AI assistants like the one demonstrated in the YouTube video - 'Deliver insights from SAP Finance data with generative AI from AWS' (https://www.youtube.com/watch?v=XKSOw1GwjlY).

- An order to cash analyst can utilize generative BI with Amazon QuickSight.

- A master data manager can access detailed product descriptions in multiple languages using an SAP Fiori app.

- The accounts payable team can automate supplier invoice processing with intelligent document processing using Amazon Extract and the AWS SDK for SAP ABAP. The manager or audit team can detect anomalies and outliers using Amazon Q and large language models via Amazon Bedrock.

- Legal or policy managers in insurance or legal firms can leverage Amazon Bedrock to gain insights from policy documents, enabling them to better serve their customers.

- Any employee can generate e-mails with real-time insights from SAP reports using the AWS SDK for SAP ABAP and Amazon Bedrock.

All these use cases can be implemented using pattern 3 except the second one, which uses generative BI using Amazon QuickSight, as discussed in the next section.

Generative BI

Generative BI, in very simple terms, is generative AI–powered business intelligence. Generative BI is all about combining Amazon Bedrock LLMs with existing QuickSight capabilities, such as natural language understanding and insight generation, to create new, easy-to-use natural language experiences with reliable analytics delivered through QuickSight's analytics engine.

Amazon QuickSight capabilities can be categorized into six main areas:

- **AI-powered dashboard authoring experience**: This feature empowers SAP business analysts to build dashboards more quickly and efficiently.

- **Generative data stories**: Business users can easily share their findings through generative data stories, facilitating collaboration and insights dissemination.

- **Building your own AI-powered data apps**: Developers can leverage QuickSight to rapidly create innovative data-driven applications.

- **Ask questions in natural language**: ML models interpret user questions and intent to generate visualizations, simplifying the data exploration process. Figure 6-39 shows an example.

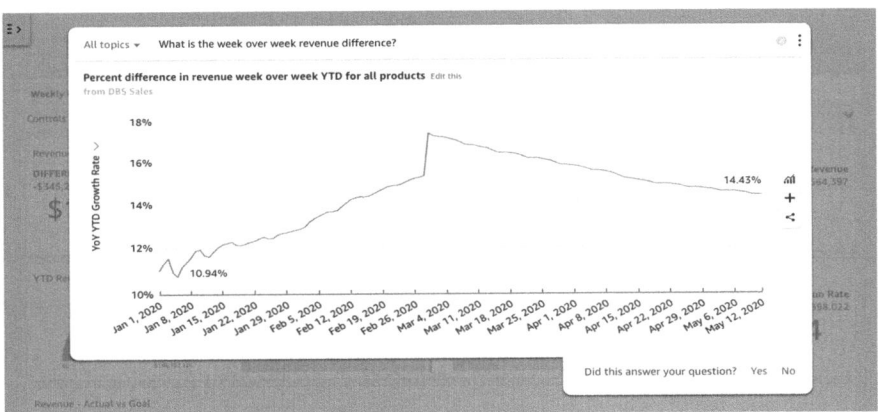

Figure 6-39. *Generating visuals based on questions asked in natural language*

- **Forecast**: Users can visualize future trends and
  trajectories for up to three measures simultaneously,
  enabling proactive decision-making based on
  predictive insights. Figure 6-40 shows an example.

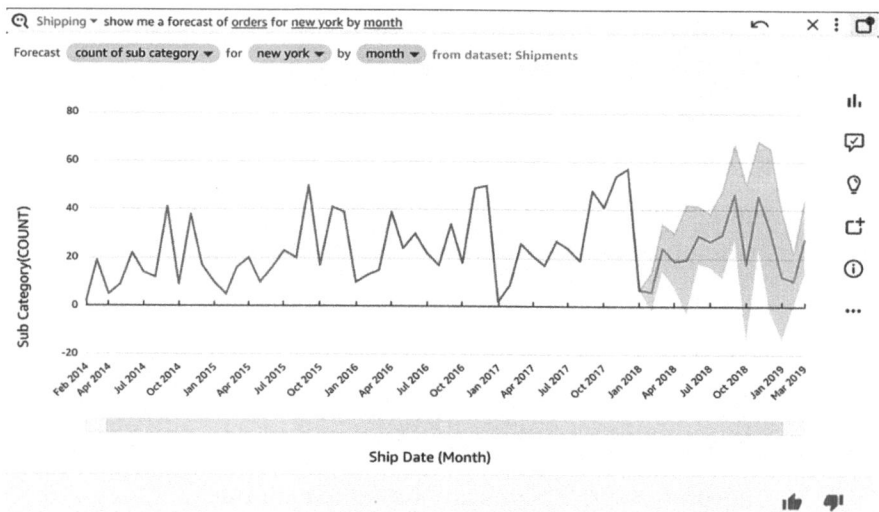

Figure 6-40. *Forecasting abilities in Amazon QuickSight*

- **Ask "why"**: This feature allows users to identify
  key drivers behind changes in the data through
  contribution analysis, quantifying the impact of each
  driver on the observed outcomes. Figure 6-41 shows an
  example.

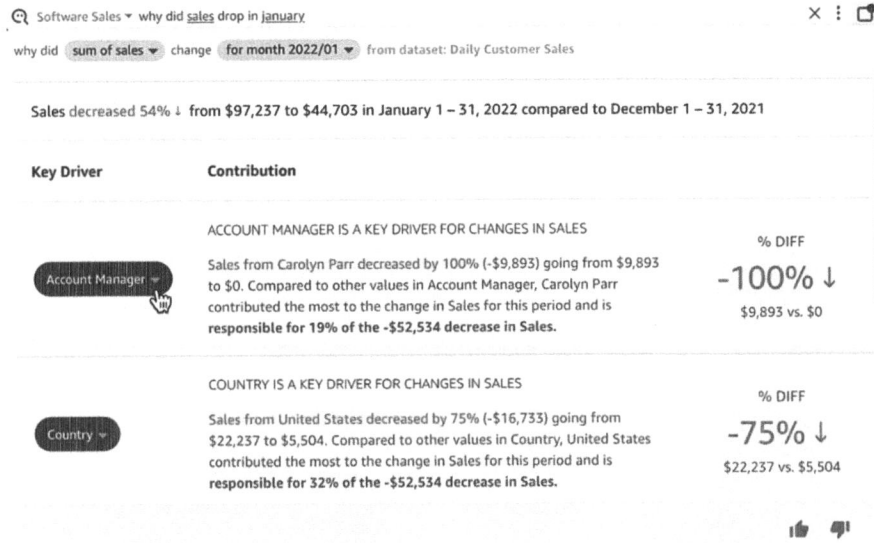

Figure 6-41. *Ask "why" feature in Amazon QuickSight*

Amazon Q in QuickSight

Amazon QuickSight Q represents a breakthrough in business intelligence by incorporating ML-powered natural language query (NLQ) capabilities. QuickSight's overarching mission is to empower end users to derive value from data swiftly and effortlessly, without the need to rely on IT or technical teams for answers to their queries. With Amazon QuickSight's Q functionality, everyday users can now ask questions about their data in plain English. QuickSight's ML models interpret these questions and intent, providing visualizations to address the queries effectively. Furthermore, Amazon Q facilitates forecasting questions, enabling users to gain insights into future trends for up to three measures concurrently. Additionally, users can pose "why" questions to identify the primary drivers behind changes in the data and quantify the contribution of each driver.

Before we dive deep into Amazon Q for QuickSight capabilities, it is very important to differentiate between Amazon Q itself and Amazon Q in QuickSight, as outlined in Figure 6-42.

	Amazon Q	Amazon Q in QuickSight
Commonalities	• Enterprise data • Data never leaves customers' account	• Enterprise data • Data never leaves customers' account

Differences	• Unstructured, semi-structured and structured data
	
	• Use Cases:
	  • Q&A on document data
	  • Document summarization
	  • Text generation (emails, abstracts, blurbs, etc.)
		• Structured data
		
		• Use Cases:
		  • Q&A on analytical data
		  • Building dashboards
		  • Data insight summarization
		  • Storytelling with data

Figure 6-42. *Amazon Q vs. Amazon Q in QuickSight*

Amazon Q brings three experiences to QuickSight customers:

- **AI-accelerated dashboard authoring**: This feature (see Figure 6-43) helps business analysts quickly build dashboards and reports:

  - **Build visuals**: Business analysts can use natural language to quickly build visuals for dashboards and reports.

  - **Build calculations**: Business analysts can easily create calculations using natural language without looking up or learning specific syntax.

  - **Refine visuals***:* Business analysts can quickly update visuals by describing desired formats using natural language.

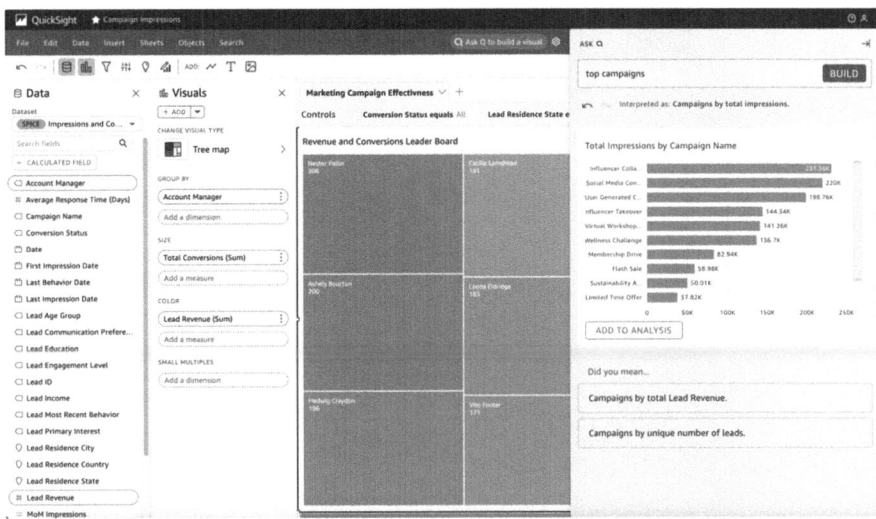

Figure 6-43. *AI-accelerated dashboard authoring*

- **AI-answered questions about data**: This feature helps users who may not be as tech-savvy to make informed business decisions.

  - **Executive summaries of dashboards**: As shown in Figure 6-44, with a click of a button, instant summaries of key dashboard insights in natural language can be created to explain top movers, outliers, and more.

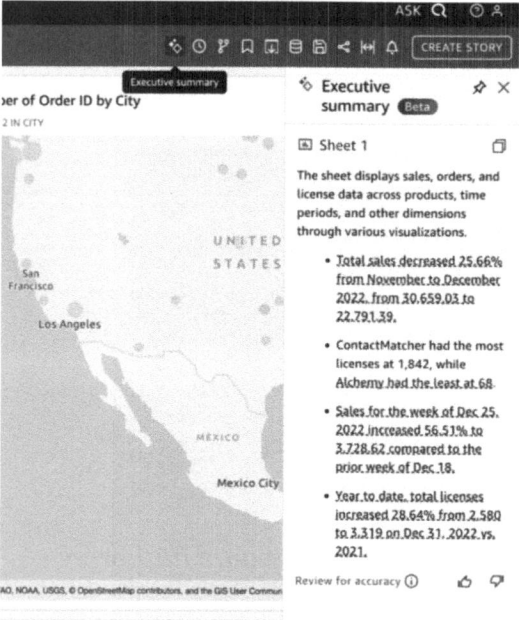

Figure 6-44. *Executive Summary feature*

- **Q Topics**: Amazon Q can insert and add semantic information to datasets automatically. This reduces the time and effort needed to start asking questions about data in natural language.

- **Amazon Q suggested questions and "What's in my data?"**: These features display potential queries that can be asked.

- **Multi-visual answers with narrative insight summaries**: This feature provides explanations of answer context.

- **Support for vague questions and "Did you mean?" alternatives**: This feature enables iterative fact-finding.

- **AI-assisted data storytelling**: This feature (see Figure 6-45) enables business users to discover and share findings from their data to drive team decisions.

  - **Interpret data for others**: Assists users in deriving meaning from data and making informed decisions to drive team actions

  - **Generate stories using AI**: Creates cohesive, impactful, and insightful narratives by analyzing data with minimal input

  - **Create refined content**: Enables users to control AI verbosity, customize narrative text, and apply captivating visual themes to enhance content presentation

  - **Governed and always up to date**: Facilitates quick and seamless sharing and updating of data at any time, ensuring governance and data currency

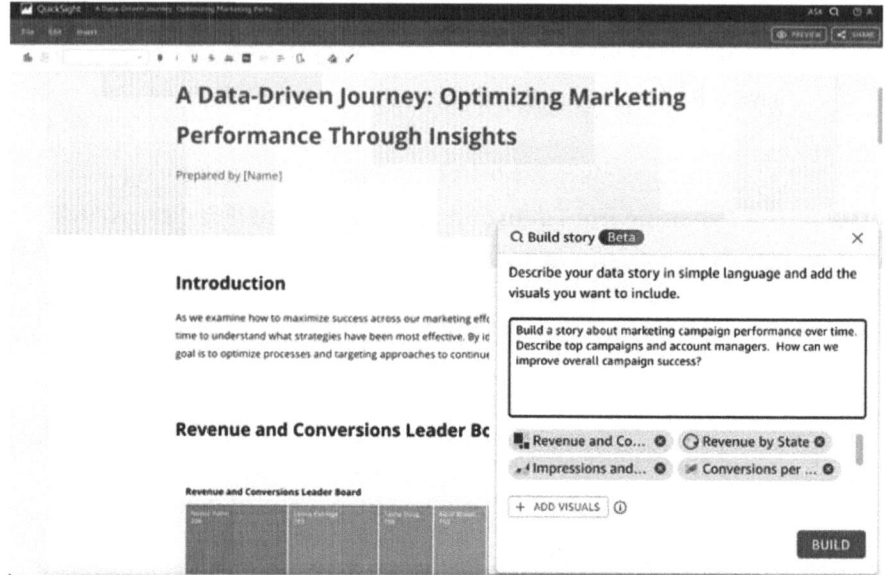

Figure 6-45. *AI-assisted data storytelling*

If you are interested in seeing a video demonstration of Amazon Q generating insights from SAP data, please refer to the YouTube video - 'Unlock procure-to-pay insights in SAP data with generative AI from AWS' (https://www.youtube.com/watch?v=Kpn-SXKuDa4).

Summary

In this chapter, you embarked on a journey into the depths of AI, ML, and the emerging frontier of generative AI. Beginning with the most common AI and ML services, you explored their applications across diverse use cases, along with customer stories. From case studies spanning from AI services like Amazon Lookout for Equipment to ML services like Amazon SageMaker, you saw that AI and ML technologies continue to revolutionize industries, driving innovation and efficiency.

Throughout this chapter, you have gained an understanding of the fundamentals of generative AI, especially how LLMs work and how you can customize them. This chapter also discussed the relevance of generative AI within the realm of SAP. You explored SAP's approach to AI and its capabilities. Additionally, this chapter provided insights into various patterns for leveraging generative AI capabilities within the SAP environment on AWS.

Furthermore, this chapter delved into a range of use cases encompassing both technical and business applications to unravel the intricacies of generative AI, examining its capabilities and potential applications across various sectors. Whether it's generating creative content, enhancing decision-making processes, or streamlining repetitive tasks, generative AI holds the promise of transforming how you interact with technology. As these technologies continue to evolve, understanding their capabilities and implications will be crucial for individuals and organizations seeking to harness their transformative potential.

Chapter 7 transitions into a totally separate topic, DevOps and SysOps for SAP.

CHAPTER 7

DevOps and SysOps for SAP

DevOps and SysOps create a seamless balance between agility and reliability in IT operations.

One of the primary reasons customers migrate their SAP landscape to AWS is to take advantage of the innovation that AWS offers while also being cost effective in running SAP. SAP customers often report benefits such as faster time to market and increased speed of innovation. This is because their staff becomes significantly more productive using AWS due to the wide range of options available to accelerate build, test, and operational activities, which are typically manual in an on-premises environment. As introduced in Chapter 1, in this chapter we will explore the end-to-end life cycle of SAP artifacts from a DevOps perspective, covering everything from provisioning infrastructure for SAP to developing and testing artifacts and managing daily operations. We will also explore how to use AWS technologies to manage and maintain the SAP landscape, ensuring optimal performance, availability, and reliability, referred to as SysOps for SAP.

The following topics will be covered in this chapter:

- DevOps in action

- Building a CI/CD pipeline with AWS

© Bidwan Baruah, Krishnakumar Ramadoss and Abarajith Vivekanandha 2024
B. Baruah et al., *Evolve from Infrastructure to Innovation with SAP on AWS*,
https://doi.org/10.1007/979-8-8688-0890-6_7

- DevOps for SAP

- SysOps for SAP

DevOps in Action

While closer cooperation between teams can be achieved with manual processes, it's much more efficient to use tools to automate as much of the work as possible. Automating the build, test, and deployment steps allows the work to be done more quickly, which in turn means that results from those stages are available much sooner. Automation is central to a DevOps approach because it enables the tight feedback loops that are essential to building quality and eliminating waste.

Putting DevOps into Practice

Building an automated continuous integration and continuous delivery/deployment (CI/CD) pipeline puts the DevOps ideals into practice. The continuous integration practice of committing frequently encourages small batch sizes, which can move through the pipeline quickly. An automated build and test system makes it possible to verify each change and provide feedback far more quickly than a manual process.

As a developer, getting rapid feedback on what you've just written is more efficient, as you are less likely to have lost the context for the change. Frequent automated testing also helps to improve quality, as catching and fixing bugs early avoids other code being built on top of them.

Automating deployment to preproduction servers makes the process consistent and reliable, providing opportunities to make more use of staging environments for testing and feedback. Deploying small changes to production regularly rather than batching them up into large, infrequent releases reduces the risk of something going wrong in live environments, as there are fewer variables to combine into unintended consequences.

If a bug does emerge, isolating and fixing it is quicker thanks to the smaller batch size. Releasing updates incrementally means you're delivering value to users regularly, and you can use the feedback on those changes to inform what you build next, thus continuing to improve the product.

Building a DevOps Culture

DevOps itself is a culture. It requires teams to think a new way and adhere to the mechanisms, tools, to establish a culture in which development teams can make small changes quite often while also maintaining the highest quality and standard of development. DevOps takes out the blame game by meaningfully tracking the code changes, allowing teams to make smaller changes with smaller blast radius and, most importantly, automate the testing flow to ensure the code quality.

Putting all these ingredients together gives you fast and regular feedback on your code. By checking that any change to the code base, be it a bug fix, refactor, or part of a new feature, at least results in a clean build that passes tests, teams can avoid building new work on top of bad foundations and the inevitable rework that follows.

Whereas in the past there may have been a handoff from your developers to testers, and then from testers to release managers, with continuous delivery, your team (which may include people from a range of disciplines) owns the entire process of building, testing, and releasing their product. This comes with several advantages:

- By avoiding the traditional silos, your development team has a better understanding of the business and operational needs for delivering their product to your users.

- In turn, this creates an opportunity to bring development practices to what was typically a manual, and often quite lengthy, process.

341

- Automating the steps involved in delivering a product to live status not only speeds up the process but also reduces the risk of errors and makes it more stable and reliable.

The exact steps involved in delivering software to your users and, therefore, the required stages in the delivery pipeline vary according to business and user needs, but it's common practice to deploy to at least one preproduction environment before software is released into the wild.

Preproduction environments include testing environments for additional layers of testing, such as security, load, and performance tests; sandbox environments for ABAP and functional teams to familiarize themselves with new features; and acceptance testing environments for QA and product professionals to verify that changes work as intended. With continuous delivery, each successful build is automatically deployed to each of the preproduction environments, with confidence in the quality increasing with every stage. This is a worthwhile goal that takes some investment to put into practice. Some guiding principles are presented next.

Build Once

Only by promoting the same artifact through each stage of the pipeline can you have confidence that it has passed all prior stages of testing.

Store Code Changes in Source-Controlled Repositories

Keeping all these code changes in source-controlled repositories helps ensure deployments are consistent and repeatable.

Automate Every Deployment

The deployment to each environment should be scripted so that it can be run automatically and in exactly the same way each time. The release to go-live should be scripted for the same reason, but with continuous delivery, this final step is not automatic.

Clean Your Runtime Environment

For a completely consistent approach, the runtime environment should be reset to the same conditions for each new deployment. Thanks to containers, this is now much easier to implement, either on local infrastructure or in the cloud.

Avoid Out-of-Sync Situations

With continuous integration, it is crucial to maintain consistency between pipelines and their targeted system states by avoiding manual changes outside of the pipeline; this ensures that all modifications are tracked and synchronized, preventing discrepancies and potential deployment issues.

With continuous delivery, your team takes responsibility for the delivery of software and benefits from the timely feedback provided throughout the process.

Continuous deployment takes the practices of continuous integration and delivery to their logical conclusion. If a build passes all previous stages in the pipeline successfully, it is automatically released into production. This means that as soon as any change to your software has passed all tests, it is delivered to your users. Continuous deployment shortens the feedback loop from code change to use in production, giving your team timely insight into how their changes perform in the real world without having to compromise on quality.

Although automating the deployment of software to production is not suitable for every product and organization, it's worth considering the steps required to get there, as each individual element is valuable on its own:

- **Have confidence in your tests**: Deploying to production automatically requires a high level of confidence in your pipeline, particularly your automated tests. A great testing culture, where your team invests in test coverage and performance and prioritizes fixing the build and the pipeline over new features, is essential.

- **Monitor production**: Even with all the preceding measures in place, continuous deployment can feel like a risky practice. What happens if a bug goes undetected in testing only to emerge in production? Time, money, and reputation are all potentially at stake. While the same problem could get through with a manual deployment, having a critical issue automatically released by "the system" can significantly erode stakeholder confidence. This is where being proactive in looking for signs of trouble rather than waiting for bug reports to come in makes all the difference. Monitoring stats for any deviation from the norm, particularly just after a release, can alert you to issues before they cause a noticeable problem for your users.

- **Choose when and how to release**: A common objection from those new to continuous deployment is that if your developers commit early and often and let things march out of the door without some manual oversight, then your users will find unfinished or half-baked features that are not ready for use. Rather than resorting to branches and missing out on the advantages of continuous integration, the solution is to use feature flags, which allows your developers to check what's visible and what's hidden from the user when they're writing the code.

- **Streamline your pipeline**: If something does go wrong in production, you want to be able to respond quickly. In some cases, it might be possible to roll back the release to the previous version. Often, however, things aren't that straightforward. Some changes, like changes to the database schema and other updates, can make rolling back impractical, necessitating a fix instead. Skipping steps in your pipeline may seem efficient but often leads to unforeseen issues that thorough testing would have prevented. Instead, it's better to invest in streamlining your pipeline, from build speed to test performance. Not only does this mean you can deploy changes quickly when you need to, but it also shortens the feedback loops throughout the process.

Control the Rollout of Changes

Although continuous deployment means releasing automatically if all previous stages pass muster, that does not mean surrendering all control. There are various deployment practices that are used to minimize risk and control the rollout. Canary releases allow you to test the waters with a small group of your users initially, while blue-green deployments can be used to manage the transition to a new version.

Continuous integration, delivery, and deployment are sequential stages of the software delivery process. They help teams release software faster, shorten the feedback loop, and automate repetitive tasks. Identifying the key differences between continuous integration, delivery, and deployment is no less important than understanding how these three steps work together to deliver software to users in a reliable and stable manner. Continuous integration is the practice of merging any new code changes into the main branch. Continuous delivery automates manual tasks that are required to build and test software (for example, by automating tests). Continuous deployment is a logical continuation of the practice of automating build and test steps, and at this stage, software is deployed automatically once it passes all the necessary checkpoints.

345

Benefits of Automation in the Build Environment

A software release can be a painful and time-consuming process. One involves weeks of manual integration, configuration, and testing, while the ever-present risk of discovering a showstopper threatens to force everyone back to square one. The time commitment involved in getting code ready for release can mean changes are delivered every few months at best. But there is another way.

CI/CD has enabled many organizations to release on a more frequent basis without compromising on quality. With CI/CD, code changes are shepherded through an automated pipeline that handles the repetitive build, test, and deployment tasks and alerts you about any issues.

If you're wondering whether the benefits of CI/CD are worth exploring or if you need help convincing your stakeholders, read on to find out what a difference a CI/CD pipeline can make to your organization.

Faster Time to Market

Enterprises may have led the way in adopting Agile and DevOps techniques to transform their development processes and deliver constant improvements to their users, but many smaller organizations are following suit, resulting in the landscape becoming increasingly competitive.

Understanding your users' needs, coming up with innovative features, and turning them into robust code is not necessarily enough if your competition is moving more quickly. With an automated CI/CD pipeline, you can ship changes weekly, daily, or even hourly.

New features can be launched faster, with deployment strategies giving you the option to experiment and collect feedback, which you can then incorporate into the next update. Being able to push changes out quickly and with confidence means you can respond to new trends and address pain points as they emerge.

Better Code Quality

Testing your code's behavior is an essential step in the software release process, but doing it thoroughly can also be extremely time consuming. A central part of any CI/CD pipeline is to run a series of automated tests on every build. Although writing automated tests requires an investment of time and expertise, doing so pays significant dividends.

As anyone who has had to follow a manual test script knows, testing is a repetitive process that demands high levels of concentration. Automating tests ensures they are performed consistently, making the results more reliable. Because automated tests are quicker to run than their manual equivalents, it becomes feasible to test much more frequently.

Testing your code regularly and thoroughly means you'll discover bugs sooner, making it easier to fix them, as less functionality has been built on top of them. Over time, this results in better-quality code.

Once you've invested in a first layer of automated tests, the time you save on running those tests manually can be spent developing additional layers of automated tests, such as end-to-end or performance tests, and on manual exploratory testing. The latter puts your quality assurance or test engineers' creative skills to use by identifying new failure modes, while their findings can be used to extend your test coverage.

Faster Bug Fixes

Even with improved code quality thanks to automated testing, bugs will still occasionally sneak their way through to production. If you're committing changes regularly and shipping frequently, each release to production will contain a relatively small number of code changes, making it much easier to identify the cause of an issue. As your commits are more granular, if you decide to back out of the change, you're less likely to make other useful changes to it. When it's urgent to get a fix out to production, it can be tempting to skimp on manual testing in order to save time, despite the risk of introducing a new failure to production. With a CI/CD pipeline, running automated tests is no longer a significant overhead, so there's less temptation to compromise on quality.

Smoother Path to Production

As we all know, practice makes perfect, and what's true of shooting hoops or mastering scales also applies to software releases. Adopting CI/CD is best done incrementally, starting with CI practices and building up your pipeline over time. As you start deploying changes more frequently, you'll identify pain points and steps in your current process that slow you down, such as refreshing data in a test environment or having to reconfigure parameters before deploying to a particular system.

Adding automation for builds, tests, environment creation, and deployments makes each step consistent and repeatable. Having broken it down, you can keep optimizing each stage to make your process more efficient. From being a significant event that occupies multiple teams for several days, CI/CD releases mature into a familiar and predictable occurrence.

Efficient Infrastructure

Automation is a central part of any CI/CD pipeline, serving to make the release process repeatable and reliable. In the early stages of implementing continuous integration, your focus will be on automating the build process and writing and running automated tests. Once you've established a solid CI foundation, the next stage is to automate the deployment of your build to test and staging environments.

Taking an infrastructure-as-code approach involves automating the creation of those environments. Rather than managing individual servers manually, their configuration is scripted and stored in version control so that new environments can be brought online quickly without the risk of inadvertent changes and inconsistencies. This not only makes the continuous delivery stage faster and more robust but also allows you to respond quickly to requests for additional preview and training environments with minimal interruption to development work.

Measurable Progress

Many of the tools available to support an automated CI/CD pipeline also instrument the process, providing you with a whole host of metrics, from build times to test coverage to defect rates to test fix times. Armed with this data, you can identify areas that might need attention so you can keep improving your pipeline. Slower builds may indicate a need to increase capacity, while an increase in mean fix times might be a sign of a process or cultural issue.

Conversely, metrics can also provide reasons to celebrate, and so they should; consistently extending your code test coverage, reducing your defect rate, or increasing your release frequency all belong on the team's collective brag sheet as signs of a great working culture. Being able to measure how your CI/CD pipeline is supporting your organization's goals is another advantage of the practice.

Tighter Feedback Loops

Rapid feedback is a key part of the DevOps approach, with applications throughout the pipeline. It starts with automated build and test steps to inform you of immediate problems, helping you to work more efficiently and effectively than if there is a long delay between the original work and the results.

Similarly, shipping updates regularly provides you with far more immediate feedback on what you've built than if you batch changes for a larger release every few months. By collecting feedback, observing user behavior, and tracking key performance indicators, you can identify what's working well and prioritize modifications and improvements.

A frequent release cadence also gives you an opportunity to experiment with alternative designs or behaviors, whether by running side-by-side comparisons with A/B testing or by deploying new versions and comparing results over time.

Feeding insights into a cycle of continuous deployment allows you to see how your changes perform soon after you've made them. That means you can keep iterating and tweaking without the loss of context that results from a long delay between coding and release.

Collaboration and Communication

DevOps is as much about building a collaborative culture as it is about new processes and tools. In order to get started with CI/CD, you need to start breaking down barriers between teams and encouraging more communication. Aligning around the overarching aim of delivering a product that meets user needs and understanding all the steps involved in reaching that goal helps everyone to focus on what needs to be achieved rather than being limited by their team's remit.

Breaking down silos between development and operations is the start of a virtuous circle. A CI/CD pipeline provides an opportunity for the many functions and specialists involved in building a product, from security experts to marketing teams, to get better visibility of the software development process and to collaborate with each other more.

Many of the tools available to help manage your CI/CD pipeline also make it easier for non-developers to see what is in train, while access to staging environments allows them to engage with and provide feedback on what is being built. Sharing details of what is being released, usage metrics, and the results of experiments opens the door to more communication, which in turn fosters innovation.

Building a CI/CD Pipeline with AWS

The AWS developer tools and services that are available for you to build a CI/CD pipeline wit AWS are shown in Figure 7-1 and briefly described in this section.

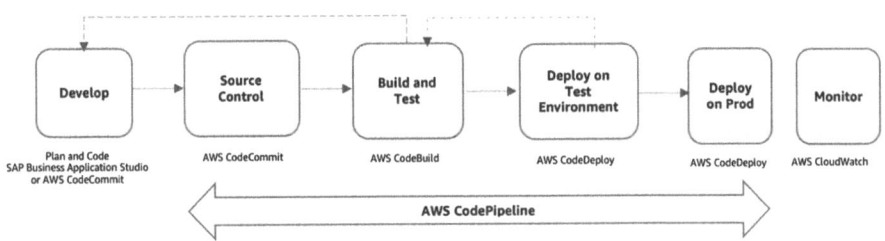

Figure 7-1. *Tools for the CI/CD process*

AWS CodeCommit

AWS CodeCommit is a secure and scalable private Git repository designed to facilitate collaborative code development. By leveraging this service, developers can eliminate the burden of hosting, maintaining, backing up, and scaling their own source control servers. AWS CodeCommit enables developers to customize user-specific access to repositories, ensuring granular control over code collaboration. The service prioritizes security by automatically encrypting files during transit, providing an additional layer of protection. With its reliable and highly available infrastructure, AWS CodeCommit is a managed service that ensures uninterrupted access to repositories. Moreover, developers can conveniently maintain their repositories in proximity to their build, staging, and production environments on AWS, streamlining the development workflow.

AWS CodeBuild

AWS CodeBuild is a comprehensive continuous integration service that streamlines the software development process. It handles tasks such as compiling source code, executing tests, and generating deployment-ready software packages. By utilizing AWS CodeBuild, there's no need for manual provisioning, management, or scaling of build servers. Simply specify the source code location and desired build settings, and AWS CodeBuild will execute the build scripts, encompassing compilation, testing, and packaging of your code. This service facilitates the creation of a fully automated software release pipeline, enabling seamless promotion of code changes across multiple deployment environments. Furthermore, AWS CodeBuild offers compatibility with existing Jenkins build jobs, eliminating the need to configure and manage Jenkins build nodes.

AWS CodeDeploy

AWS CodeDeploy is a comprehensive deployment service that streamlines the automation of software deployments across various compute services, including Amazon Elastic Compute Cloud (Amazon ECC), on-premises instances, serverless AWS Lambda functions, and Amazon Elastic Container Service (Amazon ECS) services. This service facilitates the repetition of application deployments across different groups or instances through a file and command-based installation model. With AWS CodeDeploy, managing deployments to thousands of hosts becomes seamless, thanks to advanced monitoring capabilities and traffic-shifting functionality. It supports multiple deployment types, including in-place, canary, and blue/green deployments, enabling flexibility in the deployment process. Additionally, AWS CodeDeploy empowers users to configure alarms that trigger rollbacks and halt ongoing application deployments, ensuring the stability and reliability of the deployment process.

AWS CodePipeline

AWS CodePipeline is a continuous delivery service that streamlines the software release process by providing a visualized and automated workflow. It allows you to model and automate the necessary steps to release your software, from building the code to deploying it to preproduction environments, testing the application, and finally releasing it to production. Whenever there is a code change, AWS CodePipeline automatically builds, tests, and deploys your application according to the defined workflow. The service also offers the flexibility to integrate partner tools and custom tools at any stage of the release process, allowing you to create an end-to-end continuous delivery solution tailored to your needs.

By automating the build, test, and release processes, AWS CodePipeline enhances the speed and quality of your software updates by subjecting all changes to a consistent set of quality checks.

AWS CodePipeline seamlessly integrates with various AWS services, including AWS CodeCommit, Amazon Simple Storage Service (Amazon S3), AWS CodeDeploy, AWS Elastic Beanstalk, AWS OpsWorks, and AWS Lambda. Additionally, it offers integration with numerous partner tools, enabling a wider range of capabilities. You also have the option to write your own custom actions and integrate existing tools with AWS CodePipeline, providing further flexibility and extensibility

DevOps for SAP

As introduced in the Chapter 1 section "Pillars of SAP on AWS Innovation," the fifth pillar is DevOps and SysOps for SAP. (The section "SysOps for SAP" is presented later in this chapter.) Figure 7-2 represents the end-to-end process of DevOps for SAP together with the challenges, solutions, and tools for the entire portfolio.

	Plan & Setup	Develop		Test	Monitor & Operate	
Challenges	Development and Functional Teams wait for SAP System Builds	• Distributed Development • Feature Development • Continuous Integration		Manual Testing	Lack of visibility into applications	System Refresh, Patching, etc. takes a lot of time
Solutions	Infrastructure provisioning and SAP System Builds are completed in minutes/hours using IaC	ABAP based systems • abapGit + gCTS • Git based development • Git based CI	Non-ABAP based systems Standard DevOps solutions	• AI powered Change Impact Analysis • Test Automation • Robotic Test Automation	Applications are integrated with Observability solutions	Automate tasks
Tools	• AWS LaunchWizard • AWS CloudFormation • AWS Service Catalog • Terraform • SaltStack	Git Solutions gCTS Config Management Chef, Ansible, Puppet, etc.	CI/CD Solutions AWS CodeBuild AWS CodeDeploy AWS CodePipeline	Third-party Solutions like Worksoft and Tricentis	• AWS CloudWatch • AWS CloudWatch AppInsights • AWS Cloud Trail • Third party solutions like Datadog	• AWS Systems Manager • AWS Systems Manager for SAP • AWS Serverless System Refresh Solutions from SAP • Third-party solutions like Libelle and Avantra
ALM Tools	←		SAP Cloud ALM / SAP Solution Manager			→

Figure 7-2. *Phases of DevOps, with challenges, solutions, and tools*

Plan and Setup Phase

Automation is integral to the plan and setup phase of DevOps for SAP, so you first need to understand the options for automation on AWS.

Automation As a Concept on AWS

Existing automation solutions for SAP in the on-premises world typically have three common components: an agent running on the SAP application and database servers, a centralized orchestration server to manage automation tasks and to collect logs, and integration with notification systems. This setup requires maintaining the central server, managing agent versions, and handling their life cycle. However, with AWS native services that are serverless and fully managed, like AWS Lambda functions,

AWS Step Functions (as an orchestrator), Amazon EventBridge, and Amazon CloudWatch Logs, you can create automation solutions for any SAP requirements without needing to maintain a server or infrastructure. In this section, we explore these SAP-specific automation solutions using AWS technologies.

Figure 7-3 shows the anatomy of a traditional automation solution, and Figure 7-4 shows how the anatomy changes with AWS automation tools.

Structure of a Traditional Automation Solution

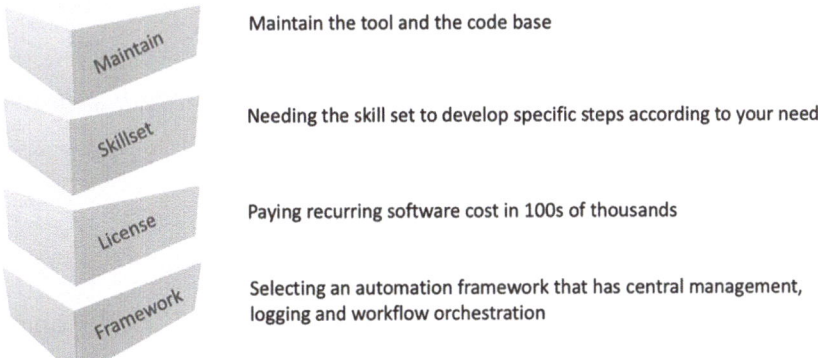

Maintain	Maintain the tool and the code base
Skillset	Needing the skill set to develop specific steps according to your need
License	Paying recurring software cost in 100s of thousands
Framework	Selecting an automation framework that has central management, logging and workflow orchestration

Figure 7-3. Structure of a traditional automation solution

With AWS Technologies

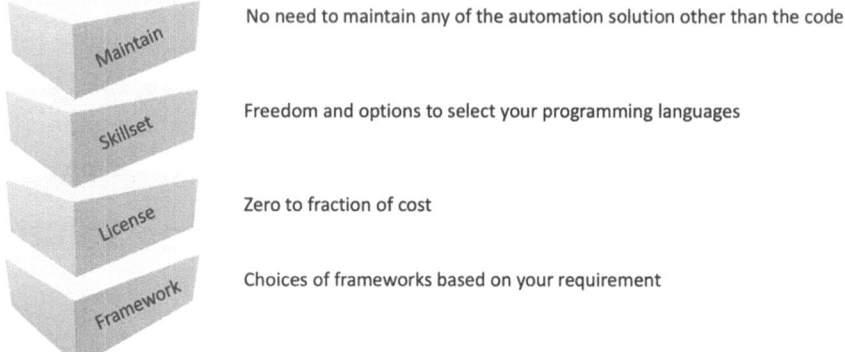

No need to maintain any of the automation solution other than the code

Freedom and options to select your programming languages

Zero to fraction of cost

Choices of frameworks based on your requirement

Figure 7-4. *Advantages of using an automation solution with AWS technologies*

Infrastructure As Code

AWS provides Infrastructure as Code (IaC) through AWS CloudFormation, enabling automated, repeatable, and scalable deployment and management of infrastructure on AWS. With AWS CloudFormation, you can define your entire SAP environment as code using JSON or YAML templates. These templates describe the AWS resources you need, such as Amazon Elastic Cloud Compute (Amazon EC2) instances, network configurations, storage volumes, and more.

Here's how you can provision SAP infrastructure using AWS CloudFormation:

1. **Define resources**: In your AWS CloudFormation template, you specify the resources required for your SAP environment, such as Amazon EC2 instances with specific CPU and memory configurations, network settings (VPCs, subnets, security groups), Amazon Elastic Block Store (Amazon EBS) volumes, and

other AWS services like Amazon Relational Database Service (Amazon RDS) databases or Amazon Elastic File System (Amazon EFS) file storage.

2. **Specify dependencies**: AWS CloudFormation allows you to define dependencies between resources, ensuring that resources are created and configured in the correct order. For example, you can specify that an Amazon EC2 instance should be created after the necessary networking resources are provisioned.

3. **Automate provisioning**: AWS CloudFormation takes care of provisioning and configuring the resources defined in your template. You can automate the deployment of SAP software components like the SAP kernel, database, and other components by including them as part of the AWS CloudFormation stack.

4. **Ensure consistency and repeatability**: Once you have defined your SAP environment in an AWS CloudFormation template, you can use the same template to consistently provision the same infrastructure across different environments (dev, test, prod) or regions, ensuring consistency and repeatability.

5. **Change management**: AWS CloudFormation keeps track of the changes made to your infrastructure over time. You can update your existing AWS CloudFormation stacks by modifying the template, and AWS CloudFormation will handle the necessary changes to the resources, making it easier to manage and track changes to your SAP environment.

357

Benefits of using AWS CloudFormation for SAP infrastructure provisioning include the following:

- Improved infrastructure consistency across environments

- Reduced manual effort for provisioning SAP software components

- Ability to track and manage changes to the SAP environment over time

- Integration with other AWS services and automation tools (e.g., AWS CodePipeline, AWS Lambda)

By defining your SAP infrastructure as code using AWS CloudFormation, you can achieve a consistent, repeatable, and automated provisioning process, while also benefiting from the scalability, reliability, and security features of the AWS Cloud.

AWS Launch Wizard for SAP

AWS Launch Wizard for SAP is a service that automates the complex process of sizing, configuring, and deploying SAP applications based on SAP HANA and SAP Sybase databases on AWS, following AWS and SAP best practices. AWS Launch Wizard provides a guided experience and simplifies and automates the deployment of SAP systems on AWS, including configuration of high-availability clusters, SAP HANA system replication, and AWS Backup. This enables organizations to rapidly provision and configure SAP systems following best practices. Key benefits include

- Accelerated deployment of production-ready SAP environments

- Automated provisioning and configuration of AWS resources

- Integration with AWS services for monitoring, logging, and governance

- Support for high-availability architectures

- Customization through pre- and post-deployment scripts

- Reusable deployment artifacts for consistent, repeatable deployments

Let's discuss some of the important benefits in this list.

SAP High-Availability Architecture

AWS Launch Wizard has the capability to configure high-availability clusters, which is a complex task. As SAP specialist Solution Architects, we have seen so many customers struggling to set up their high-availability cluster the right way so that whenever there is a disruption on one of the cluster nodes or Availability Zones, the failover is seamless. Some of the common mistakes we have seen customers making while setting up their clusters manually include

- Incorrect cluster configuration.

- Constraints are stopping the cluster from failing over.

- Source/destination check is not disabled on the cluster.

- The EC2 instances does not have the right IAM roles with access to update the route table.

- For SAP HANA, proper permissions are not set for the SAPHanaSR Python hook.

AWS Launch Wizard takes care of all these issues behind the scenes, and the customer can rest assured that the cluster has been set up correctly and will work perfectly fine.

Custom Deployment Scripts

You can provide custom pre-deployment and post-deployment configuration scripts that run on various instance tiers (SAP HANA Database, Primary Application Server, Enqueue Replication Server) during the deployment process. These scripts can be used for tasks like OS hardening, deploying security/logging software, installing monitoring tools, and updating DNS entries.

AWS Service Catalog Integration

AWS Launch Wizard can create AWS Service Catalog products from successful deployments. These products contain AWS CloudFormation templates and associated application configuration scripts stored in Amazon S3. You can use these AWS Service Catalog products, along with integrations offered by AWS Service Catalog, with third-party tools like ServiceNow, Jira, or Terraform for self-service provisioning and governance.

By leveraging AWS Launch Wizard, organizations can focus on their core SAP application management and business processes, rather than undifferentiated infrastructure tasks.

Now, let's move to the next phase, the develop phase.

Develop Phase

In terms of development in SAP, the Chapter 1 section "Pillars of SAP on AWS Innovation", we discussed the two approaches of development in SAP (depicted in Figure 1-4): classical on-stack development (tightly coupled) extensions and side-by-side (loosely coupled) extensions. Let's look at each approach and how DevOps can be implemented with each of them.

Classical ABAP Development

SAP still largely functions as a monolithic application, especially with a classical ABAP development perspective. Features like object lock in SAP ABAP during development, deploying changes through transports, and so forth are unique to SAP. ABAP development practices have traditionally been centered on waterfall-like development practices, which involve long development cycles, extensive testing phases, and sequential (rather than iterative) development, which is contrary to the fast, iterative cycles promoted by DevOps. Classical SAP development lacks native support for CI/CD pipelines, which are crucial for DevOps. As a result, ABAP teams are confined to collaborating within a single monolithic code base, making distributed development impossible.

To address this challenge, SAP introduced Git-enabled Change and Transport System (gCTS), allowing developers to manage their ABAP source code in Git repositories like AWS CodeCommit. This enables parallel work on ABAP code bases, collaboration, change monitoring, and management of ABAP development similar to non-SAP software development processes. gCTS was first introduced in SAP S/4HANA 1909 and is available for both customizing and workbench requests with SAP S/4HANA 2020. The Git client typically used for ABAP development is abapGit. This process was described in detail in Chapter 1. With SAP strategically moving to an SAP HANA–only database strategy, keep in mind that database changes are also covered by gCTS. With SAP HANA Transport for ABAP, you can move both your database and SAP changes under a single transport.

Implementing Git-Enabled Change and Transport System (gCTS)

Before you get started, you need the following:

- **Git platform**: To host your repositories. Examples include AWS Code Commit, GitHub, etc.

- **SAP ABAP system**: SAP S/4HANA 1909 (2020 for customizing support)

- **Continuous integration server (optional)**: To host your pipelines. You may use AWS CodePipeline or the Jenkins shared libraries and templates provided in Project "Piper," which is discussed in an upcoming section.

Figure 7-5 shows the overall approach. To get started, below are the high level steps:

1. **Install abapGit**: This is the Git client used for ABAP development. Install it in your SAP S/4HANA system. You can find the code at `https://github.com/abapGit/abapGit`.

2. **Create and link a Git repository**: Set up a Git repository and link it in the abapGit configuration. You can use AWS CodeCommit as your Git repository for commits.

3. **Prepare your system**: Implement the necessary Business Add-Ins (BAdIs) that are specific to gCTS, create repositories for projects, and complete other preparatory steps.

4. **Develop and commit artifacts**: Develop project artifacts and check in files using the commit-by-task approach.

5. **Use abapGit for code management**: Once
    installation and configuration are complete, use the
    abapGit transaction to push and pull code changes
    between your SAP system and your Git repository.

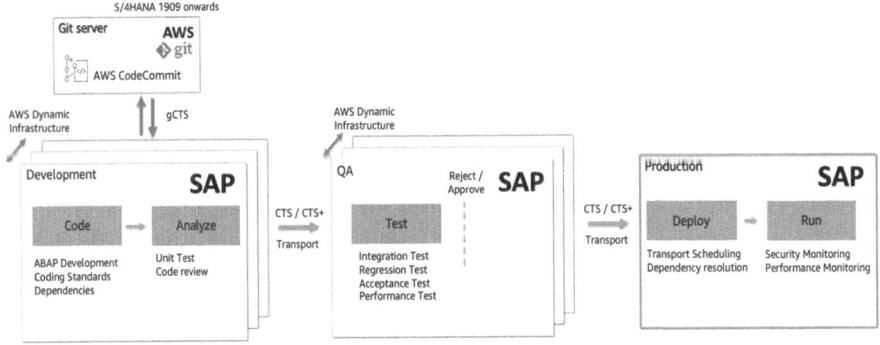

Figure 7-5. *DevOps for SAP ABAP: classical ABAP development*

CI/CD Pipelines for SAP ABAP

The following are some other tools and services available to build your CI/
CD pipelines for SAP ABAP.

SAP Continuous Integration and Delivery Service in SAP BTP

SAP Continuous Integration and Delivery (CI/CD) service in SAP Business
Technology Platform (BTP) is a set of tools and services that automate the
process of integrating code changes and delivering them to production
environments. This service enables development teams to streamline
and accelerate their processes by ensuring that new code changes are
consistently integrated, tested, and deployed with minimal manual effort.
All you need to do is connect your SAP S/4HANA system to this SAP BTP
service via the Cloud Connector.

Project "Piper"

Project "Piper" (https://www.project-piper.io/) is an open source initiative by SAP to provide a set of best practices and tools for implementing CCI/CD for SAP development projects for the SAP ABAP stack. The project offers templates, tools, and a framework for building CI/CD pipelines using either Remote Function Call (RFC) or Open Data Protocol (OData) to connect to the ABAP system.

The following are the key components of Project "Piper":

- **Pipelines as code**: Offers predefined pipelines that can be customized to suit specific project needs. These pipelines define the steps and stages involved in the CI/CD process.

- **Pipeline configuration**:

  - **Jenkins shared libraries**: Provides shared libraries for Jenkins, which can be used to configure and run CI/CD pipelines

  - **Extensibility**: Enables developers to extend and customize the libraries to add new functionalities or modify existing ones

- **Best practices**: Includes guidelines, templates, and examples to help developers implement best practices in CI/CD for SAP projects.

Use cases for Project "Piper" include the following:

- **SAP S/4HANA development**: Helps manage CI/CD for custom developments and enhancements in SAP S/4HANA.

- **SAP Fiori applications**: Supports the CI/CD process for developing and deploying SAP Fiori applications.

- **Integration projects**: Manages the CI/CD pipeline for integration projects involving multiple SAP and non-SAP systems.

Third-Party Tools

Several third-party tools are available that provide tools and features to build CI/CD pipelines for ABAP applications. By way of example, Basis Technologies (https://www.basistechnologies.com/) offers an automated DevOps and testing platform engineered for SAP, enabling customers to accelerate innovation, ensure continuous quality and delivery, and lower risk across even the most complex SAP landscapes. This platform supports the following functionalities:

- **Automated deployment**: Automatically deploy SAP transports (and even non-SAP changes) after approval.

- **Comprehensive analysis**: 60+ advanced analyzers automatically identify the impact of change, manage dependencies, enforce code quality, and perform other functions.

- **DevOps toolchain integration**: Out-of-the-box integrations with GitLab, Jira, ServiceNow, HP ALM, Eclipse IDE, and more are supported.

- **One-click deployment rollback**: The ultimate safety net against business downtime, the BackOut feature restores SAP production systems to their pre-import state following a deployment. Changes are reversed without having to redevelop, test, and approve fixes.

Side-by-Side Extensions: SAP CAP and SAP ABAP RAP

The SAP Cloud Application Programming Model (CAP) is a framework provided by SAP for building cloud-native applications and services. CAP is designed primarily for developing cloud-native applications using modern programming languages such as JavaScript (Node.js) and Java in SAP BTP by leveraging tools and libraries tailored for these languages. By leveraging CAP, developers can build modern, cloud-native applications that are aligned with SAP's strategy and ecosystem, ensuring compatibility and ease of integration with other SAP and non-SAP cloud solutions.

You can configure a CI/CD pipeline specifically designed for developing applications that adhere to the SAP Cloud Application Programming Model.

SAP ABAP RESTful Application Programming Model (RAP), on the other hand, is available with the SAP BTP ABAP environment and uses SAP ABAP development constructs for building cloud-ready applications and services. SAP RAP is applicable for uses cases either with a deep integration with the SAP S/4HANA system or those that leverage SAP ABAP's business logic.

The following list outlines considerations regarding when to use SAP CAP and when to use SAP RAP:

> **Programming language:** If you are building ABAP-based extensions and applications outside the core SAP S/4HANA system, you should use SAP RAP; if you are planning to build non-ABAP-based extensions using languages like Node.js and/or Java, you should use SAP CAP.

> **Developer skillset:** SAP CAP gives you the flexibility to use developers who are not SAP centric.

Business logic: If you need to reuse/leverage the complex business logic that SAP has provided in the core SAP S/4HANA system, you are better off using SAP RAP.

Like all architectural choices, there is no right or wrong choice; SAP CAP and SAP RAP each has its pros and cons.

Test Phase

As discussed in Chapter 1, testing in SAP is usually a prolonged and intricate process, primarily because even a minor change to a technical artifact results in a substantial overlap between various business processes within SAP. This necessitates testing across multiple business processes and obtaining sign-off from respective Business Process Owners.

Due to their expertise in SAP business processes, business users find themselves burdened with extensive testing responsibilities. However, this need not be the case, as testing can be automated. Third-party products from Tricentis, Worksoft, Basis Technologies, and others offer solutions for automating the testing process, reducing the involvement of Business Process Owners to cases where new test scenarios need to be recorded.

When dealing with changes impacting multiple business processes, it's essential to note that not all processes require testing. These third-party tools come equipped with AI-driven features that help identify and focus on the most at-risk processes, streamlining the testing effort rather than testing all impacted business processes indiscriminately.

Monitor and Operate Phase

In the Monitor and Operate phase for SAP from a DevOps perspective, continuous monitoring, tracing, and optimizing tasks like SAP System Refresh are essential for maintaining a high-performing, reliable and efficient environment.

Observability

Recall from Chapter 1 that the term *observability* was coined by R.E. Kalman, who defined it as follows: "A system is observable if its state can be inferred from measurements. If that system can be controlled and observed, it can be optimized." In modern terms, observability describes how well you can understand what is happening in a system, often by instrumenting it to collect metrics, logs, or traces. Hence, you'll often see monitoring, tracing, and logging described as the *three pillars of observability*.

Why Observability Matters?

Observability gives you the ability to efficiently detect, remediate, and investigate. This can and should improve your operational availability by reducing mean time to resolution (MTTR). This is why observability is so important. Now, while reducing MTTR is important, observability is about a lot more than resolving incidents, or even preventing them.

Observability can help you understand impact. For example, during an incident, if you can see how many people are affected by the issue, and if it is affecting any of your key business functionality, then you can make decisions about the impact of this incident. Understanding the impact lets you make decisions about how urgent the incident is, when you need to involve others, and who you need to involve. You don't want to be the one calling everyone out at 3 a.m. for something that 0% of your customers are impacted by.

Beyond incidents, observability is also about understanding the impact of changes that you make, so that you can make data-driven decisions.

On the technical side, consider a new deployment. Having good baseline measurements around the key parts of your customer experience enables you to see any impact your deployment has had, and by how much, good or bad. For example, has the deployment achieved the improvement you anticipated? On the business side, has the new customer-facing initiative been too successful and overloaded the application?

Having the right data, real time and historic, lets you make those data-driven decisions on both the technical side and business side. And when you can do this, and see the impact, you not only have happy colleagues but also can make decisions that make your customers happy.

Here are some of the drawbacks of lacking effective observability:

- Increased downtime

- High mean time to detection (MTTD)

- High mean time to resolution (MTTR)

- Challenges with scaling and consolidating multiple observability tools

- Siloed organizations and finger-pointing

- Financial repercussions

The Observability Maturity Model

AWS developed the AWS Observability Maturity Model as an essential framework because customers were asking AWS for prescriptive guidance.

Common challenges customers reported included the following:

- Having outdated tools

- Not having visibility of how systems are delivering business value

- Not getting the best from observability tools, or sometimes having no observability at all

369

Mostly, customers want to build observability practices and culture in their organization. As shown in Figure 7-6, the Observability Maturity Model helps customers evaluate where they are so that they can determine where they want to be and how to get there. As they expand their workloads, their observability is expected to mature.

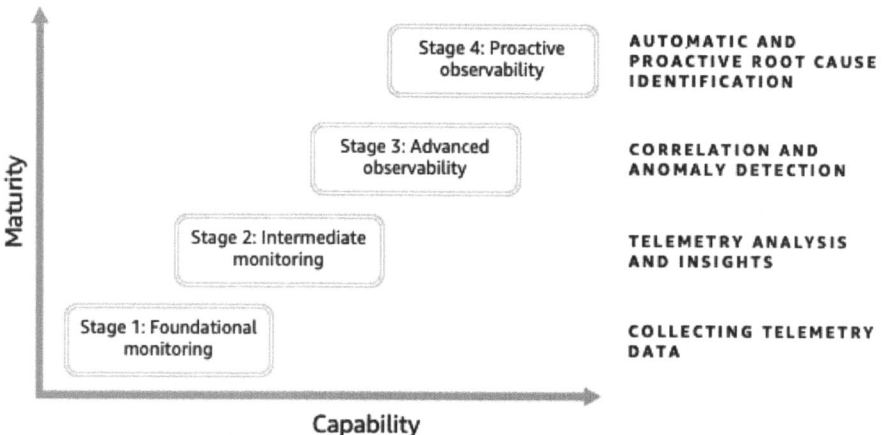

Figure 7-6. *Observability Maturity Model*

Detect, Investigate, and Remediate

Observability is all about the ability to detect, investigate, and remediate.

Detect

Often, customers don't detect issues as soon as they begin because (usually) there is a lag between when an issue starts and when they are notified. You want to reduce this lag as much as possible. Detection should be proactive and multifaceted (i.e., alarms on telemetry, synthetic testing, etc.). Anomaly detection is a key tool in the toolbox, as well as the ability to link together related alarms to reduce alarm fatigue. You can respond to failures quicker if you alert near the source of the telemetry, logs, and traces.

Investigate

Investigation is where SMEs spend the most amount of time during an operational event—it is the largest contributor to MTTR. Cutting through the chaos and understanding what to focus on is really important and remains a difficult task for many customers. Leverage logs, metrics, and tracing to help you investigate quickly to understand the root cause. Correlation across metrics, logs, and traces is key here. Your time is valuable and you need to ensure you are focusing on the information that matters during an operational event.

Remediate

Once you have identified the cause of a failure, you need to remediate it, which might be only a short-term fix. Don't forgot to do post-event analysis to determine how you could have prevented the failure in the first place. Your goal should be to ensure the same issue never happens again, but if it does, your results from post-event analysis will help you identify and remediate it automatically.

Observability with Amazon CloudWatch

Amazon CloudWatch is Amazon's default observability solution designed for scalability and availability, providing comprehensive visibility into cloud resources and applications. It enables the collection, monitoring, analysis, and action on data from AWS resources and on-premises servers. Amazon CloudWatch allows you to collect default metrics without action on your part. For example, Amazon EC2 instances automatically publish CPU utilization, data transfer, and disk usage metrics to help you understand changes in state.

How Amazon CloudWatch Works

Figure 7-7 shows how the Amazon CloudWatch agent collects information and sends it to CloudWatch logs and metrics. The agent collects log files specified as an individual file or group of files and then sends them to a log stream (by default, this is named using the instance ID) in a log group that you have configured. Metrics are collected using data from the operating system. These metrics are sent to a CloudWatch metrics Namespace (by default, this is named CWAgent) and you can configure dimensions to organize and aggregate your metrics.

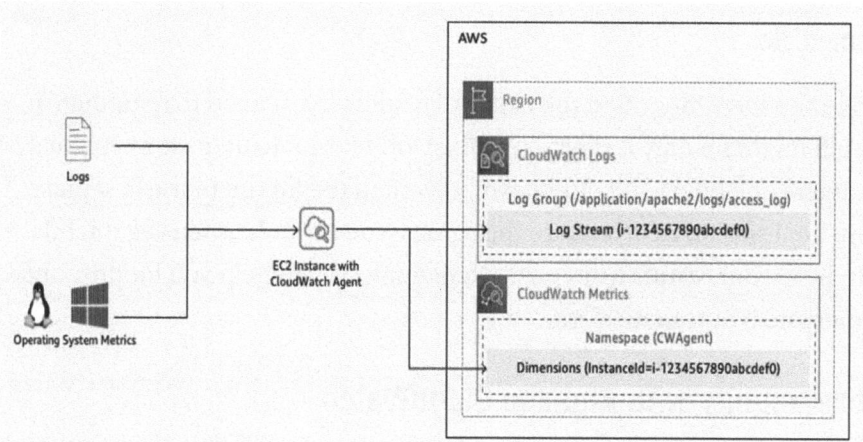

Figure 7-7. *How Amazon CloudWatch works*

Metrics are data about the performance of your systems. By default, many services provide free metrics for resources (such as Amazon EC2 instances, Amazon EBS volumes, and Amazon RDS DB instances). You can also enable detailed monitoring for some resources, such as your Amazon EC2 instances, or publish your own application metrics. Amazon CloudWatch can load all the metrics in your account (both AWS resource metrics and application metrics that you provide) for search, graphing, and alarms.

The Amazon CloudWatch agent allows you to collect additional metrics that are not available to the hypervisor, such as memory. On a server running Windows Server, installing the CloudWatch agent enables you to collect the metrics associated with the counters in Windows Performance Monitor. On a server running Linux or macOS, you can collect metrics about the CPU, disk space, disk IO, network throughput, memory, processes, swap usage, and more.

As shown in Figure 7-8, here are some of the ways you can visualize the data, enabling faster troubleshooting and performance analysis as well as take automated actions for faster response to events.

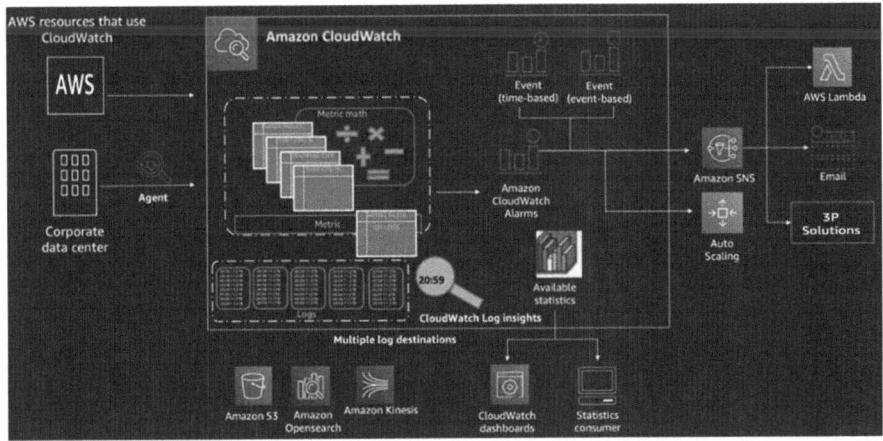

Figure 7-8. *End-to-end observability with Amazon CloudWatch*

- **CloudWatch Log Insights**: CloudWatch Log Insights is a feature within Amazon CloudWatch that allows users to query, analyze, and visualize log data from various AWS services and applications.

  - Explore, analyze, and visualize logs with ease, troubleshooting operational problems efficiently.

  - CloudWatch Logs Insights scales with log volume and query complexity, providing fast answers and billing based on queries run.

- Publish log-based metrics, create alarms, and
  correlate logs and metrics in CloudWatch
  Dashboards for comprehensive operational visibility.

- **Automate response:** You can automate response
  using CloudWatch Alarms and Actions, which is a
  feature in Amazon CloudWatch that enables users
  to automatically take specific actions in response to
  changes in metrics or log data.

  - You may set alarms and automate actions based
    on predefined thresholds or machine learning
    algorithms that detect anomalous behavior. For
    example, you can start Amazon EC2 Auto Scaling
    automatically or stop an instance to reduce billing.

  - Use Amazon EventBridge for serverless workflows,
    triggering services like AWS Lambda, Amazon
    Simple Notification Service (SNS), and AWS
    CloudFormation.

- **Operational visibility and insight:** Once all the
  statistics are available, you may use Amazon
  CloudWatch dashboards to visualize applications
  and infrastructure, correlate logs and metrics side
  by side, troubleshoot issues, and set alerts with
  CloudWatch Alarms.

  - Amazon CloudWatch provides automatic dashboards,
    real-time data with 1-second granularity, and up to 15
    months of metrics storage and retention.

  - Amazon CloudWatch performs metric math to
    derive operational and utilization insights, such as
    aggregating usage across an entire fleet of Amazon
    EC2 instances.

You can collect custom metrics from your own applications to monitor operational performance, troubleshoot issues, and spot trends. Amazon CloudWatch Application Insights uses these custom metrics to provide deeper insights into application performance.

Observability with Amazon CloudWatch AppInsights for SAP

Amazon CloudWatch Application Insights for SAP provides a centralized and unified view of your SAP application health across the operating system, database, and application layers. It consolidates logs, metrics, and events from various components, enabling you to monitor and troubleshoot issues from a single place. This comprehensive approach streamlines the identification and resolution of problems, enhancing operational efficiency and reducing MTTR.

SAP HANA Monitoring

Application Insights offers robust monitoring capabilities for SAP HANA databases. It collects and analyzes SAP HANA–specific metrics and logs, providing insights into database performance, health, and potential issues. Key features include

- Monitoring SAP HANA database performance metrics (e.g., memory usage, CPU utilization, disk I/O)

- Tracking SAP HANA database errors, alerts, and warnings

- Visualizing SAP HANA–specific metrics and logs in customizable dashboards

- Setting alarms and notifications for SAP HANA–related events and anomalies

SAP NetWeaver Monitoring

In addition to SAP HANA monitoring, Amazon CloudWatch Application Insights supports comprehensive monitoring of SAP NetWeaver application servers. It integrates with SAP NetWeaver to collect and analyze performance data, logs, and events, enabling you to

- Monitor SAP NetWeaver application server performance (e.g., work processes, queue lengths, response times)

- Analyze SAP NetWeaver-specific logs and error messages

- Visualize SAP NetWeaver metrics and logs in dedicated dashboards

- Set alarms and notifications for SAP NetWeaver-related issues and anomalies

Anomaly Detection Using AI/ML

Amazon CloudWatch Application Insights leverages advanced machine learning algorithms to detect anomalies in your SAP environment. It continuously analyzes metric patterns, logs, and events to identify deviations from normal behavior, enabling proactive issue detection and resolution. Key capabilities include

- Automatic baselining and anomaly detection for SAP-related metrics

- Identification of unusual patterns in logs and events

- Correlation of anomalies across multiple data sources (metrics, logs, events)

- Customizable sensitivity settings for anomaly detection

How Amazon CloudWatch Application Insights Works

At a high level, the operation of Amazon CloudWatch Application Insights can be divided into the following four distinct phases, as illustrated in Figure 7-9:

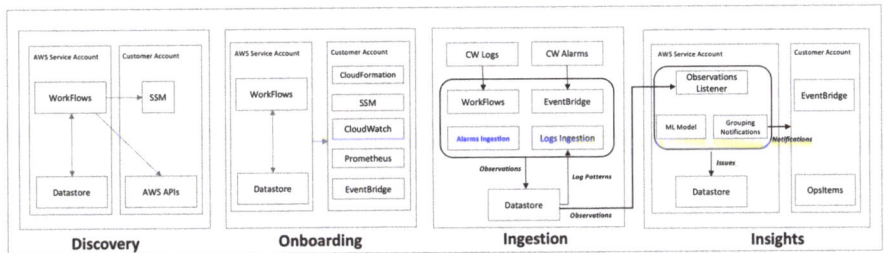

Figure 7-9. *How Amazon CloudWatch Application Insights works*

1. **Discovery**: In this phase, information about the workload hosted on the customer's AWS account is retrieved. The system discovers details about the workload, including the type of deployment, the resources used by the application, and more. This is achieved through the execution of AWS Systems Manager documents (SSM documents) and API calls.

2. **Onboarding**: During the onboarding phase, a series of workflows is executed to set up and configure monitoring agents, such as Prometheus and CWAgent, via AWS SSM Documents and AWS CloudFormation. This phase also involves configuring fixed thresholds and anomaly detection alarms based on recommended settings and setting up Amazon EventBridge rules to listen for relevant monitoring events.

3. **Ingestion**: This phase involves workflows that ingest logs and listen to alarms via Amazon EventBridge rules. Patterns are identified and captured as observations, which are then streamed to generate insights.

4. **Insights**: In this final phase, a listener monitors the observation stream from the ingestion engine. Using an ML model, the system groups observations into issues. These detected issues are then sent as SSM OpsItems, which are essentially records or tickets that represent issues or events that needs to be addressed. Additionally, Amazon EventBridge integration is enabled, to send notifications, remediation actions, and other automated responses. This is what we are going to discuss next.

Up to this point, from an observability perspective, we've discussed instrumenting and collecting data, but gathering data is merely the first step toward achieving observability. The ultimate goal is to improve your troubleshooting capabilities, make informed decisions, and accelerate innovation. Obtaining telemetry data is just the beginning. You need to comprehend the architecture, involve the human element to interpret the data, and use tools that analyze logs, metrics, and traces. Ultimately, your aim should be to transform data into information, gain knowledge and insights, and take action based on those insights. The next section discusses how to taken actions on the alerts.

Automated Actions on Amazon CloudWatch Alarms

As depicted in Figure 7-10, Amazon CloudWatch Alarms can send a notification to Amazon SNS, from which you can trigger an AWS Lambda function or push a message to Slack (https://slack.com/) or Amazon Chime via AWS Chatbot. This allows you to do almost anything, including

- Trigger an AWS Systems Manager Automation Runbook

- Resize an instance

- Send a message to Amazon Chime or Slack

- Respond with CLI commands

- Invoke disaster recovery

- Update security groups

- Automate deployments

- Instigate backups and snapshots

- Responding to security events

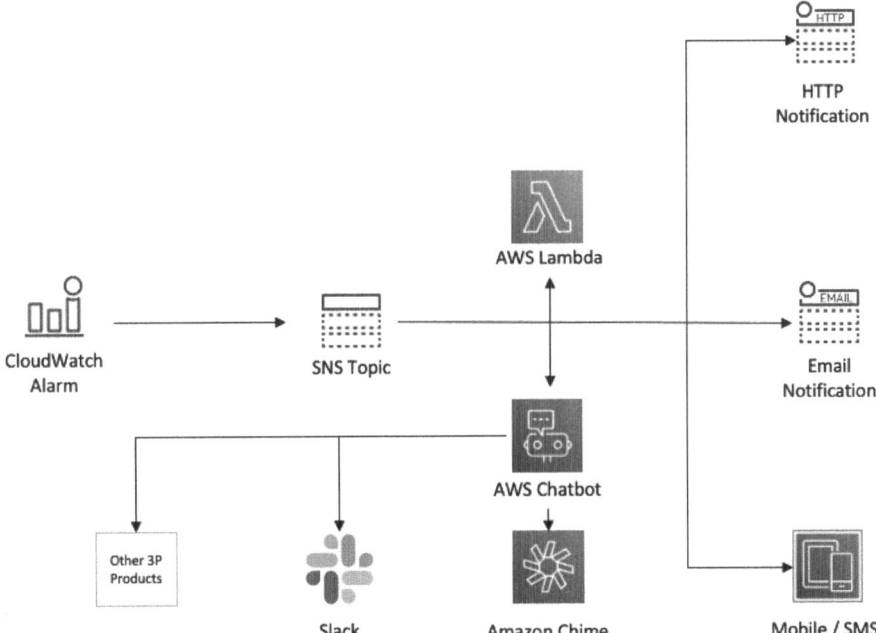

Figure 7-10. *Setting up automated actions and notifications with Amazon CloudWatch*

Integrating with Amazon EventBridge for Automated Incident Reactions

Amazon CloudWatch Application Insights seamlessly integrates with Amazon EventBridge, allowing you to automate incident response and remediation actions. When an issue is detected, Amazon EventBridge can trigger predefined workflows, such as:

- Invoking AWS Lambda functions for automated remediation tasks

- Sending notifications to incident management systems (e.g., PagerDuty, OpsGenie, and ServiceNow)

- Executing AWS Systems Manager Automation documents for incident resolution

- Scaling resources up or down based on detected anomalies

This integration enables you to respond to incidents quickly and efficiently, minimizing downtime and ensuring the reliability and performance of your SAP environment.

SAP Traces

The topic of observability would not be complete without an introduction to SAP traces. From an SAP on AWS perspective, you would use standard offerings from SAP to analyze traces. The following are some of them:

- **System Trace (ST01):** Used to trace authorizations, kernel functions, and SQL statements.

- **SQL Trace (ST05):** Used primarily for analyzing SQL statements and their performance.

- **Work Process and System-wide Process Monitoring (SM50/SM66):** Used to monitor and analyze the traces of work processes and system resources.

- **Web Service Runtime Trace (SRT_UTIL)**: Used to trace web service calls in SAP.

- **HTTP Trace (SMICM)**: Used to trace HTTP requests and responses in the SAP system.

- **ABAP Runtime Trace (SE30)**: Used to trace ABAP program execution and analyze performance.

- **Advanced End-to-End (E2E) Trace Analysis**: Used as a part of the SAP Focused Run offering.

- **SAP Cloud ALM Raw Data Outbound Traces API**: Used to read traces produced by the different SAP Cloud ALM monitoring use cases with the OpenTelemetry format. This API can be used to integrate data from an open source, an in-house tool, or a third-party tool by exporting SAP Cloud ALM data in the OpenTelemetry format. This is a good segue to our next topic.

Working with the Open Source Community

AWS has always collaborated with the open source community, and it is no different in the area of observability. At AWS, open source is a key part of innovation, with hundreds of engineers making ongoing contributions to open source projects. In today's cloud-enabled architectures, applications are typically an interconnected mesh of services—some of which are new cloud-native code and others of which are legacy back ends, third party, and open source—and understanding system performance across multiple components and tools becomes far more challenging.

AWS's open source–managed observability services can help you:

- Collect and query metrics using Amazon Managed Service for Prometheus (AMP), a fully managed Prometheus-compatible monitoring service for securely monitoring metrics at scale. You would typically install the Prometheus agents, which act as collectors, in the Amazon EC2 instances hosting SAP applications. Next, create AMP workspaces, which are logical spaces dedicated to storing and querying Prometheus metrics. Set up the ingestion of Prometheus metrics via collectors into these workspaces. Afterward, you can query the metrics using a service like Grafana or leverage AMP APIs.

- Improve application monitoring with interactive data visualization using AWS's prebuilt dashboards with Amazon Managed Grafana (AMG), based on the open source Grafana project. It's a managed Grafana service with rich, interactive data visualizations that makes it easy for developers, operators, and business leaders to monitor their SAP applications. Amazon Managed Grafana can connect with Amazon Managed Service for Prometheus to enable you to query, visualize, and alert on your metrics, logs, and traces.

- AWS Distro for OpenTelemetry (ADOT) is a secure, production-ready, AWS-supported distribution of the OpenTelemetry project, which is part of the Cloud Native Computing Foundation. With AWS Distro for OpenTelemetry, you can easily instrument your

applications to send correlated metrics and traces
to multiple monitoring solutions, including Amazon
CloudWatch and AWS X-Ray, and partner solutions,
including Datadog, New Relic, Splunk, and others.

How is open source applicable to SAP applications? SAP BTP, Kyma
runtime offers a fully managed cloud-native Kubernetes application
runtime based on the open source Kyma project (`https://kyma-project.`
`io/#/`). Utilizing modular building blocks, SAP BTP, Kyma runtime
includes all the essential capabilities to simplify the development and
operation of enterprise-grade cloud-native applications. It provides a
platform for extending SAP applications with custom functionalities and
services, and includes built-in connectors and APIs for integration with
other SAP and non-SAP systems. AWS Distro for OpenTelemetry can
be integrated with SAP BTP, Kyma runtime to enhance observability in
cloud-native applications. As previously mentioned, AWS ADOT provides
a secure, production-ready distribution of the OpenTelemetry project,
enabling the collection of metrics and traces from applications. The
ADOT Collector gathers telemetry data from applications, processes it,
and exports it to AWS services such as AWS X-Ray for traces and Amazon
CloudWatch for metrics. AWS X-Ray can be used to visualize traces and
understand the flow of requests through your applications, while Amazon
CloudWatch monitors metrics, sets up dashboards, and creates alarms to
notify you of issues.

Now, let's dive into some of the typical challenges of the operate phase
in an SAP environment.

SAP System Refresh

The SAP system refresh process is a highly complex, time-consuming,
error-prone, and downtime-inducing activity. Traditional refresh methods
often require taking the SAP QA system offline, leading to disruptions

in projects and productivity losses. To address these challenges, AWS provides a serverless solution that automates the refresh process while achieving near-zero downtime.

As illustrated in Figure 7-11, the solution leverages AWS Step Functions, AWS Lambda, and other AWS services to orchestrate and execute the SAP system refresh process in a seamless and automated manner. Here's how it works:

1. **Orchestration**: AWS Step Functions acts as the orchestrator, managing the entire refresh workflow. It has the ability to execute complex activities and invoke AWS Lambda functions in a sequential order, which aligns perfectly with the sequential nature of the refresh process.

2. **Automation:** An AWS Lambda function starts AWS Systems Manager Documents that launch an Amazon EC2 instance from an Amazon Machine Image (AMI). Another Lambda function restores the backup by copying the backup files from an Amazon S3 bucket to the Amazon EC2 instance that hosts the SAP database. After the database restoration is complete, an Amazon Lambda function performs post-copy tasks, like importing transports, running BDLS, and so on.

3. **Network isolation**: The refresh process takes place in a new, isolated QA environment within a network bubble. This approach ensures that the existing SAP QA system remains operational and unaffected during the refresh.

4. **Data storage and messaging**: Various AWS services are utilized for data storage, messaging, and configuration management:

- Amazon DynamoDB stores the progress of each step related to the refresh process.

- Amazon SNS sends notifications during the refresh workflow.

- An Amazon S3 bucket stores any necessary files or artifacts.

- AWS Systems Manager Parameter Store securely stores values of required variables, such as the target hostname, target SAP SID, and so on.

5. **Network/DNS switch**: Finally, the IP addresses are switched using DNS, redirecting traffic to the new SAP QA environment, effectively achieving zero downtime during the refresh process.

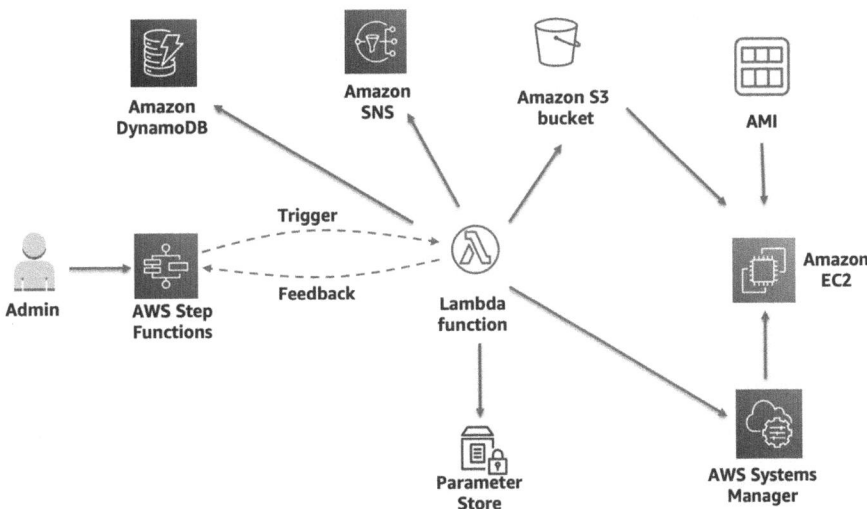

Figure 7-11. *SAP system refresh automation*

By leveraging AWS serverless services, this solution automates the complex SAP system refresh process, reducing manual effort, minimizing errors, and eliminating downtime. It provides a scalable, reliable, and cost-effective approach to managing SAP system refreshes in the cloud.

SysOps for SAP

SysOps for SAP on AWS refers to the tasks and responsibilities related to managing and operating SAP applications and underlying infrastructure on AWS. It involves various administrative and operational activities to ensure that the systems are running efficiently, securely, and reliably. AWS emphasizes creating automatable and repeatable processes, which is a key principle in SysOps on AWS.

Key focus areas include

- Automated deployments

- Observability

- Automated operations

- Security and compliance

- Business continuity

We have already discussed automated deployments and observability in great detail in the previous section, "DevOps for SAP," so in this section we'll focus on the latter three topics.

Automated Operations

Figure 7-12 illustrates all the services available to automate operational activities across the stack. Some of these services have already been discussed, and others will be covered later. In this section, we focus on AWS Systems Manager, which is the most powerful and essential service for automating operations.

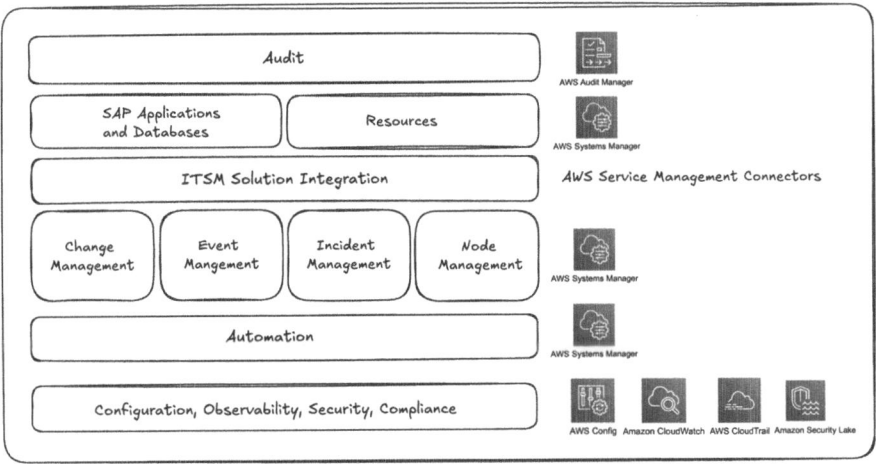

Figure 7-12. *Automated operations*

AWS Systems Manager

AWS Systems Manager provides visibility and control over your AWS infrastructure through a unified user interface. It consolidates operational data from various AWS services, allowing you to monitor and troubleshoot effectively. You can automate operational tasks across your AWS resources, manage resource and application operations, and quickly address issues without the operational overhead.

Key features of AWS Systems Manager include the following:

- Reduces the time needed to detect problems

- Includes native AWS automation commands and documents (some predefined by AWS)

- Executes commands and scripts remotely on Amazon EC2 instances

- Enhances visibility and control over resources

- Manages hybrid environments effectively

387

- Ensures security and compliance

- Facilitates patching with or without scheduled maintenance windows

- Offers AWS Systems Manager Parameter Store for secure configuration storage

- Functions as a state manager to maintain the desired state of systems

AWS Systems Manager for SAP (AWS SSM4SAP)

AWS Systems Manager for SAP helps customers manage AWS resources that are used for hosting SAP applications with SAP application awareness. It provides seamless integration between AWS services and your SAP applications, enabling efficient management and operations.

If you navigate to the AWS Console and then to AWS Systems Manager ➤ Documents and search for AWS SSM documents using search criteria **SAP**, you'll find the SSM documents specific to SAP, ready for your use. The following are some examples:

- **AWSSystemsManagerSAP-Discovery**: Discovers SAP metadata on an EC2 instance

- **AWSSystemsManagerSAP-CheckComponentStatus**: Checks the status (running, stopped, etc.) of an SSM for SAP component

- **AWSSystemsManagerSAP-CreateDLMSnapshotForSAPHANA**: AWS Dynamic Lifecycle Manager (DLM) script for SAP HANA databases

- **AWSSAP-InstallBackintForAWSBackup**: A composite document for installing and configuring AWS Backint Agent

- **AWSSystemsManagerSAP-BackupHanaDatabase**: Backs up the specified SAP HANA database

- **AWSSystemsManagerSAP-BackupJobIdRenewal**: Renews continuous job IDs on an EC2 instance

- **AWSSystemsManagerSAP-DeleteHanaBackupFromCatalog**: Deletes the specified SAP HANA backup from the backup catalog

Now let's cover a few use cases wherein you may use AWS Systems Manager to build solutions:

- Automating the start and stop of SAP instances

- Patching SAP HANA databases and OS

- Automating SAP configuration health checks on AWS

- Autoscaling for SAP

Use Case 1: SAP Start and Stop Automation Using AWS Systems Manager

Managing the start and stop processes for distributed SAP landscapes can be a complex and time-consuming task, especially when dealing with multiple instances and components. Manually starting and stopping these systems can be error-prone and lead to potential downtime or inconsistencies.

Solution Overview

AWS Systems Manager for SAP provides a centralized and automated approach to starting and stopping SAP instances. AWS Systems Manager contains multiple features. The ones applicable in this use case are Run Command and State Manager. Run Command lets you run documents against a managed server. Documents are a simple way of running administrative tasks such as executing scripts, installing patches, or performing software installations. State Manager is a secure and scalable configuration management service that automates the process of maintaining the configuration state for your managed resources.

The solution works by defining automation documents (runbooks) that encapsulate the necessary steps and sequences for starting or stopping the SAP landscape. These runbooks can be executed on demand or scheduled based on your requirements, ensuring that the instances are started and stopped in the correct order and with the appropriate dependencies handled.

AWS Systems Manager communicates with the Amazon EC2 instances hosting the SAP components through secure channels, eliminating the need for manual intervention. This automation not only saves time and reduces operational overhead but also minimizes the risk of human errors during the start and stop processes.

Use Case 2: SAP HANA Database and OS Patching Using AWS Systems Manager

Keeping SAP HANA databases up to date with the latest patches and security updates is crucial for maintaining system stability, security, and compliance. However, the patching process can be complex, time-consuming, and prone to errors, especially in distributed environments with multiple SAP HANA instances.

Solution Overview

AWS Systems Manager for SAP provides automation capabilities to streamline the SAP HANA database patching process. It leverages features like patch baselines and automation documents to automate the patching workflow, reducing manual effort and minimizing potential errors.

The solution involves defining automation documents that encapsulate the necessary steps for patching the SAP HANA database, such as stopping dependent applications, taking backups, applying patches, and validating the patched system. These automation documents can be executed on demand or scheduled based on your patching requirements.

AWS Systems Manager communicates with the Amazon EC2 instances hosting the SAP HANA databases, ensuring that the patching process is carried out consistently and in the correct order across all instances. It also provides logging and reporting capabilities, allowing you to monitor the patching progress and troubleshoot any issues that may arise. Detailed information, along the with step-by-step process, is available at `https://docs.aws.amazon.com/sap/latest/sap-hana/automated-patching.html`.

Similarly, the OS patching automation solution is available at `https://docs.aws.amazon.com/sap/latest/sap-netweaver/automation-os-patching.html`.

By automating the SAP HANA database and OS patching process with AWS Systems Manager, you can significantly reduce the time and effort required for patching, minimize downtime, and ensure that your SAP HANA databases and operating systems are always up to date with the latest patches and security updates.

Use Case 3: Automating SAP Configuration Health Checks on AWS

AWS has published a Solution Guidance designed to help customers track over 100 configuration checks and identify any drifts from AWS best practices (see https://aws.amazon.com/solutions/guidance/automating-sap-configuration-health-checks-on-aws/). This solution can evaluate SAP configurations against AWS best practices. It empowers organizations to identify any deviations, ensuring that their SAP workloads run on secure, high-performing, resilient, and cost-efficient infrastructure. With this Guidance, SAP provides a framework to add new checks, regularly assess their SAP workload configurations, identify high-risk configuration drifts, and quickly resolve issues. Furthermore, the solution maintains a comprehensive history of changes, enabling organizations to audit the configurations changes over a period of time.

Solution Overview

This solution is published as a framework using services like AWS Lambda and AWS Systems Manager, based on the recommendations from SAP Lens for AWS Well-Architected Framework, which is a collection of best practices and design principles for running SAP on AWS.

- Supports 100+ checks on SAP HANA database and application servers

- Identifies any best practices that are not compliant and sends a notification via e-mail

- Provides customers the option to deploy an Amazon QuickSight dashboard

AWS has published an SAP systems inventory template and an AWS CloudFormation template, along with detailed instructions, at the following GitHub repository: https://github.com/aws-solutions-library-samples/guidance-for-automating-sap-configuration-health-checks-on-aws.

Customers can take advantage of these templates and follow the below steps to implement. If they intend to make changes in the architecture of the solution, which is shown in Figure 7-13, they would need to customize the AWS CloudFormation template.

1. Create an Amazon S3 bucket. Download the SAP systems inventory template from the GitHub repository and update it with the provided example SAP workload inventory.

2. Launch the AWS CloudFormation template from the GitHub repository with the input as the Amazon S3 bucket. AWS CloudFormation will deploy the AWS resources. Upload the SAP inventory template to the Amazon S3 bucket.

3. Run SAP health checks by executing the Amazon Lambda function on-demand, or schedule it periodically using Amazon EventBridge.

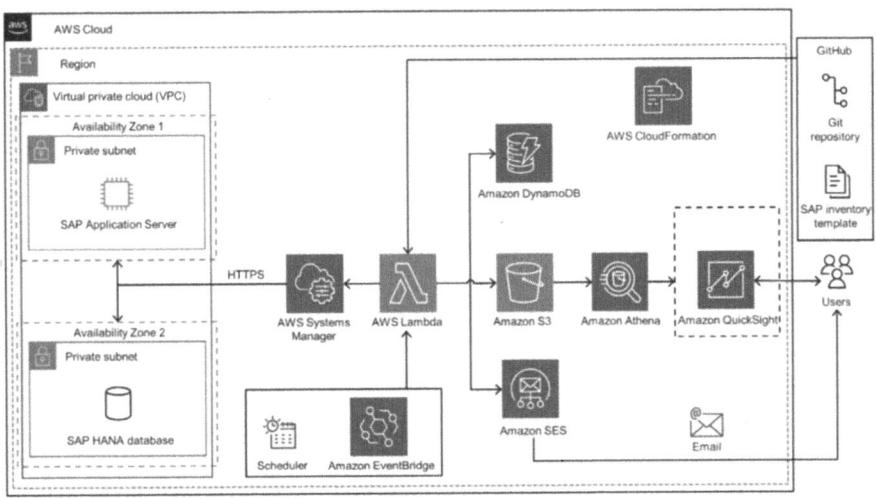

Figure 7-13. *Automation of SAP health checks*

The Amazon QuickSight dashboard based on the health checks
provided in the template looks like the image shown in Figure 7-14.

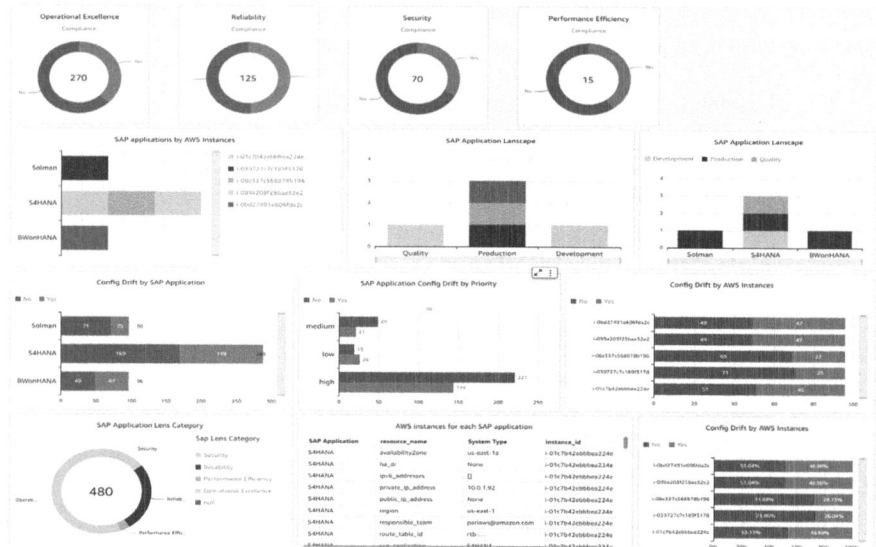

Figure 7-14. *Dashboard for SAP health checks*

This solution provides a single pane of view of the entire SAP
landscape with aggregated information on configuration drifts and
anomalies against SAP and AWS best practices for all SAP systems across
multiple AWS accounts.

Use Case 4: Autoscaling for SAP

In an SAP system, there is a need to add or remove SAP application servers
dynamically based on the system load. For example, during month-end
or year-end operations, the system load is typically higher, and more
application servers may be required to handle the increased workload.
Conversely, when the load decreases, the additional application servers
can be shut down to optimize resource utilization and costs.

Solution Overview

The solution, which is shown in Figure 7-15, leverages various AWS services to monitor the SAP system's workload and automatically scale the application servers accordingly. Here's how it works:

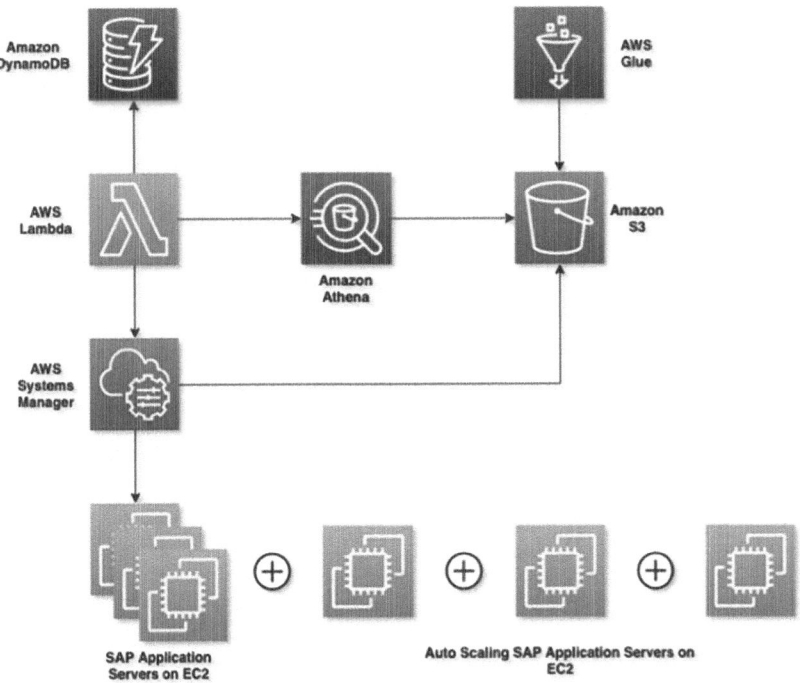

Figure 7-15. *Autoscaling for SAP*

1. **Monitoring:** Every minute or so, start an AWS Lambda function, whose first task is to check the number of idle and occupied work processes across the SAP system by using an AWS SSM document. The collected data is then pushed to Amazon S3. You may use Amazon EventBridge to start this time-based event.

2. **Data processing**: The AWS Lambda function determines whether an additional application server is needed to be spun up or scaled down. This is done using AWS Glue and Amazon Athena. AWS Glue, a fully managed extract, transform, and load (ETL) service, transforms the data and then using Amazon Athena, the number of idle work processes is determined.

3. **Scaling decision**: Based on the above data, if the number of idle work processes falls below a predefined threshold, the solution using AWS Systems Manager triggers the launch of an additional SAP application server instance to handle the increased workload.

4. **Scale down**: When the load decreases and there are enough idle work processes available, the solution initiates a soft shutdown of the additional application server instances. This ensures that user activities are not disrupted during the scale-down process. The function of the DynamoDB is to maintain the names and current states of all SAP application servers. For example, once a particular SAP application server is shutdown, the state is recorded back in the Amazon DynamoDB; ensuring that the next time AWS Lambda function runs, it is aware that not all SAP Application Servers are running and if needed which SAP application server needs to be started back up.

By leveraging AWS services such as AWS SSM, Amazon DynamoDB, AWS Lambda, Amazon S3, and AWS Glue, this custom-built solution enables dynamic scaling of SAP application servers based on the system's workload. It helps optimize resource utilization, reduce costs, and maintain performance during periods of high demand. Now, let's move on to the next focus area of SysOps for SAP on AWS, security and compliance.

Security and Compliance

Ensuring security and compliance for SAP workloads on AWS is of great importance due to the critical nature of the data and processes managed by these systems. SAP environments often handle sensitive business information and financial data, making them prime targets for cyber threats. Robust security measures and compliance frameworks are essential to protect this data from unauthorized access, breaches, and other security incidents. Leveraging AWS services like AWS Config, AWS CloudTrail, and AWS Security Lake provides continuous monitoring, detailed logging, and centralized security management, which are vital for maintaining a secure and compliant infrastructure.

Let's cover each of these services in further detail.

AWS Config

Manual monitoring of the configurations of AWS resources, such as EC2 instances, security groups, and Amazon S3 buckets, to understand configurations and changes, is impractical. This task can be automated using AWS Config, which provides automated compliance checks and remediation. As shown in Figure 7-16, AWS Config stores a comprehensive view of resource configurations and their changes over time by capturing configuration snapshots. AWS Config allows users to define rules that represent these best practices, compliance requirements, or operational policies, and can also help identify and automatically remediate resources that don't comply with these rules.

397

With AWS Config, you can review configuration changes and relationships between AWS resources, access detailed configuration histories, and assess overall compliance with your internal guidelines. This simplifies compliance auditing, security analysis, change management, and operational troubleshooting. AWS Config can retain this data for up to seven years and integrate with your IT service management (ITSM) systems.

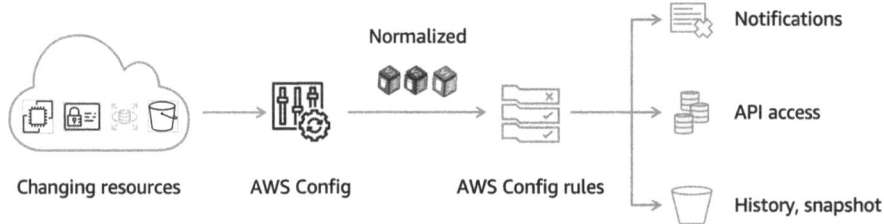

Figure 7-16. *How AWS Config works*

Now, if you are wondering how you are going to build the AWS Config rules, don't be concerned, because AWS delivers hundreds of predefined AWS Config Managed Rules. Here are some examples of AWS Config Managed Rules relevant for SAP:

- Making sure Instance Detailed Monitoring is enabled to comply with SAP Support requirements

- Confirming that all Amazon EBS volumes are encrypted

- Making sure Amazon EBS snapshots are not publicly accessible

However, you also can build your own AWS Config custom rules. Examples of these custom rules which are relevant for SAP are,

- Verifying that all SAP instances are utilizing approved Amazon Machine Images (AMIs)

- Ensuring AWS SSM Agent is running on all SAP instances

- Ensuring the Amazon S3 bucket holding SAP production backups has S3 Object Lock enabled, encryption enabled, cross-region replication enabled, and active lifecycle configuration rules

You may also go a step further and use other AWS services to build a solution that can audit your SAP profile parameters as well as SAP HANA and OS parameters. Let's look at a possible solution you can build.

Use Case: Auditing Your SAP System on a Continuous Basis to Stay Compliant

In this solution, you can build an AWS Lambda function that will access the existing bastion host, which in turn will talk to the SAP instances in a private subnet to collect the required parameter value. This value is evaluated with AWS Config to be either compliant or noncompliant. Figure 7-17 depicts the high-level architecture. Please note this diagram doesn't cover all aspects of the solution but only a representation of essential elements within it.

This solution is described in detail in the blog https://aws.amazon.com/blogs/awsforsap/audit-your-sap-systems-with-aws-config-part-ii/.

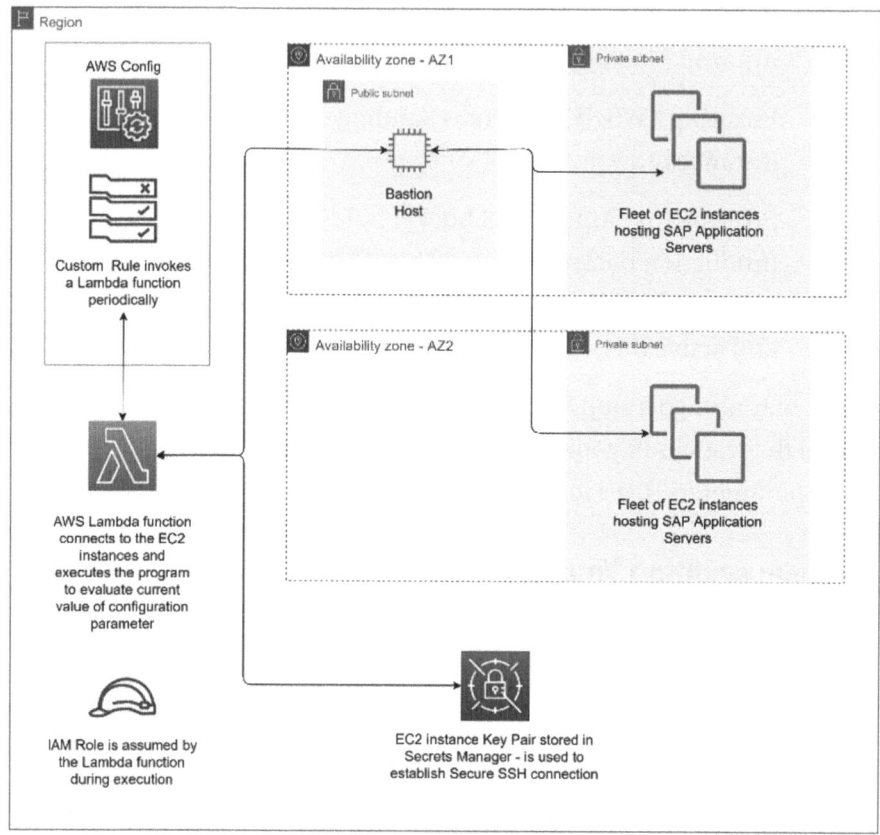

Figure 7-17. *Automation of SAP audit checks*

This solution is just an example; you may build your own based on your own preferences. Now, let's look at AWS CloudTrail, another service that helps to ensure security and compliance for SAP workloads on AWS.

AWS CloudTrail

AWS CloudTrail is a service offering governance, compliance, and auditing capabilities for your AWS account. It is primarily used to track user and resource activity across your AWS infrastructure, supporting governance and auditing efforts. AWS CloudTrail can also identify and respond to unusual usage through automated analysis. AWS CloudTrail can also track critical changes such as pipeline updates, deletions, and creations, and sends notifications to the DevOps team for auditing purposes.

AWS CloudTrail is used for

- Compliance aid

- Visibility into activity

- Anomaly detection

- Data exfiltration detection

- Automated security analysis

- Analysis of permissions

- Detection of unusual activity

Primary use cases include

- Identifying users logged in during an incident, and all actions taken during that period

- Notifying when admin privileges are granted or console logins occur

- Detecting access from authorized networks or IP addresses

- Identifying when a file or object has changes to public access, and who made the change

- Understanding top user, role, or service callers of API calls or Amazon Lambda functions

- Complying with internal and regulatory compliance requirements with the immutable history of all AWS activities

Typically, in a large organization, there are many activities across various AWS accounts that are recorded in multiple locations, such as AWS CloudTrail, AWS Config, and even non-AWS sources. This fragmentation makes it challenging for humans to correlate and analyze the data. This is where AWS CloudTrail Lake comes into play, discussed next.

AWS CloudTrail Lake

AWS CloudTrail Lake provides an out-of-the-box turnkey solution that, with just a few clicks in the console, sets up a centralized store for viewing and querying all the data. It is a managed audit and security lake with several benefits: it has immutable storage; it has advanced analytical capabilities built in, like SQL querying and dashboards for visualization; it is multisource, meaning you can consolidate activity events from AWS and sources outside AWS; and it supports AWS multi-region and multi-accounts.

There is another AWS service called Amazon Security Lake. It automates the sourcing, aggregation, normalization, and data management of security data across your organization into a security data lake stored in your account. It makes your organizational security data broadly accessible to your preferred security analytics solutions, supporting use cases such as detection, investigation, and forensics. This service is particularly beneficial for security operations teams or SOC (Security Operations Center) teams, as it allows them to analyze a wide range of data, including infrastructure and networking logs, alongside Amazon CloudTrail logs. However, for audit and compliance teams focusing specifically on user activity, API call details, and resource activity from a compliance perspective, Amazon CloudTrail Lake is more appropriate. Table 7-1 summarizes the difference in use cases.

Table 7-1. *Differences Between AWS CloudTrail Lake and AWS Security Lake*

	CloudTrail Lake	Security Lake
Use cases	• Meet Audit and compliance needs for immutable storage for up to 7 years and proof of immutability • DevOps/ application/ support teams for operational troubleshooting on individual member account logs or organization level logs • Streamlined security/compliance investigations with built-in SQL query engine of lake	• Central security teams using partner SIEM solutions (using OCSF schema for security monitoring and investigations
Data sources available	• Auditable data sources such as CloudTrail, Config, Audit Manager evidence, 3rd party audit events(Beta)	• Security logs from AWS (CloudTrail, VPC flow logs, Route53, WAF, Security Hub findings etc.) and participating partner solutions (Okta, Crowdstrike etc.)
Who can create the Lake?	• Central teams such as security or compliance operations on organization level data using management/delegated admin account • Individual applications teams on data specific to their accounts from member accounts	• Central security team at an organization level using delegated admin account
How to analyze logs?	• Using built-in SQL query interface of CloudTrail Lake , and CloudTrail Lake dashboard	• Using security data analysis tools of your choice (Amazon Athena, Splunk, Sumologic etc.)

Business Continuity

AWS CTO Werner Vogels famously said, "Everything fails all the time." To safeguard our SAP systems from various types of failures—whether physical, logical, or other disasters—we need to implement robust protection measures.

Before diving deep into disaster recovery (DR), it's crucial to understand that high availability (HA) and DR take on different meanings in the context of on-premises versus cloud environments. Historically, HA and DR have specific connotations, but cloud services, especially AWS, have redefined these terms. Typically, on-premises HA/DR is not equivalent to cloud availability, as explained next.

Traditionally, on-premises HA meant protection against hardware failure within the same data center. However, AWS inherently provides features like Auto Recovery, which automatically shifts workloads to another host if there are hardware, software, or network issues. Therefore, setting up HA for these types of events is unnecessary on AWS. Similarly, traditional DR involves switching to another data center in a different geographic location in case of a primary site disaster, focusing on protection against a single data center or geographical location failure. AWS Availability Zones (AZs) inherently provide these protections. By setting up your SAP systems across multiple AZs, you ensure geographic protection and protection against a single data center failure.

So, what are Availability Zones? AZs are clusters of data centers with redundant power, networking, and connectivity, housed in separate facilities and different flood plains. Many AWS services are architected in multi-AZ mode by default.

What does this mean for you? It means you need to think broadly about what you're protecting your systems from—whether physical errors, logical errors, or disasters. You must consider the agreed-upon recovery point objective (RPO) and recovery time objective (RTO) around which you need to architect your systems, ensuring it is cost-effective. For instance, does it make sense to run a hot standby database if you can tolerate a few hours of outage?

Many customers set up systems within a single region across multiple AZs, viewing this as both HA and DR due to the separate physical locations and power grids, and multiple data centers protecting their data. Other customers may have compliance requirements to set up their DR site a certain distance away, such as 200 miles, or to protect against an entire AWS region's disruption. In such scenarios, they use a separate AWS region as the DR site.

We will not talk about the architecture patterns available for HA and DR because there is a lot of documentation available online on the patterns, particularly at `https://docs.aws.amazon.com/sap/latest/sap-hana/hana-ops-patterns.html`.

Instead, we'll focus on how to modernize and automate your DR drills. High availability is already an automated process managed by existing HA clustering software, so there is no need for modernization in this area. For DR, you may end up using a combination of technologies like AWS Elastic

Disaster Recovery (AWS DRS), native database replication, and so forth. With the power of AWS, you can automate the orchestration of different steps across technologies.

Automation of DR Runbook

In terms of execution of the overall disaster recovery, it can get complex, with a lot of dependencies, steps, and technologies, because DR is not just about bringing up the SAP systems. AWS DRS provides inherent automation in terms of recovery of the instances, and you may also anchor additional scripts to added requirements. However, when examining an existing DR runbook or attempting to create one, you'll find it involves multiple steps across various technologies. For instance, a section of the DR runbook for conducting DR drills might include the following steps:

- Disable monitoring of SAP production and other non-SAP/'Friends of SAP' applications to prevent alert notifications.

- Stop SAP production.

- Stop other non-SAP applications.

- Initiate recovery of SAP application servers using AWS DRS.

- Initiate takeover of SAP HANA.

- Initiate recovery of non-SAP servers using AWS DRS.

- Start SAP.

- Start non-SAP applications.

- Establish integration/connectivity between SAP and non-SAP applications.

- Perform verification steps.

405

The following are some challenges of the overall process:

- **Complexity**: SAP DR typically involves intricate steps across multiple layers (database, application, OS, network, storage).

- **Human error**: Manual execution of complex procedures is prone to mistakes, potentially leading to business downtime.

- **Time efficiency**: Manual processes can be time-consuming, impacting overall recovery time.

One of the SAP on AWS customers began with a simple automation process and then continued to expand it. They started by automating the manual steps they were performing in the AWS console to initiate the AWS DRS recovery and start SAP in the launched instances. They created a workflow in AWS Step Functions with two basic steps: calling the AWS DRS StartRecovery API and calling a SSM Document to mount Amazon EFS and start the SAP application. Once they accomplished this, they began adding other steps. We recommend taking a similar incremental approach.

Figure 7-18 presents the high-level architecture of a solution that you may use to orchestrate the complex disaster recovery process. This solution will help you to manage some of the DR challenges previously discussed. Note that this is one possible solution we've developed; you can build your own based on what makes sense for you. Figure 7-19 shows the details of the DR automation solution. The solution might seem overwhelming at first, so it's advisable to start small and gradually add steps to build out the entire solution over time.

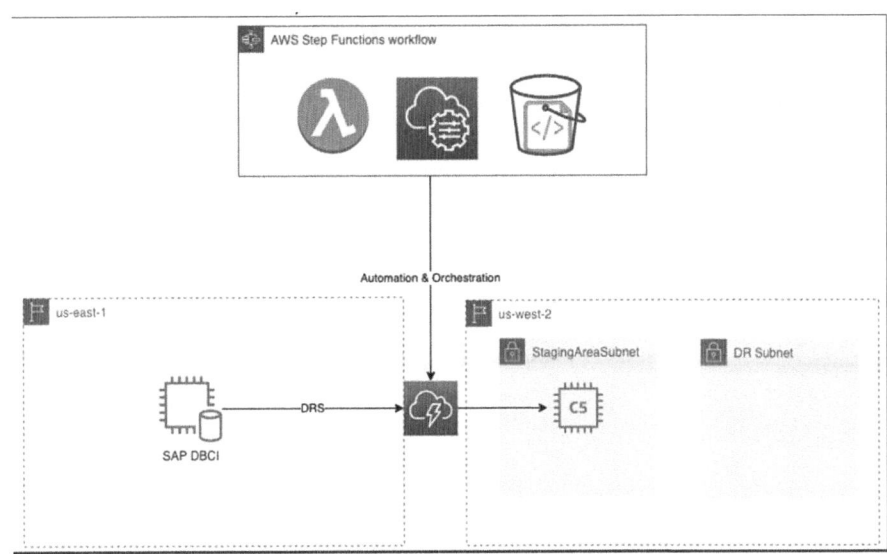

Figure 7-18. *Architecture of DR automation solution*

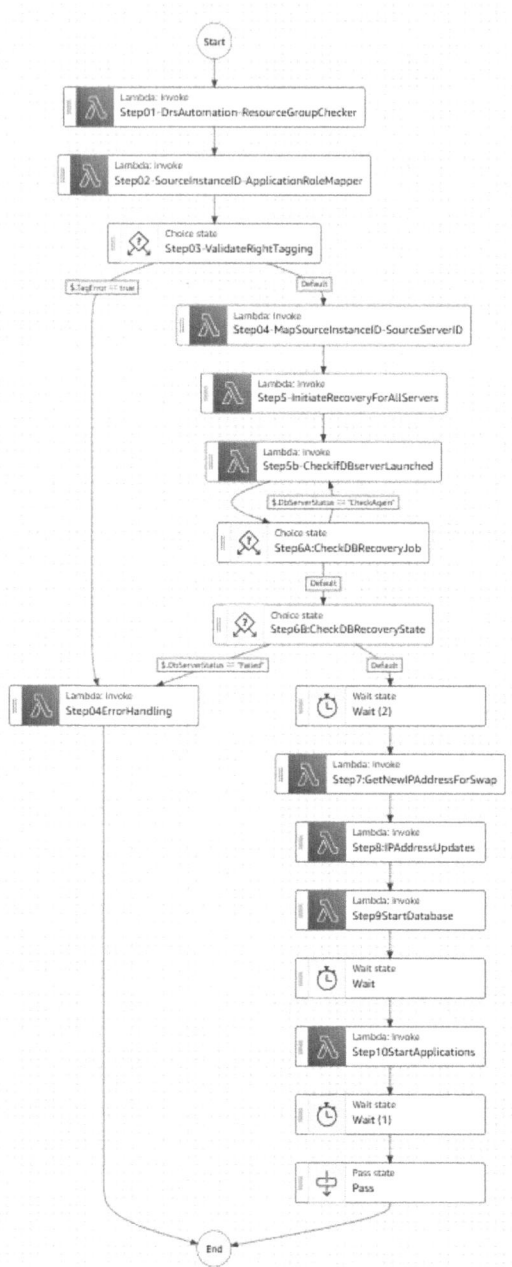

Figure 7-19. *Details of the DR automation solution*

Solution Overview

The automation solution architecture presented here leverages AWS Step Functions as the central orchestration service, coordinating a series of AWS Lambda functions and AWS Systems Manager commands to execute various DR tasks:

1. **Workflow orchestration**: AWS Step Functions manages the complex flow of DR processes.

2. **Task execution**: AWS Lambda functions handle specific DR steps, including:

   - SMM Documents and OS-level scripts

   - SAP application-level commands

   - Database operations

   - AWS API actions (e.g., DNS updates, AWS DRS actions via APIs like Elastic Disaster Recovery: StartRecovery)

   - Systems Manager facilitates execution of commands on target systems

The following are the prerequisites for this automation solution:

1. Create a resource group that consists of your primary SAP system. Make sure to add all the servers that belong to a single SID to the resource group name. If you have already created a resource group as a best practice or if you have used the AWS Launch Wizard for SAP and the resource group already exists, you may use the same instead of creating a new one.

2. Assign a tag to each Amazon EC2 instance in your resource group to indicate the role of that server in within that SID. For example, you may assign "DRSKEY":"DBSERVER" to your database server, "DRSKEY":"MSSERVER" to your SAP ABAP SAP Central Services (ASCS), and so on. This will help the automation to execute the respective start and stop commands later.

3. While invoking the automation, you will pass this resource group name that you have created in the first step.

How the Automation Workflow Works

A typical DR automation workflow for an AWS DRS bubble test might include

1. Precheck validations

2. AWS DRS initiation in an isolated staging area

3. Monitoring of DR process progression

4. SAP database and application startup sequence

5. Network and DNS adjustments

6. Verification of system availability

The workflow includes error handling and notification mechanisms, ensuring administrators are alerted to any issues during the process. If any of the steps fail, execution is stopped and notification is sent to the administrators.

Now, lets go through the ten steps in the workflow, which is depicted in Figure 7-19.

Step 1: This step is a prerequisites check step. It's an AWS Lambda function built with Python and boto3 (a Python package to call AWS APIs). The automation works by getting the resource group name as an input (from the input presented in the previous section). It gets the resource group name and checks if the given group name exists. If the group exists, it proceeds or it fails.

Step 2: This AW Lambda + Python + boto3 step checks each server in the group and sees if it has the right server role mapped based on the DRSKEY tag. If any of the servers are missing the mapping, it assigns an error message in the output.

Step 3: This step checks if the resources are correctly tagged. If there is no error, it passes to step 4; otherwise, the error is handled.

Step 4: This AWS Lambda + Python + boto3 step does the heavy lifting of checking the AWS DRS replication status and mapping the source and replication servers. It identifies the respective source and target ID and ensures AWS DRS is in ready state to do a DR takeover. if anything fails, it stops or continues to step5.

Steps 5 and 6: These AWS Lambda + Python + boto3 and AWS Step Functions logic steps initiate the recovery, which means the AWS DRS servers start to build the servers in the secondary site. These steps also monitor the AWS DRS job to completion. Once the initiation is done, it monitors two things: AWS DRS job execution and DB server launch status.

If any issue with either recovery or launching the DB and applications servers, it fails or goes to the next step.

Steps 7 and 8: These two steps are responsible for updating the network settings. They vary customer to customer. Overall, they update the DNS records and make sure the newly launched servers are accessed by end users.

Step 9: This step uses AWS Lambda with Python, boto3, and AWS SSM Agent to call OS-level commands to start the database. If it is successful, it proceeds to step 10. Otherwise, it fails.

Step 10: Using a similar type of AWS Lambda function as in the step 9, it starts the application server.

Benefits of This Automation Solution

The automation solution presented offers the following benefits:

- **Reduced human error**: Automation minimizes the risk of mistakes during complex DR procedures.

- **Improved RTO**: Streamlined, automated processes significantly reduce overall recovery time.

- **Consistency**: Automation ensures DR processes are executed consistently across multiple systems and scenarios.

- **24/7 readiness**: Automated systems can initiate DR processes at any time, improving overall resilience.

- **Cost efficiency**: Automation reduces the need for extensive manual intervention, lowering operational costs.

- **Enhanced testing capabilities**: Automation facilitates more frequent and comprehensive DR testing without disrupting production.

Summary

In this chapter you learned how the integration of DevOps and SysOps practices for SAP on AWS offers a transformative approach to managing and optimizing SAP landscapes. By leveraging AWS's robust suite of tools and services, organizations can automate daily operations, enhance system reliability, and achieve significant cost savings. The use of infrastructure as code (IaC), CI/CD pipelines, and advanced monitoring solutions like CloudWatch Application Insights for SAP ensures a proactive and efficient management of SAP environments. These capabilities not only streamline operations but also enable faster time-to-market, improved collaboration, and increased agility.

To fully realize these benefits, organizations should prioritize the following:

1. **Adopt IaC**: Implement AWS CloudFormation or AWS Cloud Development Kit (AWS CDK) to automate the provisioning and management of SAP infrastructure, ensuring consistency and repeatability across environments.

413

2. **Implement CI/CD pipelines**: Utilize AWS
    CodePipeline, AWS CodeBuild, and AWS
    CodeDeploy to establish automated build, test, and
    deployment processes, reducing manual effort and
    minimizing the risk of errors.

3. **Enhance observability**: Integrate AWS CloudWatch
    and CloudWatch Application Insights for
    comprehensive monitoring of SAP systems,
    enabling proactive issue detection and resolution.
    Start using AWS CloudTrail and AWS CloudTrail
    Lake for enhancing the security, compliance, and
    operational transparency of your AWS environment.
    By providing detailed logs of account activity and
    API calls, it enables you to monitor, audit, and
    respond to changes in your AWS infrastructure,
    where your SAP systems runs, effectively.

4. **Automate routine tasks**: Leverage AWS Systems
    Manager for SAP to automate start/stop procedures,
    patch management, and other routine operational
    tasks, freeing up valuable time for strategic
    initiatives.

5. **Scale SAP environments dynamically**: Implement
    custom-built auto-scaling solutions using AWS
    services to optimize resource utilization and reduce
    costs during peak and off-peak periods.

6. **Modernize DR procedures**: The combination of AWS DRS and serverless automation technologies represents a powerful solution for modern SAP disaster recovery. By leveraging these advanced capabilities, organizations can ensure their critical SAP systems remain protected and recoverable in the face of diverse threats, while simultaneously optimizing costs and operational efficiency. This approach not only meets the stringent requirements of today's digital business landscape but also provides a foundation for future scalability and innovation in disaster recovery strategies.

By focusing on these strategic initiatives, organizations can effectively optimize their SAP development and system operations on AWS, ensuring high performance, reliability, and cost-efficiency.

CHAPTER 8

Starting your SAP on AWS Modernization Initiatives

Modernization begins with small but focused initiatives that build momentum towards a bigger and a transformative vision.

This final chapter focuses on topics related to how your organization can get started. It begins by identifying which business processes are worth modernizing and presenting ready-to-use solutions. It then introduces SAP CAF in more detail, which is followed by an explanation of how to develop an SAP on AWS modernization roadmap. This roadmap will enable your organization to start small, think big, and scale over time. Once the roadmap is established, we will discuss strategies to secure business buy-in for your innovation projects and how to effectively navigate internally within your organization until you have a plan for a smooth transition to operations after go-live of these projects.

The following topics will be covered in this chapter:

- What's worth solving?

- Ready-to-use solutions

- AWS CAF in action

© Bidwan Baruah, Krishnakumar Ramadoss and Abarajith Vivekanandha 2024
B. Baruah et al., *Evolve from Infrastructure to Innovation with SAP on AWS*,
https://doi.org/10.1007/979-8-8688-0890-6_8

- Creating an SAP on AWS modernization roadmap

- How to get business buy-in

- Governance

- Transition to operations

What's Worth Solving?

Determining which business challenges are worth solving in an SAP on AWS modernization initiative involves a thorough evaluation of the organization's strategic goals, current pain points, and potential opportunities for improvement. Start by conducting a comprehensive assessment of existing processes and systems to identify inefficiencies, bottlenecks, and areas of high cost or risk. Engage with stakeholders across various departments to understand their needs and gather insights into which challenges most significantly impact productivity and growth. Prioritize challenges that align closely with the organization's strategic objectives and have the potential for substantial return on investment (ROI). Consider factors such as the feasibility of implementation, especially for the first use case. By focusing on these high-impact areas, organizations can ensure that their modernization efforts drive meaningful and measurable business outcomes.

Strategy Is Everything

A well-defined strategy is everything in a modernization initiative. It helps manage risks by anticipating challenges and developing mitigation plans, ensuring a smooth transition. Additionally, a strategic approach aligns modernization efforts with industry best practices, enhances decision-making, and measures progress through defined metrics, ultimately leading to cost efficiency and competitive advantage. However, everything starts with the evaluation of which business challenges need prioritization, as discussed next. Identifying and prioritizing these challenges ensures that the modernization efforts focus on areas that will deliver the most significant impact, thereby maximizing the value derived from the cloud transformation.

Evaluation Framework

An effective evaluation framework is instrumental in determining whether a business process warrants modernization. The Chapter 4 section "Strategic Innovation Approach" discussed an evaluation framework to identify if a business process is worth modernizing. Figure 8-1 provides a recap of this framework at a high level.

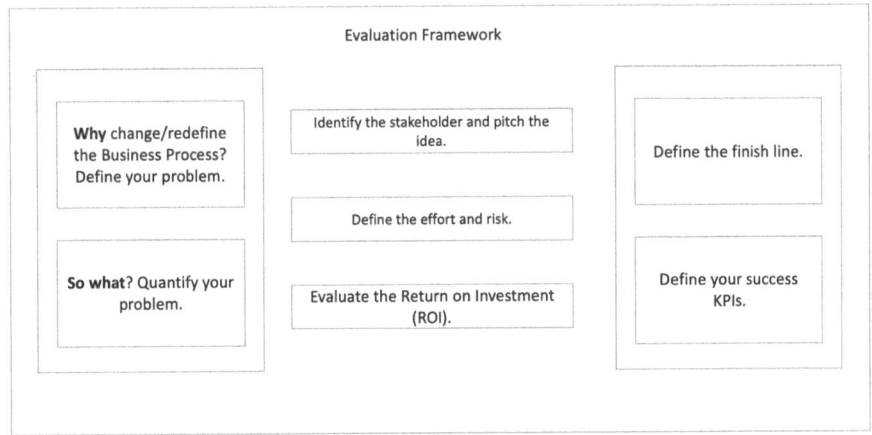

Figure 8-1. Evaluation framework

This framework encompasses several key features, starting with a thorough definition and quantification of the problem at hand. Understanding the magnitude and impact of the problem allows for informed decision-making. Next, the framework outlines the associated risks and efforts involved in modernizing the process, providing a holistic view of potential challenges. Evaluating the ROI helps gauge the financial viability of the modernization effort. Defining the target state based on business outcomes establishes clear objectives and goals for the modernized process, while specifying key performance indicators (KPIs) ensures that progress and success can be measured effectively. Finally, if the analysis indicates that modernization is indeed warranted, the idea can be pitched to stakeholders, backed by data-driven insights and a compelling business case. This comprehensive framework enables organizations to make informed decisions about modernization initiatives, maximizing value and driving positive outcomes.

Now, let's explore methods for identifying business processes that can be optimized or reimagined. Usually, the best resource for identifying pain points within the business are the Business Process Owners, as they have an in-depth understanding of their specific processes. But, there are some challenges. First of all, the Business Process Owners might identify an area of improvement based on gut feeling. Also, often complications arise due to overlaps with other business processes, necessitating collaboration with other Business Process Owners. Additionally, reliance on tribal knowledge can be a challenge. To address these issues, tools like SAP Signavio are available to assist. SAP Signavio will give you data-driven visibility of your business processes.

SAP Signavio

SAP Signavio, a process mining and business process management tool from SAP, can help you identify and prioritize the right initiative. It can provide insights into areas of productivity improvement by offering a comprehensive view of an organization's business processes. It captures and analyzes data from various sources, such as transaction logs and system records, to visualize how processes are executed in real time. By mapping out the end-to-end process flows, SAP Signavio highlights bottlenecks, inefficiencies, manual activities, and areas of deviation from the intended process. An example of this analysis is shown in Figure 8-2.

This visibility allows organizations to identify opportunities for streamlining workflows, reducing cycle times, and optimizing resource utilization. Additionally, SAP Signavio's analytics capabilities enable users to quantify process performance metrics, such as throughput times, wait times, and error rates, providing a clear understanding of where improvements can be made. By leveraging these insights, organizations can prioritize initiatives aimed at enhancing productivity, driving operational excellence, and, ultimately, achieving their business objectives.

Potential areas for productivity improvements per SAP Signavio Process Insights - analysis

Figure 8-2. *Example analysis of business processes by SAP Signavio*

Once the business processes needing improvement are identified using SAP Signavio, you can use SAP LeanIX to inventory your applications and capabilities. This tool helps assess the current applications to determine if they can be leveraged for modernizing business processes. If new applications and capabilities are required, SAP LeanIX can identify them. Based on the modernization goals set by the business process owners, SAP LeanIX aids in pinpointing the necessary capabilities and applications for the transformation. SAP Signavio and SAP LeanIX together can help you build a culture of continuous improvement, not just a one-off activity to determine inefficient business processes.

Now that we have discussed how to identify uses cases that are worth innovating, let's explore ready-to-use solutions that can help you solve your business challenges.

Ready-to-Use Solutions for Your Business Challenges

Once you have identified your use case, the next step is to build the solution. This solution might be built using AWS services, SAP BTP services, or a partner tool, or it might involve leveraging a combination of AWS services, partner products, and SAP BTP services to create a robust and scalable solution.

AWS Solutions and SAP Missions

By now, you are familiar with all the AWS services that can help you innovate. But you might not necessarily know which combination of AWS services or partner products can help you solve your business problem.

What Are AWS Solutions?

An AWS Solution is a combination of AWS services, partner products, and open source technologies that is designed to address a specific business problem. AWS Solutions (https://aws.amazon.com/solutions/) are tailored for particular use cases, which may be industry-specific, technology-specific, or line-of-business (LoB)-oriented. AWS Solutions are built by AWS Solutions Architects and partners and often come with AWS CloudFormation templates for easy deployment. AWS Solutions for SAP-specific use cases are available at https://aws.amazon.com/solutions/enterprise-resource-planning/.

One type of AWS Solution is *Guidance*. Guidance consists of prescriptive technical advice designed to assist customers in independently building solutions using AWS services, along with relevant AWS and AWS Partner Solutions. Guidance includes thoroughly vetted architecture diagrams, optional sample code, white papers outlining architectural considerations and technical requirements, and reference materials.

Introduced in Chapter 1, a great example is "Guidance for SAP Data Integration and Management on AWS" (`https://aws.amazon.com/solutions/guidance/sap-data-integration-and-management-on-aws/`), which is a very popular topic in almost all customer conversations. It's popular because there are many challenges and many approaches for SAP data extraction, thereby making the pros and cons of each choice critical. There are both strategic and technical challenges associated with SAP data extraction. Strategically, organizations face difficulties due to the mission-critical nature of SAP applications, which limits the use of invasive replication methods. Additionally, SAP's evolving requirements for supported integration platforms and protocols, as well as licensing restrictions, present significant obstacles. On the technical side, challenges include managing multiple SAP ERP code bases and data extraction options, dealing with proprietary data formats and objects, meeting incremental data capture and latency requirements, and navigating the complexities of different solution types, whether embedded within SAP, hosted on dedicated instances, or provided as SaaS.

This Solutions Guidance assists customers in navigating the various options available for SAP data integration and management. Whether their SAP system is hosted on SAP RISE, on AWS, or on premises, the Guidance helps them determine the best approach for data extraction—whether in real time or batch mode, full or incremental. It also provides direction based on the type of source SAP system, such as SAP ERP Central Component (SAP ECC), SAP S/4HANA, SAP Business Warehouse (SAP BW), or SAP HANA Database. Most importantly, the Guidance offers detailed architectural patterns for various approaches, whether you prefer to utilize AWS native services, an SAP Solution, or an AWS Partner Solution.

Why AWS Solutions?

AWS Solutions offer several advantages:

Build and deploy faster: Preconfigured solutions speed up the development and deployment process.

Gain fast time to value: Quickly realize the benefits of your investment with proven solutions.

Lower or Mitigate Risks: Utilize architectures that have been tested and validated to reduce potential risks.

Automatic deployment: Deploy the right services automatically, ensuring optimal performance and integration.

AWS best practices: Implement solutions that follow AWS best practices for security, efficiency, and scalability.

What Are SAP Missions?

Similar to AWS and its AWS Solutions, SAP publishes SAP Missions in the SAP Discovery Center (`https://discovery-center.cloud.sap/missionCatalog/`). SAP Missions are structured, goal-oriented guides that help organizations achieve specific outcomes using SAP solutions. A lot of these Missions include AWS services in their guides. They are designed to simplify the implementation and adoption of SAP technologies by providing a clear, step-by-step approach tailored to various business challenges and opportunities. The following are some of the SAP Missions (`https://discovery-center.cloud.sap/missionCatalog/?search=aws`) that include AWS services in their solutions:

- Predict Inventory Allocation with Amazon SageMaker and FedML

- Accessing data in Amazon S3 from SAP Datasphere

- Access, Share and Monetize Data with SAP Datasphere

- Integrate Amazon Athena with SAP Datasphere

- Integrate Events from Amazon Monitron with SAP S/4HANA using SAP BTP

- Data Federation from Amazon Redshift through SAP Datasphere

- Integrate Amazon Rekognition and SAP EHS for PPE Detection

- Include All Users by Enabling Accessibility in Your Digital Experience

- Build Events-to-Business Actions Apps with SAP BTP and MS Azure/AWS

- Explore your Hyperscaler data with SAP Datasphere

- Route Multi-Region Traffic to SAP BTP Services Intelligently

The following are key characteristics of SAP Missions in the SAP Discovery Center:

- **Outcome-focused**: Each Mission is designed with a specific business outcome in mind, driving innovation.

- **Comprehensive guidance**: Missions offer detailed, step-by-step instructions to help users implement SAP solutions.

- **Best practices and templates**: Missions incorporate SAP best practices and often come with reusable templates, tools, and resources to streamline the implementation process.

- **Interactive and engaging**: Missions include interactive elements like tutorials, videos, and hands-on exercises to make the learning and implementation process engaging and practical.

SAP Missions in the SAP Discovery Center are invaluable resources that guide organizations through the successful implementation of SAP solutions, ensuring they can achieve specific business outcomes efficiently and effectively.

Why AWS Solutions and SAP Missions?

AWS Solutions and SAP Missions offer several advantages:

- **Accelerated implementation**: By following a structured and well-defined path, organizations can accelerate the implementation of the solutions, reducing the time required to achieve desired outcomes.

- **Reduced risk**: Utilizing proven best practices and guidance minimizes the risk of errors and issues during implementation, leading to smoother deployments.

- **Cost efficiency**: AWS Solutions and SAP Missions help optimize resource utilization by providing clear guidance and tools, potentially reducing costs associated with trial-and-error approaches.

- **Strategic alignment**: AWS Solutions and SAP Missions ensure that the solution implementations are closely aligned with the organization's strategic goals and objectives, driving meaningful and relevant business results.

AWS CAF in Action

Now that we have discussed the right strategies to identify the right business challenge and also possibly discover the right solution, let's see how enterprises can use AWS Cloud Adoption Framework (CAF) to overcome the challenges to implement these modernization projects and get maximum value out of AWS. Some challenges include

- Cloud initiatives jammed up in long IT queues

- No automation or reference architectures for cloud services

- Limited value from a few Greenfield cloud projects

- Difficult for tech leaders to make a business case

- No aligned vision, guidance or management oversight

As introduced in Chapter 1, AWS CAF leverages AWS experience and best practices to help you digitally transform and accelerate your business outcomes through innovative use of cloud services. Leveraged by customers and partner teams, it helps derive, prioritize, evolve, and communicate a strategy for transformation. Figure 8-3 depicts the AWS CAF 3.0 value chain.

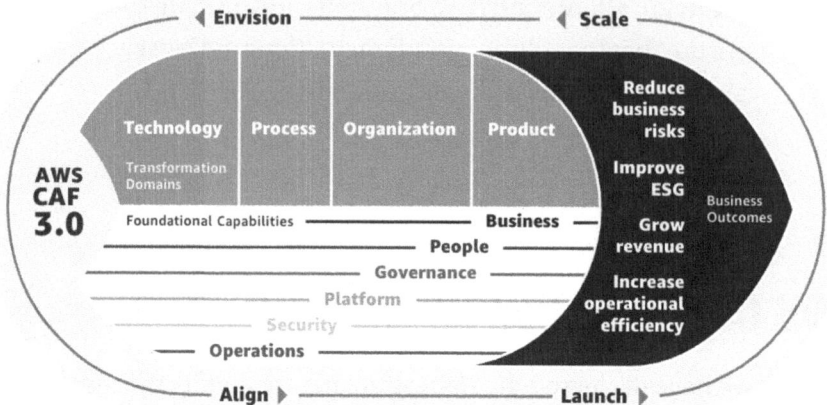

Figure 8-3. *AWS CAF value chain*

To demonstrate the value of AWS CAF, let's take a simple example in the realm of modernization of SAP on AWS. Let's say your organization wants to take advantage of AWS AI/ML technologies to drive a business outcome of increased operational efficiency. You can use AWS CAF to figure out the framework. Creating this business outcome relies on your capability to adopt AI/ML technologies. To adopt AI/ML, your organization needs to transform along at least these three domains:

- **Technology**: Establishing the capability and then enabling the usage and adoption of AI/ML

- **Process**: Focuses on modernizing your business operations via automation through the power of AI/ML

- **Organization**: Collaboration between different teams to meet your business outcomes

Transforming these domains and enabling them to use AI/ML depends on at least five of your foundational capabilities: people, governance, platform, security, and operations.

When embarking on this journey, it's essential to focus on iterative and incremental improvements. The process begins with *assessment*, which involves evaluating your AI/ML readiness in terms of integrating with SAP to determine the starting point. Collaborating with your AWS account team can be highly beneficial, as they can provide workshops with SAP and AI/ML specialists. Following this initial assessment, the adoption cycle unfolds through four key stages, as depicted in Figure 8-4:

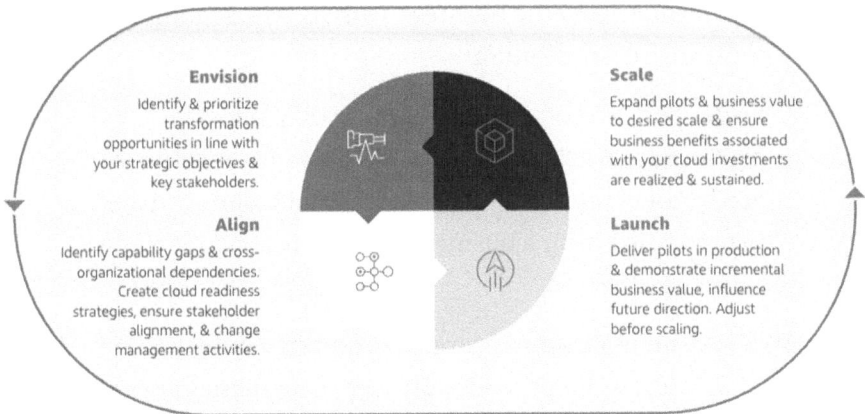

Figure 8-4. *AWS CAF adoption cycle phases*

- **Envision**: This initial phase centers on envisioning how AI/ML can accelerate your business outcomes. It involves identifying and prioritizing operations that align with your business objectives. Ensure that your transformation initiatives are tied to key stakeholders and measurable business outcomes.

- **Align:** In the second phase, the focus shifts to building foundational capabilities. This involves identifying cross-organizational dependencies and challenges, creating strategies to enhance AI/ML readiness,

ensuring stakeholder alignment, securing future buy-
in, and facilitating relevant organizational change
management activities.

- **Launch**: During this phase, the emphasis is on
  executing pilot initiatives, from early proofs of concept
  to production, demonstrating incremental business
  value. These pilots should significantly impact both the
  organization and the business, with AI/ML making a
  meaningful contribution.

- **Scale**: The final phase focuses on scaling pilots in
  production to achieve the desired business outcomes.
  Scaling here refers not only to technical scaling but also
  to expanding from a business perspective and reaching
  your customers.

So, AWS CAF helps customers organize and structure their
modernization journey, helps to identify what capabilities are needed
to successfully implement different technologies, and provides a mental
model for iteration over them. Next, we will discuss how to develop a
modernization roadmap from an organizational perspective, considering
the entire SAP on AWS cloud journey.

Creating an SAP on AWS Modernization Roadmap

There is not a single strategy or roadmap for your SAP on AWS
modernization initiative. Thinking big and starting small is an approach
to avoid analysis-paralysis. However, Figure 8-5 depicts a modernization
roadmap that is based on working with hundreds of customers who have

embarked on this journey. But, keep in mind that modernization is a journey, not just a sequence of steps. The steps depicted in Figure 8-5 will help you make incremental progress.

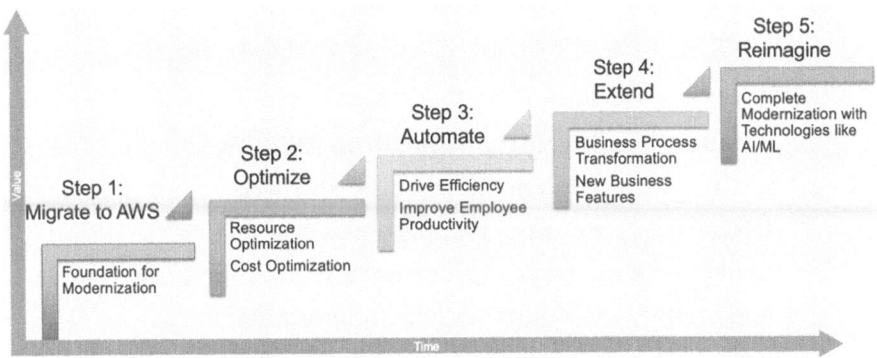

Figure 8-5. *Sample SAP on AWS modernization roadmap*

Creating a roadmap for modernizing prioritized business services involves a systematic approach that balances value with time. In this context (and throughout the chapter), *business services* refers to the activities, processes, and functions that support the operation and management of a business. These services can encompass a wide range of activities, including IT services, operational activities, and business processes within the IT systems.

The key to success is to break down the modernization journey into manageable steps. This involves mapping the modernization steps to business *objectives and key results (OKRs)* and technical KPIs, with before and after values, to validate the impact of modernization. To determine the modernization steps or activities required for each phase, it's crucial to consider factors such as complexity, dependencies, and resource availability. Referring again to the Chapter 4 section "Strategic Innovation Approach," we discussed an evaluation framework to identify if a business process is worth modernizing. This framework helps in assessing the current state of the process and its alignment with strategic

431

objectives, enabling informed decision-making regarding modernization efforts. Regular feedback from stakeholders is also crucial, allowing for adjustments to the roadmap based on lessons learned and changing requirements throughout the modernization process.

Depending on the current state of business services, the steps may include

> **Step 1: Migrating SAP workloads to AWS**: This is usually, although not necessarily, the first step as it lays down the foundation to use AWS native services. Assess the current SAP landscape and identify workloads suitable for migration to AWS. Plan and execute the migration process, considering factors like data transfer, application compatibility, and downtime minimization.
>
> **Step 2: Optimizing by adopting cloud-native architectures**: Evaluate existing architectures and redesign them to leverage cloud-native technologies and principles to maximize resource and cost optimization.
>
> **Step 3: Automating manual tasks and processes**: Key considerations to make include the following:
>
> - Automate everything you can!
>
> - Identify repetitive and time-consuming manual tasks and processes across the organization.
>
> - Foster a culture of collaboration between development and operations teams.
>
> - Implement DevOps practices such as continuous integration, continuous delivery, and infrastructure as code.

- Adopt SysOps practices for proactive monitoring, incident management, and performance optimization of IT systems.

Step 4: Extending business processes: Target the low-hanging fruit. Business process owners usually can identify inefficiencies and areas of improvement in the business processes that might be easily solved. Once these pain points and bottlenecks are known, identify the AWS services that can be used to extend the business process to close the inefficiencies. It can be as simple as sending an e-mail notification via Amazon SNS or storing and accessing documents in Amazon S3.

Step 5: Reimagining with AI/ML: Reimaging a business process requires greater effort than just getting rid of bottlenecks. Conducting a comprehensive analysis of current business processes is needed to evaluate if the process can be simplified. By reimagining business processes, leveraging innovative technologies like AI and ML, organizations can not only improve internal operations but also elevates the overall customer experience, resulting in increased satisfaction and loyalty. By leveraging Amazon Bedrock to access the latest foundation models (discussed in depth in Chapter 6), organizations can automate repetitive tasks, gain actionable insights from data, and enhance decision-making processes within their SAP ecosystems, ultimately driving efficiency, agility, and competitive advantage.

Modernization Process: An Iterative Approach

Aligning the modernization roadmap with business priorities, resource constraints, and organizational capabilities is essential for successful modernization. This involves breaking down the modernization journey into manageable phases or iterations, taking into account the complexity of each step, its dependencies on other activities, and the availability of resources such as budget, personnel, and technology.

This iterative approach to modernization is recommended because each iteration delivers incremental value while progressing toward the target state. This approach is shown in Figure 8-6 and described in the following list. By prioritizing high-impact, low-effort changes for early iterations, organizations can demonstrate quick wins and build momentum for further modernization efforts.

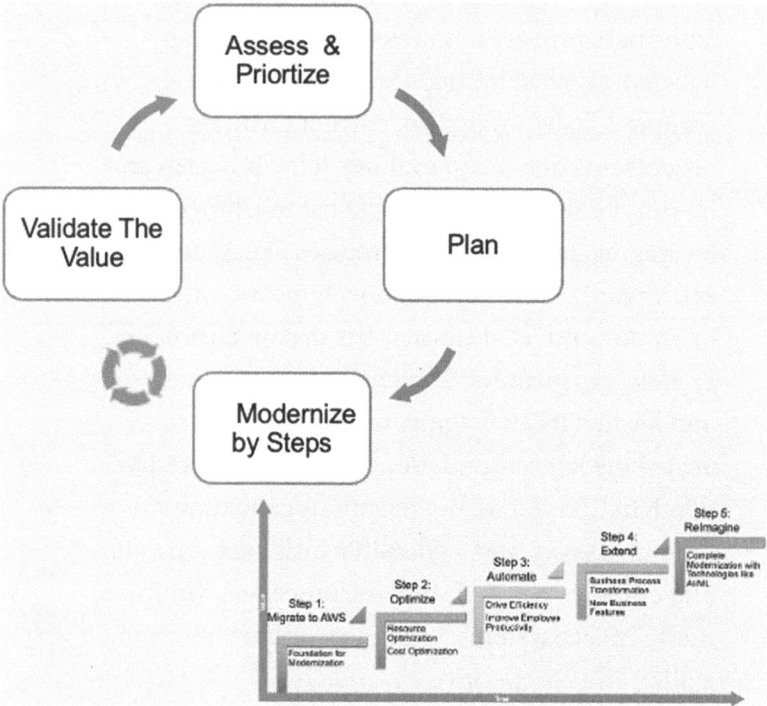

Figure 8-6. *Modernization process: an iterative approach*

- **Assess and prioritize**: Evaluate and prioritize business services based on factors like strategic importance, impact on operations, customer value, and technical debt. Identify the current stage of each prioritized business service in terms of technology, processes, and capabilities. Assessment and prioritization is done from business to technical components to determine the target strategy and roadmap (plan phase).

- **Plan**: In the plan phase, define the target modernization strategy. Establish clear goals and objectives for modernizing each business service. Determine the desired end state, considering factors like improved efficiency, scalability, agility, and alignment with business objectives. Define OKRs to measure the success of the modernization efforts.

  Modernize by steps: The modernization starts and iterates through modernization steps. Measure OKRs as a baseline. As previously presented, modernization steps may include the following, depending on the state of the business services:

  - Migrating SAP workloads to AWS as the first step.

  - Adopting cloud-native architectures for resource and cost optimization.

  - Automating manual tasks and processes to improve efficiency. Implementing DevOps & SysOps practices for faster delivery and continuous improvement.

- Transforming business processes for better agility and customer experience.

- Integrating data analytics and machine learning for insights and decision-making.

After each step, the achieved business and technical benefits are measured (validate the value phase).

- **Validate the value**: Measure OKRs after each modernization step. Compare to the baseline to measure the value provided by the modernization efforts. Assess the need for another step. If the modernization step achieves the targeted value from a business perspective, there is no need for further modernization. If not, the cycle is repeated for the next step. After all, modernization is a journey. But to start the journey, you need to get buy-in from the business stakeholders, as discussed in the next section.

How to Get Business Buy-in

Once you have developed an SAP on AWS modernization roadmap, the next crucial step is to secure business buy-in by engaging key stakeholders, communicating the benefits, and demonstrating the value this transformation will bring to the organization.

Securing business buy-in for SAP innovation projects demands a strategic approach harmonized with organizational objectives. To achieve this, attention should be directed toward the following key areas:

- Building a strong business case

- Presenting your case for innovation

- Creating a stakeholder engagement plan

- Demonstrating proof of concept

Building a Strong Business Case

Based on the modernization framework discussed in the previous section, start by understanding the organization's overarching business objectives and pain points, develop a compelling business case that outlines the benefits of the SAP innovation project in terms of cost savings, revenue generation, competitive advantage, or risk mitigation. Highlight the potential ROI and how the project aligns with strategic business objectives.

Presenting Your Case for Innovation

Craft a persuasive presentation that effectively communicates the value proposition of the SAP innovation project to key stakeholders. Focus on the following key points when you prepare your pitch:

- Storytelling—open with a captivating story that resonates with stakeholders.

- Context is king. Use context.

- Keep it simple.

- Stop focus on technology; focus on business outcomes.

- Utilize data-driven insights, case studies, and success stories to illustrate the potential impact on operational efficiency, customer satisfaction, and overall business outcomes.

Creating a Stakeholder Engagement Plan

Most modernization strategies either overlook stakeholders or refer vaguely to a group of managers needing information and decision support. Who are the specific stakeholders your modernization and innovation project targets? What value does the project deliver to each stakeholder? It's crucial for stakeholders to recognize the value delivered by your project; otherwise, securing buy-in for future initiatives will be challenging. A *stakeholder engagement plan* is a vital part of the overall organizational change management strategy for any modernization project, as it drives many of the communications, training, and change activities. The purpose of the stakeholder engagement plan is to describe the key stakeholder groups identified for the modernization project and their respective levels of influence and impact, as well as the strategy for how to engage each group to facilitate a successful transformation.

Who Are the Stakeholders?

Stakeholders are individuals, teams, or business units who

- May be impacted by project changes

- May be held accountable for project success or failure

- Have influence on the level of commitment to the change

- Have a vested interest in the outcome of the changes

Why Is Engaging Stakeholders Important?

Stakeholder engagement can

- Accelerate employee adoption and change ownership

- Prevent resistance to change proactively

- Minimize productivity impacts during transition

Effective stakeholder engagement yields the following results:

- Assurance that the right people are engaged and receive the right information, at the right time, in the right way

- Proactive management of the pace and amount of change that each stakeholder group must undergo to prevent "change saturation"

- Organizational buy-in, commitment, and capability for change

- Mitigation of risks associated with implementation of the project

Stakeholders from a Cloud Center of Excellence Perspective

Chapter 1 briefly introduced the Cloud Center of Excellence (CCoE). The CCoE is a cross-organizational leadership function that facilitates successful cloud adoption across the enterprise through alignment, enablement, and automation. As shown in Figure 8-7, the CCoE gathers cloud requirements from its customers and stakeholders, iteratively developing the right cloud solutions for the organization. Customers include business, engineering, and operations teams, while stakeholders encompass infrastructure, security, compliance, finance, and HR. Depending on the organizational structure, there may be a single central CCoE or multiple federated CCoEs.

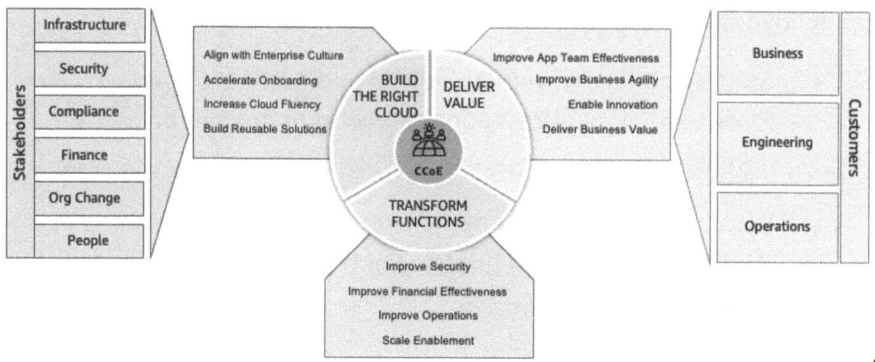

Figure 8-7. *How a CCoE works*

CCoE Objectives

The CCoE has three critical objectives:

- **Build the right cloud**: The CCoE aligns with the needs of internal customers and stakeholders, enhances their cloud literacy and fluency, creates effective reusable solutions based on collected requirements, and accelerates customer onboarding with necessary enablement.

- **Deliver value**: The CCoE improves the effectiveness of application teams, enhances business agility, boosts the organization's ability to innovate, and ultimately delivers the business value expected from cloud initiatives.

- **Transform functions**: The CCoE transforms and automates many critical functions, including security and compliance, finance, operations, and governance.

CCoE Innovation Process

As a customer-obsessed organization, you should work closely with customers to gather feedback and derive insights. These insights will inform product ideas, from which you will select the most relevant to experiment with. You will release product experiments to select customers, gather their feedback, and learn from it. This learning will be leveraged to create new or improved products and release them to the market.

The CCoE will promote innovation by

- Increasing the quantity of candidate innovations in the pipeline

- Providing the cloud foundation to speed up innovation by reducing the time it takes for an idea to progress through the innovation funnel

- Providing the cloud foundation to reduce the cost of innovation and, more importantly, the cost of failure

These initiatives will improve downstream innovation metrics. In essence, the CCoE will provide the cloud foundation to allow the organization to innovate like an industry disrupter.

Demonstrating Proof of Concept

Conduct a proof of concept (POC) or pilot project to demonstrate the feasibility and potential impact of the SAP innovation initiative. Showcase tangible results, such as improved process efficiency, cost savings, or enhanced user experience, to build confidence and garner support. Additionally, a POC allows for accurate cost estimation and resource allocation, preventing overcommitment and ensuring budgetary control.

Figure 8-8 shows a pathway to strategize a POC, divided into four phases:

1. Scoping

2. Pre-build

3. Build

4. Evaluate

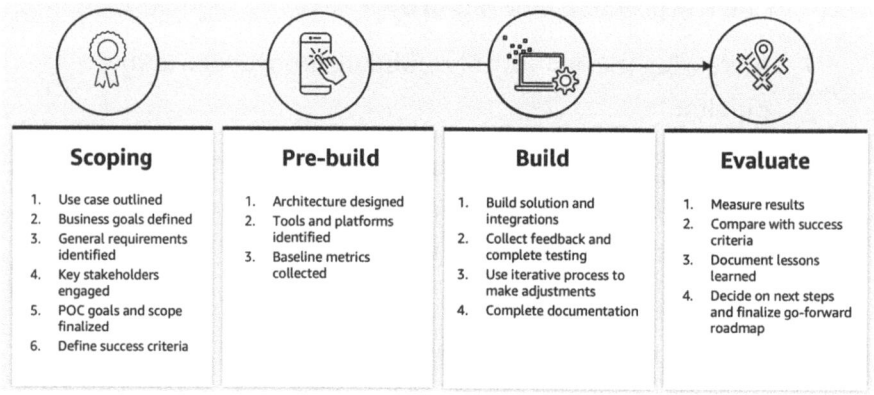

Scoping
1. Use case outlined
2. Business goals defined
3. General requirements identified
4. Key stakeholders engaged
5. POC goals and scope finalized
6. Define success criteria

Pre-build
1. Architecture designed
2. Tools and platforms identified
3. Baseline metrics collected

Build
1. Build solution and integrations
2. Collect feedback and complete testing
3. Use iterative process to make adjustments
4. Complete documentation

Evaluate
1. Measure results
2. Compare with success criteria
3. Document lessons learned
4. Decide on next steps and finalize go-forward roadmap

Figure 8-8. *Phases of a PoC*

1. **Scoping**

   Objective: Define the POC's scope, goals, and success criteria.

   Identify objectives: Clearly outline what you aim to achieve with the POC. This ideally includes defining business goals.

   Define scope: Determine the boundaries of the POC, specifying which systems, applications, and processes will be included.

Stakeholder engagement: Identify and engage key stakeholders, including business leaders, IT staff, and end users, to understand their expectations and concerns.

Success criteria: Establish measurable success criteria and KPIs that will be used to evaluate the POC's outcomes.

Resource allocation: Plan and allocate the necessary resources, including budget, personnel, and tools required for the POC.

2. **Pre-build**

Objective: Prepare the environment and necessary resources for the POC.

Infrastructure setup: Set up the necessary infrastructure as per the POC requirements.

Tool selection: Choose the appropriate tools and platforms to execute the POC.

Security and compliance: Ensure that all security measures and compliance requirements are addressed and implemented in the POC environment.

Baseline metrics: Collect baseline metrics of the current business process's performance, cost, and other relevant parameters for later comparison.

3. **Build**

   Objective: Develop and deploy the POC solution.

   Development: Begin the development and configuration of the cloud solution based on the defined scope and objectives.

   Integration: Integrate the POC solution with existing systems and workflows, ensuring compatibility and interoperability.

   Testing: Conduct thorough testing, including functionality, performance, and security tests, to ensure the solution meets the defined success criteria.

   Iteration: Use an iterative approach to development, incorporating feedback and making necessary adjustments throughout the build phase.

   Documentation: Document the development process, configurations, and any issues encountered along with their resolutions.

4. **Evaluate**

   Objective: Assess the outcomes of the POC and make informed decisions.

   Performance analysis: Compare the POC results with the baseline metrics to evaluate performance improvements, cost savings, and other benefits.

   Stakeholder feedback: Gather feedback from stakeholders, including end users, to assess satisfaction and identify any areas for improvement.

Success criteria review: Measure the results against the predefined success criteria and KPIs to determine if the POC objectives were met.

Lessons learned: Identify lessons learned, best practices, and any challenges faced during the POC to refine the strategy for future phases.

Decision-making: Based on the evaluation, decide whether to proceed with a full-scale implementation, make modifications, or explore alternative solutions.

By systematically addressing each phase with detailed planning and execution, you can ensure that the POC effectively demonstrates the viability and benefits of the cloud modernization project, providing a solid foundation for broader adoption and implementation.

Once you have secured a business buy-in and strategized for a POC, you would need to work with procurement, as discussed briefly next together with broader governance topics.

Governance

Balancing speed and agility with safety and security while investing in management and governance capabilities is crucial for the success of modernization projects. One of the most common conflicts that arise in organizations is between procurement and technology teams, particularly in projects that require agility and rapid innovation. To address this, a clear line of demarcation between these two teams must be established, ensuring both can operate effectively while aligning with the overall project goals.

Procurement leaders should focus on

- **Software and data governance and visibility**: Ensure comprehensive oversight and management of software and data to maintain security and compliance.

- **Productivity and shortened procurement cycle time**: Streamline procurement processes to enhance efficiency and reduce the time required to acquire necessary resources.

- **Cost savings and eliminating unused licensing costs**: Implement strategies to optimize spending, avoid unnecessary expenditures, and eliminate costs associated with unused licenses.

Technology users should be responsible for

- **Working with approved vendors, in compliance with policies**: Adhere to organizational policies by selecting and collaborating with vetted and approved vendors to ensure compliance and security.

- **Wide selection and variety of solutions**: Ensure a diverse range of solutions is available to meet various needs and preferences.

- **Easy access and ease of deployment**: Prioritize solutions that are user-friendly and can be deployed swiftly and effortlessly, enhancing overall productivity and user satisfaction.

In short, a clear demarcation between procurement and IT leaders is crucial for ensuring that each team can operate efficiently and effectively within their area of expertise. It enhances decision-making and overall project success by leveraging the strengths and expertise of both teams. Talking about overall project success, it is crucial to establish comprehensive governance that extends beyond procurement to include areas such as security, cost, and licensing. AWS offers various services to assist in these areas, ensuring effective management and compliance.

Enable Governance with AWS Services

Enabling governance with AWS services involves implementing a robust framework to ensure secure, compliant, and efficient operations across cloud environments. This includes

- **Secure access using AWS IAM**: Utilize AWS Identity and Access Management (IAM) to manage user access and permissions to AWS resources securely. Implement best practices such as least privilege access and multifactor authentication (MFA) to enhance security.

- **Manage governance across multiple accounts**: Employ AWS Control Tower and AWS Organizations to centrally manage governance policies and enforce security standards across multiple AWS accounts. This allows for consistent enforcement of policies and compliance requirements.

- **Automate deployment and policies with AWS CloudFormation and AWS Service Catalog**: Leverage AWS CloudFormation to automate the deployment of infrastructure resources using code templates. Utilize AWS Service Catalog to define and provision approved resources and services while enforcing governance and compliance policies.

- **Cloud financial management**: Implement cloud financial management practices using AWS Budgets, AWS Cost and Usage Reports (CUR), and AWS Cost Explorer. Monitor and optimize costs, set budgetary controls, and gain insights into cost drivers to ensure efficient resource utilization.

- **Entitlement management with AWS License Manager**: Visibility to what's being purchased is very important. Manage software licenses and entitlements effectively using AWS License Manager. Ensure compliance with licensing agreements and optimize license usage across your AWS environment.

- **AWS Marketplace**: If you decide to use third-party products, utilize AWS Marketplace for procurement, leveraging private offer policies for negotiated pricing and terms. Integrate AWS Marketplace with your procurement system for seamless purchasing and vendor management. Gain insights into vendor performance and usage metrics for informed decision-making.

Figure 8-9 illustrates all these services that are available at your disposal.

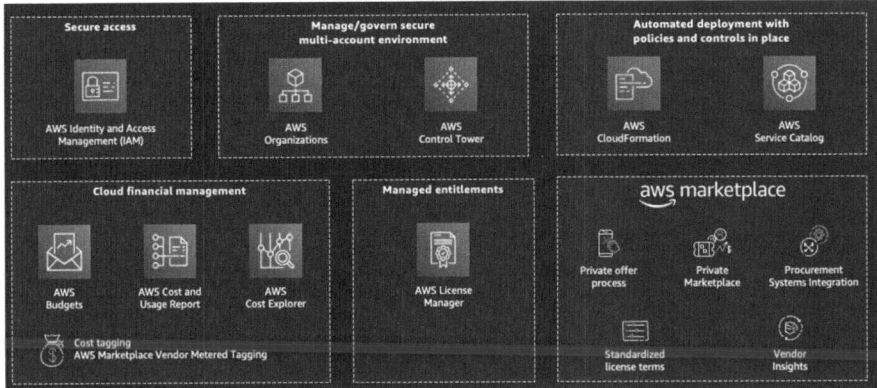

Figure 8-9. *AWS services for governance*

By incorporating these AWS services into your governance strategy, you can establish a secure, compliant, and cost-effective cloud environment while enabling efficient resource management and procurement processes.

Now, let's pivot to another important topic that is important to consider as you plan and execute your SAP on AWS modernization project.

Transition to Operations

After completing the modernization project, transitioning to operations becomes a critical consideration. It's essential to understand the implications of the new solution or product on IT operations by examining the target support framework for the implemented solution. This framework outlines a series of operational implementation activities that must be integrated into the overall project plan to ensure operational readiness. To effectively address any gaps and issues that may arise due to the modernization initiative, the following focus areas should be carefully addressed:

- **IT operations management:** Evaluate the potential impact on business operations stemming from the modernization initiative due to any missing documentation or knowledge gaps. Based on the evaluation, take steps to close the gaps. Establish robust processes and procedures for managing IT operations effectively, ensuring smooth day-to-day functioning of the IT environment.

- **Application management and operations:** Update the strategies for maintaining and supporting the applications post-modernization, including monitoring, troubleshooting, and performance optimization.

- **Business process operations:** Ensure that modernized business processes are well defined and documented for future reference.

- **End-user support:** Implement mechanisms for providing end-user support for all applications impacted by the modernization project, especially by training the end-user support team so that they can handle incidents and address any issues or concerns promptly.

- **End-to-end change and release management:** As applicable, incorporate changes to existing change management processes to manage changes to the modified business processes effectively.

- **Development and development guidelines:** Establish guidelines and best practices based on decisions taken during the modernization project to ensure consistency and to make sure developers don't go back to old ways.

- **Test management**: Update test cases to reflect the changes to the business processes. Test these updated test cases to validate the functionality and performance of the new solution before deployment into the production environment.

- **Support tools**: Identify and implement new support tools and technologies, if needed, to streamline operational tasks and enhance productivity.

By addressing these focus areas comprehensively, organizations can ensure a smooth transition to operations post-modernization, mitigating risks and maximizing the value of the investment in the new modernization initiatives.

Summary

This chapter prepared you to effectively lead innovation conversations, linking business needs with technological capabilities. The chapter offered a clear grasp of the organizational aspects of these innovations, equipping you for a transformative journey. This combination of skills should help position you as a key player in steering your organization toward innovative triumphs. By reading about how to create an SAP modernization roadmap for your organization, you should have gained valuable insights into strategic planning and execution. You learned how to assess current organizational objectives and challenges, prioritize initiatives, and develop a structured plan to achieve long-term goals. Additionally, you learned how to break down the innovation journey into manageable phases or iterations, ensuring incremental progress while minimizing risks and disruptions.

By delving into the intricacies of preparing an organization for transitioning into operations post-modernization project, you gained insights into assessing the impact of new solutions on IT operations and how to develop a robust support framework to ensure smooth transition and ongoing maintenance. You should now be equipped with the knowledge to effectively manage the transition process, mitigate risks, and maximize the value of the modernization investment. You also gained valuable insights into the critical role of stakeholder engagement to drive success in your innovation projects.

By now, you should be well equipped to elevate your roles within your organizations, assuming the mantle of innovative pathfinder, influencer, and leader. You should have gained the ability to find optimal opportunities for business process innovation, thereby enhancing your value as a proactive change agent. Success in digital transformation hinges on striking a delicate balance between ambition, possibilities, and efficiency. Through mastering the concepts outlined in this book, you learned how to identify and quantify problem statements, gaining insight into the true impact of your innovations. Furthermore, you should be well prepared to select the appropriate strategy and technology for each solution, based on the real-life use cases explained throughout this book that illustrate how leading enterprises are modernizing and innovating. Inspired by Amazon CTO Dr. Werner Vogels' famous saying, we would like to conclude by saying, "Now go innovate!"

Correction to: Evolve from Infrastructure to Innovation with SAP on AWS

Correction to:

Bidwan Baruah, Krishnakumar Ramadoss and Abarajith Vivekanandha, *Evolve from Infrastructure to Innovation with SAP on AWS,* **https://doi.org/10.1007/979-8-8688-0890-6**

This title was published inadvertently with incorrect order of authors which has now been corrected as

Bidwan Baruah

Krishnakumar Ramadoss

Abarajith Vivekanandha

The updated version of this book can be found at
https://doi.org/10.1007/979-8-8688-0890-6

C1

Index

A

Advanced Business Application
    Programming (ABAP)
  builder community, 187
  CAP/RAP, 366
  CI/CD pipelines, 363
  CI/CD service, 363
  classical development, 363
  data lakes, 142
  document processing
    check processing, 190
    connectivity complexity, 192
    formats, 192
    high-level architecture, 188
    leading online
      automobile, 187
    manual processing
      checks, 187
    point-to-point
      integration, 192
    score evaluation, 189
    security and
      authentication, 192
    solution, 188
    working process, 188–190
  financial consolidation, 193
    advantages, 195
    General Ledger (GL)
      accounts, 193
    high-level architecture, 193
    insurance company, 193
    integration, 195
    preparation, 194
    training data, 195
  location accuracy
    advantages, 198
    customer satisfaction, 196
    high-level architecture, 196
    service, 196
    workflow process, 197
  monolithic application, 361
  Piper project, 364
  third party tools, 365
  title, 362
Advance Message Queuing
    Protocol (AMQP), 115
Amazon Machine Images (AMIs),
    384, 399
Amazon Managed Grafana
    (AMG), 382
Amazon Managed Service for
    Prometheus (AMP), 382
Amazon Q, 34, 35, 307–309,
    315–320, 331–336

© Bidwan Baruah, Krishnakumar Ramadoss and Abarajith Vivekanandha 2024
B. Baruah et al., *Evolve from Infrastructure to Innovation with SAP on AWS*,
https://doi.org/10.1007/979-8-8688-0890-6

Amazon Web Services (AWS), 1, 73
    CI/CD (*see* Continuous
        integration/Continuous
        deployment (CI/CD))
    data lakes, 132–152
    financial management, 448
    governance, 447–449
    launch wizard
        catalog integration, 360
        custom deployment
            scripts, 360
        high-availability clusters, 359
        key benefits, 358
    Solutions Library, 69, 70
    vendor management, 448
Americas' SAP Users' Group
        (ASUG), 5
Application Link Enabling (ALE), 79
Application programming
        interfaces (APIs)
    BTP (*see* Business Technology
        Platform (BTP))
    connected experience, 76, 77
    customers/partners/
        employees, 75
    enterprise apps, 75
    innovation services, 16–20
    integration, 77–79
    microservices, 74
    OData protocol, 78
    pivotal role, 18
    protocols, 78, 79
    statefulness *vs.*
        statelessness, 80–84

Artificial intelligence/machine
        learning (AI/ML), 4
    Apollo Tyres, 276
    AWS CAF, 428
    Bedrock working, 48–50
    business
        transformation, 271
    cloud computing, 272
    computer vision (CV), 274
    deep learning, 44–46
    foundation models, 277–281
    Gen AI (*see* Generative AI
        (Gen AI))
    LLMs (*see* Large language
        models (LLMs))
    predictive analysis, 51–54
    quality inspection, 50
    relationship, 42, 43
    self-attention, 277
    tools/services, 50
    use cases mapping, 274–277
Asset Performance Management
        (APM), 252
Asynchronous point-to-
        point model
    Amazon
        EventBridge, 109, 110
    Amazon SQS, 106
    architectural style, 109
    dead-letter queue, 105
    disadvantage, 106
    message bus pattern, 108–110
    message queues, 104–106
    queue pattern, 104

router pattern, 108

synchronous (*see* Synchronous (request-response) model)

Automated Predictive Library (APL), 51

AWS Distro for OpenTelemetry (ADOT), 382, 383

B

Bedrock architecture, 50–52

Bidirectional encoder representations from transformers (BERT), 282

Business Application Programming Interfaces (BAPIs), 77, 145

Business intelligence (Gen BI), 34
  ask/why feature, 330, 331
  forecasting abilities, 330
  natural language, 329
  QuickSight capabilities, 328
  QuickSight Q
    accelerated dashboard authoring, 332, 333
    Amazon Q, 332
    customers, 332
    data storytelling, 335, 336
    executive summary feature, 334
    natural language query, 331

Business process innovation
  ABAP (*see* Advanced Business Application Programming (ABAP))

data interchange, 198–205

JRAs (*see* Joint Reference Architectures (JRAs))

real-world application, 164

SAP BTP services, 168–170

strategic innovation approach
  capabilities, 164
  current/target state, 166
  evaluation framework, 164–167
  POC, 167
  problem statement, 166
  target state, 166

Business process management (BPM), 13, 15, 206, 420

Business Technology Platform (BTP), 2, 19, 52, 64, 65, 168, 309, 363
  API management, 85
  asynchronous model, 104–110
  AWS SDK/SAP ABAP, 171
  cloud strategy alignment, 85
  cost/pricing models, 87
  coupling, 100, 101
  data format coupling, 100
  decoupling, 89
  discovery portal, 168
  event-driven architecture, 111–116
  hybrid approach, 88
  integration architecture, 101
  integration patterns, 101
  key considerations, 85
  management capabilities, 86

Business Technology Platform
(BTP) (*cont.*)
microservices
architecture, 95–98
monolithic architecture, 91, 92
native services, 172
S/4HANA system, 170
scalability/performance, 86
security features/mechanisms, 87
serverless computing, 98, 99
SOA, 93, 94
software architecture, 90, 91
strategic points, 169, 170
synchronous model, 101–103
teams, 88
technology coupling, 100
temporal coupling, 100
traditional monolithic
systems, 89

C

Chain-of-thought (CoT)
prompting, 292, 293
Change Request Management
(ChaRM), 68
Cloud Adoption Framework (AWS
CAF), 12, 13, 417
adoption cycle phases, 429
challenges, 427
domains, 428
initial assessment, 429
scaling, 430
value chain, 427, 428

Cloud Adoption Office (CAO), 9, 10
Cloud Application Programming
Model (CAP), 182, 312, 366
Cloud Center of Excellence
(CCoE), 10–12
cloud creation, 440
definition, 439
deliver value, 440
innovation process, 441
objectives, 440
transform functions, 440
working process, 440
Cloud Integration Automation
Service (CIAS), 260
Cloud service providers
(CSPs), 39, 169
CloudWatch application, 371
AppInsights, 375
automated actions/
notifications, 378, 379
detection, 376
discovery, 377
EventBridge, 380
incident response/
remediation actions, 380
ingestion, 378
monitoring capabilities, 375
NetWeaver application, 376
onboarding phase, 377
working process, 377
automate response, 374
EC2 instances, 372
end-to-end observability, 373
log insights, 373

operational visibility/insight, 374
working process, 372
CloudWatch Application Insights
     (CWAI), 60, 375–377, 380
Continuous integration/continuous
     delivery (CI/CD), 61, 62
Continuous integration/
     Continuous deployment
     (CI/CD), 55, 340
  ABAP development, 361–365
  AWS process
     CodeBuild, 352
     CodeCommit, 351
     CodeDeploy, 352
     CodePipeline, 353
  DevOps
     automation, 348
     bugs, 347
     code quality, 347
     collaboration/
        communication, 350
     deployment strategies, 346
     feedback loops, 349
     infrastructure-as-code
        approach, 348
     measurable progress, 349
     pipeline, 346
     shooting hoops/mastering
        scales, 348
     testing, 347
  pipeline, 351
Cost and Usage Reports (CUR), 448
Create, Read, Update, and Delete
     (CRUD), 81

D

Data analytics
  analytics tools, 26
  AppFlow, 32, 33
  AWS Lake Formation, 38, 39
  business insights, 26
  capabilities, 34
  components, 29
  data exchange, 37, 38
  Datasphere SAP, 39, 40
  end-to-end architecture
        pattern, 36
  general availability (GA), 28
  ingest data, 31
  intelligent assets, 26
  Invista, 26
  journey of, 28–31
  Order-to-Cash (O2C), 38
  Simple Storage Service (S3), 26
  simplification/
        democratization, 36
  third-party solutions, 40
  transfer data, 31
  visualization, 34, 35
  warehouse, 27, 28
Data archiving strategies
  benefits, 120
  definition, 119
  legal/regulatory
        requirements, 120
  management evolution, 122
     aging, 123, 124
     build/buy solution, 126

Data archiving strategies (*cont.*)
    context, 127
    data tiering, 124
    hot/warm data, 123
    ILM capabilities, 123
    regulatory requirements, 122
    total cost of ownership
      (TCO), 124, 126
    volume/performance, 125
  migration processes, 120
  online business, 119
  storage gateway, 127, 128
  types of, 121
Database as a
    service (DBaaS), 39
Database Migration Service
    (DMS), 134
Data lakes
  application-level extraction
    application level, 144, 145
    data intelligence, 145
    higher-level extraction
      methods, 149
    key consideration, 146, 147
    orchestration tools, 145
    semantic layer, 147–150
  AWS services
    catalog design, 135
    commercial solution, 140
    data extraction
      consideration, 137
    data sources, 132
    ingestion, 134
    integrating SAP data, 136

    licensing/compliance/cost
      considerations, 139
    processing/analytics, 135
    scalable data, 133–136
    security/compliance, 136
    storage classes, 133
    storage layer, 135
    volume/velocity/variety
      (VVV), 137–140
  concept of, 131
  diversity/complexity, 130
  electronic data interchange,
    202, 203
  end-to-end enterprise
    analytics, 155
  extraction patterns
    application stack, 141, 142
    database level, 142
    details, 140
    Glue/Lambda functions, 143
    packages, 143, 144
    third-party adapters, 142
  structured data, 131
  traditional data, 131, 132
  traditional data warehouses, 130
Data manipulation language
    (DML), 151
Data Mesh
  access control, 160
  Amazon DataZone, 160
  architecture principle, 158
  business domains, 158
  domain-centric data
    architecture, 159

limitations, 158

paradigm shift, 161

SAPdata/AWS services, 159

traditional data lakes, 158, 159

Data Provisioning Agent (DPA),
177, 178

Deep learning (DL), 44–46

definition, 42

Development/Operations team
(DevOps), 339

ABAP-based SAP systems, 60–62

automation, 340

challenges/solutions/tools, 354

continuous integration, 340–350

culture building, 341

advantages, 341

automation, 346–350

consideration, 343

continuous
deployment, 343–345

continuous integration, 345

deployment/delivering
process, 342, 345

monitor production, 344

preproduction
environments, 342

principles, 342

runtime environment, 343

source-controlled
repositories, 342

stages, 342

streamlining, 345

data storage/messaging, 384

development phase, 360–367

ABAP development, 361–365

side-by-side extension, 366

implementation, 54

monitor/operate
phase, 368–383

observability

automation, 384

CloudWatch, 371–375

definition, 368

detect issues, 370

investigation, 371

maturity model, 369, 370

network isolation, 384

open source
community, 381–383

operational availability,
368, 369

orchestration, 384

remediate, 371

traces, 380, 381

overview, 56

plan/setup phase, 354–360

automation solutions,
354, 356

AWS Launch
Wizard, 358–360

benefits, 358

change management, 357

CloudFormation, 356

consistency and
repeatability, 357

dependencies, 357

IaC, 356–358

provisioning, 357

Development/Operations team
(DevOps) (*cont.*)
traditional automation
solution, 355
SAP artifacts
CloudWatch, 60
IaC, 57
monitoring/observability, 59
solutions/tools, 56, 57
testing, 59
system refresh
process, 383–386
testing phase, 367
traditional process, 54, 55
Digital Manufacturing Cloud
(DMC), 264, 265
Digital Manufacturing Cloud for
Execution (DMCe), 264
Digital Manufacturing Insights
(DMCi), 264
Digital Production Platform
(DPP), 230
Disaster recovery (DR), 66, 67, 403,
405, 409, 415
Disaster recovery plans (DRPs), 66

E

Elastic Block Store (Amazon
EBS), 356
Elastic Cloud Compute (Amazon
EC2), 6, 356
Elastic Compute Cloud (Amazon
ECC), 352

Elastic Container Service (Amazon
ECS), 352
Elastic Disaster Recovery (AWS
DRS), 66, 404
Elastic File System (Amazon EFS),
6, 128, 357
Elastic MapReduce (Amazon
EMR), 135
Electronic data interchange (EDI)
architecture pattern, 205, 206
businesses exchange
documents, 198
configurations/maintenance, 200
core features, 201
costs, 200
data lake, 202, 203
errors, 200
functions, 199, 200
inbound transformation, 201
insightful analytics, 202
insights, 201
integration, 204
primary methods, 199
protocols, 199
trading partner management, 201
transaction visibility, 202
translation/mapping, 199
Enterprise Architects (EAs)
CAF value chain, 12, 13
framework, 13–15
methodology, 14
multiple frameworks, 11
solution capabilities/technical
components, 11

Enterprise Architecture
    Management (EAM), 15
Enterprise Asset Management
    (EAM), 260
Enterprise resource planning
    (ERP), 73
    implementations, 74
    migrations, 75
    monolithic applications, 76
    statefulness *vs.*
        statelessness, 81–84
Enterprise Services (ES), 77,
    79, 93, 95
Enterprise service bus (ESB), 95
Environmental, social, and
    governance (ESG), 12
ERP Central Component (ECC), 2,
    141, 236, 423
Estimated times of arrival
    (ETAs), 203
Event-driven architecture (EDA)
    architecture works, 111
    event-driven architectures, 116
    full-state events, 113
    key properties, 112
    SAP S/4HANA, 114–116
    serverless functions, 116
    source/context, 112
    sparse events, 113
    time-sensitive processing, 112
Event-driven integrations, 7,
    73, 74, 182
Extended Warehouse Management
    (EWM), 259

Extensible Markup Language
    (XML), 77
Extract, transform, and load (ETL),
    29, 31, 133, 316, 396

F

Federation
    Amazon Athena, 152, 153
    virtual database, 152
    virtualization (*see* Virtualization)
Fine-tuning process
    definition, 294
    detailed architecture, 295
    epochs, 296
    full fine-tuning, 298
    high-level overview, 297
    instruction, 300, 301
    large language model, 297
    PEFT techniques, 299–301
    pretraining, 295, 297
    reinforcement learning from
        human feedback, 301–304
    sentiment analysis, 296
    SFT, 298
Foundation models (FMs), 42, 45–47
    Amazon Bedrock, 280, 281
    Amazon Bedrock Agents,
        280, 281
    embedding process, 278
    evaluation, 277
    multimodal text, 279
    text-to-text, 278
    types of, 278

G

General Data Protection Regulation
(GDPR), 123, 139, 144
Generative AI (Gen AI)
Amazon Bedrock
ABAP code, 321, 322
code generation, 321
eclipse IDE, 324, 325
editor mode, 323, 324
preferences, 326
prompt, 326
response, 327
Amazon Q
administration/operational
tasks, 317
automation script, 317
benefits, 316
code suggestions, 319
coding assistants, 317
final output, 320
instructions, 319
operational activities, 316
scripting process, 318
business applications/
processes, 327–336
business knowledge *vs.*
process, 309
capabilities, 310, 312–314
categorization, 314
enterprise transformation, 311
functional business process, 310
Joule, 310
primary patterns, 311

primary technical case
Amazon Bedrock, 321–327
Amazon Q, 316–321
code generation, 316
transformative/innovative
technologies, 276, 277
Generative pretrained transformers
(GPTs), 282
Git-enabled Change and Transport
System (gCTS), 61, 62,
361, 362
Graphics processing units
(GPUs), 272

H

HANA, 60, 141–143, 177, 390
Hypertext Transfer Protocol
(HTTP), 78

I

Identity and Access Management
(IAM), 87, 136, 447
In-context learning (ICL), 290,
293, 294
Industrial Data Fabric (IDF), 255
Industrial data lake/analysis
AWS IoT Core/data ingestion
architecture pattern, 227
comprehensive solution, 223
end-to-end architecture
patterns, 226
message delivery/
scalability, 223

OEE serves, 228–232
processing data, 225
several options, 226
creation, 215
data ingestion, 221–223
data lake, 215
digital model assets, 218
high-level overview, 216
ingestion patterns, 217
IoT SiteWise Edge, 219–221
SiteWise, 217–219
Information lifecycle management (ILM), 122, 123
Infrastructure as code (IaC), 57, 348, 356–358
Infrastructure hosting/innovation, 1
AI/ML technologies, 42–54
back-office process, 4
business process, 4
challenges, 5
clean core, 16
components, 3
data (*see* Data analytics)
DevOps, 54
enterprise architects, 11
extensions/customizations, 16
Industrial customers, 20–25
integration options, 20–24
intersections shape, 3
mindset/culture, 5
adopting microservices, 7
blast radius, 6
cloud native, 6
digital decoupling, 6, 7

fail fast, 8, 9
learning process, 8
operational perspective, 8
organizations, 9–11
rehosting, 5
mobile application, 16
on-stack extension, 18
SAP BTP integration, 19
SysOps, 64–69
traditional data center, 3
Instruction
fine-tuning, 300, 301
Integrated development environment (IDE), 318
Intelligent Scenario Lifecycle Management (ISLM), 52
Intermediate Document (IDoc), 79, 145
Internet of Things (IoT), 4, 210
categories, 20
core serves, 22, 23
end-to-end architecture, 24
industrial data, 20
Monitron, 25
physical device/logical entity, 21, 22
predictive maintenance, 24, 25
real-world things, 22
SiteWise Edge operates, 219–221
smart factory technologies, 266–268
stages, 21
IT service management (ITSM), 58, 398

J

Java Connector (JCO), 79
Joint Reference
        Architectures (JRAs)
  business process, 174
  data-to-value architecture, 175
    architecture, 176
    business semantic
      models, 179
    community blog series, 176
    Datasphere, 175, 178
    data-to-value
      architecture, 176
    federation/queries data, 179
    ingestion process, 178
    problem/solution, 175
    results, 179
  event notifications, 180
    architecture, 181
    community blog, 181
    flow control, 181, 182
    implementation, 182
    problem, 180
    proposed solution, 180
    user notification, 181
  integration/development, 180
  pillars, 174
  platform foundation, 183
  SAP Build Work Zone
    Amazon Route 53, 185
    architecture flow, 184
    BTP, 185
    end user access, 185
    high availability, 183
    implementation, 186
    SAP S/4HANA system, 186
    services, 183
    solution, 184

K

Key Management Service
    (KMS), 136
Key performance indicators (KPIs),
    23, 216, 420
Koch Ag & Energy Solutions
    (KAES), 236

L

Large language models
    (LLMs), 35, 277
  advantage, 282
  architecture, 286
  components, 285
  customization
    Amazon Q, 307–309
    associated costs, 290
    fine-tuning, 294–304
    organizational development
      maturity, 289
    patterns/required skills, 289
    primary techniques, 288
    prompt
      engineering, 290–294
    retrieval-augmented
      generation, 304–307

Generative AI (Gen AI), 326, 327
hallucination, 282
high-level
    architecture, 283, 284
methods, 284
natural languages, 281
prompt, 283
three-dimensional space, 288
transformer, 285
vector database, 286
vector embeddings, 282–284
Line-of-business (LoB), 422
Long Range Wide Area Network
    (LoRaWAN), 22, 222, 223
Low-rank adaptation (LoRA), 299

M

Machine learning (ML), 4
    ABAP financial
        consolidation, 194
    AI (*see* Artificial intelligence/
        machine learning (AI/ML))
    definition, 42
    end-to-end enterprise
        analytics, 156
    engineering techniques, 45–47
    IoT, 24
    SageMaker, 45–47
    SAP ERP solution, 84
Manufacturing Execution (ME),
    259, 265
Manufacturing Execution System
    (Cloud MES), 21, 253, 254

Manufacturing Integration and
    Intelligence (MII), 258, 265
Mean time to resolution (MTTR),
    368, 369
Microservices
    architecture, 95–98
    decoupling process, 89
    delivery status tracking, 97
    integration service, 96
    Lambda functions, 96, 97
    notification service, 96
    statefulness *vs.* statelessness, 81
    supply chain operation, 95
    transformative journey, 98
Mobile Application Integration
    Framework (MAIF), 262
Modern data strategies, 117
    analytics/insights, 118
    archiving (*see* Data archiving
        strategies)
    data lakes, 130–150
    end-to-end enterprise
        analytics, 154
        business insights/
            innovation, 155
        enterprise analytics, 157
        formation, 156
        machine learning, 156
        non-SAP applications, 155
        performing analysis, 156
        user access, 157
    key components, 118
    virtualization/
        federation, 151–155

Modernization projects
  AWS solution/SAP missions
    accelerate
      implementation, 426
    strategic approach
    SAP LeanIX, 421
Modernization Roadmap
  AWS CAF (*see* Cloud Adoption
    Framework (AWS CAF))
  AWS solution/SAP missions
    advantages, 424, 426
    business process, 422
    characteristics, 425
    guidance, 422
    missions, 424
    strategic goals/
      objectives, 427
    strategic/technical
      challenges, 423
  business process, 422, 433
  buy-in business
    business case, 437
    innovation projects, 436
    presentation, 437
    stakeholder engagement
      plan, 438–441
  evaluation, 418
  evaluation
    framework, 431
  evaluation/prioritization, 435
  goals/objectives, 435
  governance
    AWS services, 447–449
    capabilities, 445

    procurement leaders, 446
    technology, 446
  innovative
    technologies, 433
  iterative approach, 434
  key considerations, 432
  migration, 432
  operational
    implementation, 449–451
  operation/management, 431
  proof of concept, 441–445
  resource/cost optimization, 432
  roadmap, 417
  stakeholder engagement plan
    cross-organizational
      leadership
      function, 439–441
    implementation, 439
    individuals/teams/business
      units, 438
    meaning, 438
    ownership adoption, 438
    proactive management, 439
  steps, 435
  strategic approach
    business process, 420–422
    evaluation
      framework, 419
    features, 419
    prioritization, 418
    Signavio, 420, 421
  validation, 436
Multifactor authentication
  (MFA), 447

N

Natural language processing
(NLP), 42, 52, 279, 281
Natural language
query (NLQ), 331

O

Objectives and key results
(OKRs), 431
One-shot inference
technique, 292
Open Data Protocol (OData),
149, 364
Operational Data Provisioning
(ODP), 149
Operational Technology
(OT), 21, 257
Optical character recognition
(OCR)-, 165
Overall equipment
effectiveness (OEE)
calculation, 228
formula, 228
Georgia-Pacific process, 229
high-level
architecture, 231, 232
manufacturing facilities, 230
plant maintenance (PM)/
quality maintenance (QM)
module, 231
production efficiency, 227
production processes, 230

P, Q

Parameter-efficient fine-tuning
(PEFT), 298, 299
Predictive Analysis Library
(PAL), 51–54
Predictive maintenance
Amazon Monitron, 24, 25
reactive/preventive, 232
Predictive maintenance system
Amazon Monitron, 235–237
architecture, 237
condition-based
maintenance, 237
end-to-end architecture, 235
KAES implementation, 236
operations technology
(OT), 235
sensor data/inference, 236
implementation, 233
lookout/equipment
architecture pattern, 240
historical equipment
data, 239
Toyota Motors, 239
workflow, 238, 239
machine data, 233
maturity model, 234
Programmable logic controllers
(PLCs), 21, 215
Prompt engineering
CoT prompting, 292, 293
definition, 290
ICL, 290, 293, 294

Prompt engineering (*cont.*)
  one-shot
    techniques, 292
  zero-shot inference, 291
Proof of concept (POC), 441
  cloud configuration, 444
  division, 442
  evaluation, 444, 445
  phases, 442
  prebuild process, 443
  scoping/objective, 442

R

Recovery point objective
  (RPO), 66, 404
Recovery time objective
  (RTO), 66, 404
Reinforcement learning from
  human feedback (RLHF),
  297, 298, 301–304
Relational Database Service
  (Amazon RDS), 6, 28, 357
Remote Function Call (RFC), 77,
  79, 80, 146, 364
RESTful Application Programming
  Model (RAP), 366
Retrieval-augmented generation
  (RAG), 280
  advantages, 306, 307
  Amazon
    Bedrock, 305, 306
  approach, 304
  foundation model, 304

verb forms, 304
working process, 304, 305
Return on investment (ROI), 418

S, T, U

SAP Cloud ALM
  operation, 68, 69
SAP Landscape Transformation
  (SLT), 148
Security Assertion Markup
  Language (SAML), 83
Security Operations Center
  (SOC), 402
Service-oriented architecture
  (SOA), 74, 77, 93, 94
Session-based web services
  (SOAP), 80
Simple Notification Service (SNS),
  110, 180, 189, 374
Simple Object Access Protocol
  (SOAP), 77, 79
Simple Queue Service (SQS),
  106, 170
Simple Storage Service (Amazon
  S3), 6, 26, 127, 170, 224, 353
Smart Data Access (SDA), 151
Smart factory technologies
  concepts, 209
  definition, 210
  digital manufacturing, 212
    architecture pattern, 263
    asset performance
    management, 260

business network asset
  collaboration, 261
components, 258
different solutions, 262
DMCe/DMCi serves, 264
end-to-end integration, 264
field service
  management, 262
integration, 260
integration capabilities, 258
key tenets, 214
Plant Connectivity
  (PCo), 259
product/process design,
  212, 213
research and development
  (R&D), 213
SAP solution, 258
service and asset
  manager, 261
shop-floor applications, 258
strategic solution, 213
supply chain
  management, 213
sustainability goals, 214
digital twin
  implementation, 247
  industrial scenario, 244
  physical system, 244
  prerequisites, 245
  stages, 246, 247
  working process, 246
edge and edge
  device, 210–212

far edge/near edge, 211
industrial data lake/
    analytics, 215–232
Industry 4.0, 210
IoT project, 266–268
manufacturing/industrial
    page, 251–257
  automation software
    management, 252
  business challenges, 251
  Cloud MES, 253
  composable enterprise,
    254, 255
  Computer Vision for
    Quality, 257
  connected worker, 253
  IDF, 255
  lean daily
    management, 256
  OT Cybersecurity, 257
  performance
    management, 252
  Predictive Quality, 257
  solution packages, 252
operational metrics, 267
POC, 266
predictive maintenance (*see*
    Predictive
    maintenance system)
quality control
  Amazon Lookout Vision,
    241, 244
  architecture pattern, 243
  cost-effective solution, 242

Smart factory technologies (*cont.*)
  cutting-edge technology, 240
  first pass yield/rolled
    throughput yield, 242
  measurement
    documents, 243
  rapid inference, 242
  sampling data, 241
  user-friendly/accessible, 242
  stage process, 266
  third-party solutions, 267
  voice technology, 248–251
Standard operating procedures
  (SOPs), 253
Statefulness *vs.* statelessness
  application layer system, 82–85
  microservices, 81
  OData services, 80, 81
  SAP/API Gateway, 82
  SAP ecosystem, 81
  session/client
    interaction, 80
Supervised fine-tuning (SFT), 297,
  298, 302
Supervisory control and data
  acquisition (SCADA), 21,
  23, 221, 258
Synchronous (request-
  response) model
  advantages/disadvantages,
    102, 103
  components, 101–103
  integration pattern, 102

Systems Applications and
  Products (SAP)
  AWS (*see* Amazon Web
    Services (AWS))
  digital transformation, 1
  infrastructure (*see* Infrastructure
    hosting/innovation)
Systems Manager (AWS SSM)
  advantage, 393
  automatic configurations, 392
  autoscaling, 394, 395
  build solutions, 389
  database process, 391
  data processing, 396
  documents/search, 388
  EC2 instances, 391
  HANA database, 390
  health checks, 393, 394
  key features, 387
  monitoring, 395
  recommendations, 392
  scale-down process, 396
  scaling decision, 396
  solution overview, 390
  start/stop processes, 389
Systems Manager for SAP
  (SSM4SAP), 65, 388–389
Systems Operations (SysOps), 386
  Amazon Security Lake, 402
  automated operations, 386, 387
    systems manager, 387–397
  business continuity
    availability zones (AZs), 404

disaster recovery (DR), 403
DR runbook, 405–413
high availability (HA), 403
CloudTrail, 400–402
CloudTrail Lake, 402, 403
configurations (AWS Config)
audit checks, 400, 401
resource configuration, 397
rules, 398
working process, 398
DevOps, *see* Development/
Operations team (DevOps)
DR runbook
architecture, 407
automation solution, 408,
409, 412
automation workflow, 410
challenges, 406
high level architecture, 406
notification, 410, 412
steps, 405
task execution, 409
infrastructure hosting/
innovation, 64–69
principles, 386
security/compliance, 397–400

V

Value-added
networks (VANs), 199
Virtualization
comparison, 153, 154
meaning, 150
SAP HANA SDA, 151
virtual database, 151
Voice technology, 248–251
Alexa Skills Kit, 249, 250
architecture pattern, 249
benefits, 248
fundamental principles, 251
intent schema, 249
Lambda function, 250
resources, 249

W, X, Y

Web Application
Firewall (WAF), 87

Z

Zero-shot inference
technique, 291